Security and Stability in the New Space Age

This book examines the drivers behind great power security competition in space to determine whether realistic strategic alternatives exist to further militarization.

Space is an area of increasing economic and military competition. This book offers an analysis of actions and events indicative of a growing security dilemma in space, which is generating an intensifying arms race between the US, China, and Russia. It explores the dynamics behind a potential future war in space and investigates methods of preventing an arms race from an international relations theory and military-strategy standpoint. The book is divided into three parts: the first section offers a broad discussion of the applicability of international relations theory to current conditions in space; the second is a direct application of theory to the space environment to determine whether competition or cooperation is the optimal strategic choice; the third section focuses on testing the hypotheses against reality, by analyzing novel alternatives to three major categories of space systems. The volume concludes with a study of the practical limitations of applying a strategy centered on commercialization as a method of defusing the orbital security dilemma.

This book will be of interest to students of space power, strategic studies, and international relations.

Brad Townsend is currently an Army Space Officer assigned as a space policy advisor on the Pentagon Joint Staff, and has a Ph.D. in Military Strategy from the Air University.

Space Power and Politics

Series Editors: Everett C. Dolman, *School of Advanced Air and Space Studies, USAF Air, Maxwell, USA,* and Thomas Hoerber, *ESSCA, France.*

The Space Power and Politics series will provide a forum where space policy and historical issues can be explored and examined in-depth. The series will produce works that examine civil, commercial, and military uses of space and their implications for international politics, strategy, and political economy. This will include works on government and private space programs, technological developments, conflict and cooperation, security issues, and history.

Space and Defense Policy
Edited by Damon Coletta and Frances T. Pilch

Space Policy in Developing Countries
The Search for Security and Development on the Final Frontier
Robert C. Harding

Space Strategy in the 21st Century
Theory and Policy
Edited by Eligar Sadeh

Transatlantic Space Politics
Competition and Cooperation Above the Clouds
Sheng-Chih Wang

Understanding Space Strategy
The Art of War in Space
John J. Klein

A European Space Policy
Past Consolidation, Present Challenges and Future Perspectives
Edited by Thomas Hoerber and Sarah Lieberman

Security and Stability in the New Space Age
The Orbital Security Dilemma
Brad Townsend

Security and Stability in the New Space Age
The Orbital Security Dilemma

Brad Townsend

Routledge
Taylor & Francis Group

LONDON AND NEW YORK

First published 2020
by Routledge
2 Park Square, Milton Park, Abingdon, Oxon OX14 4RN

and by Routledge
52 Vanderbilt Avenue, New York, NY 10017

Routledge is an imprint of the Taylor & Francis Group, an informa business

© 2020 Brad Townsend

British Library Cataloguing-in-Publication Data
A catalogue record for this book is available from the British Library

Library of Congress Cataloging-in-Publication Data
Names: Townsend, Brad, 1980- author.
Title: Security and stability in the new space age : alternatives to arming / Brad Townsend.
Description: Abingdon, Oxon ; New York : Routledge, 2020. | Series: Space power and politics | Includes bibliographical references and index.
Identifiers: LCCN 2019052696 (print) | LCCN 2019052697 (ebook) | ISBN 9780367432072 ; (hardback) | ISBN 9781003001843 ; (ebook)
Subjects: LCSH: Space security. | Space warfare–Prevention. | Space weapons–United States. | Space industrialization–United States. | United States–Military policy.
Classification: LCC UG1520 .T695 2020 (print) | LCC UG1520 (ebook) | DDC 358/.8–dc23
LC record available at https://lccn.loc.gov/2019052696
LC ebook record available at https://lccn.loc.gov/2019052697

ISBN: 978-0-367-43207-2 (hbk)
ISBN: 978-1-003-00184-3 (ebk)

Typeset in Times New Roman
by Wearset Ltd, Boldon, Tyne and Wear

Contents

	List of figures	vi
	Preface	vii
1	Introduction	1
2	The orbital security dilemma	12
3	Power, motive, intent, and information	42
4	Deterrence and reassurance	87
5	Satellite communications	116
6	Remote sensing	140
7	Missile warning	170
8	Competition or cooperation?	200
9	Conclusion	221
	Bibliography	224
	Index	246

Figures

2.1	The classic prisoner's dilemma and stag hunt	14
2.2	Chicken	16
2.3	State motivations	18
2.4	Intention and motivation	19
2.5	Severity of the security dilemma	20
2.6	Quality of arming decisions	21
2.7	Offense–defense, reassurance, and vulnerability	27
2.8	Approach to measuring the security dilemma	35
3.1	Number of space launches per year	55
3.2	Gross domestic product adjusted for purchasing power parity in constant 2011 US$	58
3.3	Manufacturing value added	59
3.4	Military expenditure in billions of US$	59
3.5	Percent of population urbanized	60
3.6	State motives and information	76
4.1	Orbital offense–defense balance	97
4.2	Degree of militarization and threat to space stability	99
4.3	Relative deterrence capabilities of different space systems at various conflict levels	106
5.1	President Kennedy signs the Communications Satellite Act, August 31, 1962	119
5.2	Percentage of DOD fixed satellite services acquired by DISA, 2003–2016	121
5.3	DOD fixed satellite service bandwidth cost and usage (excludes MSS)	122
5.4	Alternative approaches to COMSAT acquisition	124
5.5	HTS footprint compared to a traditional transponder	129
6.1	Percentage of revenue for DigitalGlobe and GeoEye provided by government and NGA 2007–2016	153
6.2	DigitalGlobe image showing Russian military units within Ukraine on August 21, 2014	154
6.3	Synthetic-aperture radar image of several fields and buildings at Kirtland AFB	157
7.1	DSP 16 deployed from STS-44 Shuttle Bay	177

Preface

Less than a decade ago, as the National Aeronautics and Space Administration (NASA) approver for external payloads launched from the International Space Station (ISS), I was asked to inspect a group of small CubeSats that were seeking permission to use the soon-to-be installed NanoRacks CubeSat mechanism on the ISS. I was given an address in Houston to meet a representative from the small startup company and I arrived to find myself at a U-Haul storage facility. Confused, I double-checked the address before calling my contact at the startup to ensure I was in the right place. Used to well-funded labs with cleanrooms operated by Boeing, Lockheed, and others I found it hard to believe I was in the right place, surely no spacecraft of any type was being stored or assembled at a U-Haul? I was wrong. One of the startup founders met me and escorted me into a small rented space at the U-Haul facility to inspect a group of satellites that were not much larger than elongated shoe-boxes. I left that day thinking that these small CubeSats were interesting toys that would never amount to much. Again, I was wrong. Today that company, Planet, has a multi-billion-dollar valuation and operates hundreds of small imaging satellites that, working together in a constellation, image the entire Earth each day, something that was inconceivable a decade ago. Planet is just one example of the commercial innovators that are driving an accelerating pace of change that is overturning long-standing paradigms for how we get to space and how we utilize it.

The long-term implications of this commercial revolution are as yet unclear, but what is clear is that the great space powers are in a race to leverage these new technologies to ensure their own security in the space domain. This competition shows worrying signs of developing into an arms race with uncertain outcomes. These observations led to several questions that form the basis of this book. Is an arms race in space necessary to ensure security in space or is there another possibility? Can competition in space be turned away from a security competition and toward a competition focused more on economic gains and scientific prestige? While conducting that research and completing this book I have found it difficult to keep up with the constant drumbeat of change as first the United States Space Command was resurrected and then a separate US military space service. These events and their strategic implications have only reinforced the relevance of the questions that drove me to write this book. A book that I hope will give its readers a lens through which to view the events occurring in space and contemplate possible alternatives.

This book is not meant to be all-encompassing despite its scope. To address every concern, possibility, and category of space system would try the patience of all but the most dedicated of readers. For these perceived oversights, I apologize in advance. I must also apologize for the assumed knowledge of the reader. This book assumes that the reader is familiar with basic astrodynamics and the fundamental limitations of space systems. That is not to say that the reader needs to be a space expert, though a brief review of the fundamentals of orbits and the space environment would be helpful. Everett Dolman's book *Astropolitik* includes an excellent summary chapter of these fundamentals as well as a contrasting viewpoint to that found in this book.[1]

Research of the scope found in this book is not possible without opportunity and assistance. I would like to thank both the US Air Force's School of Advanced Air and Space Studies (SAASS) and the US Army's Advanced Strategic Planning and Policy Program's Goodpaster Fellowship for giving me the opportunity to conduct this research. There are, or at least were, very few graduate programs and sponsors that would accept inter-disciplinary research on this topic area. In particular, I would like to thank Dr. Derrick Frazier who served as my advisor and advocate throughout this work. To the members of my committee, Andrea Harrington, Robert Davis, and Everett Dolman, I would also like to express my thanks.

Most importantly, I want to express my gratitude to my wife for supporting my decision to pursue my chosen career and the obligations, separations, and hardships for both of us that have come with it. She has been, and remains, a source of sanity and perspective. To my children, who have had difficulty understanding why I needed to write so much, I apologize for being occasionally distracted during the course of this work.

Note

1 See Chapter 3 in Everett C. Dolman, *Astropolitik: Classical Geopolitics in the Space Age*, 1st edition (London; Portland, OR: Routledge, 2001).

1 Introduction

Competition in space is inevitable; the form that it will take is not. Will it be violent, peaceful, or somewhere in between? What are the options for shaping it? These are questions worth asking as we enter a new space age driven by an ongoing technological revolution in launch and satellite design. This technological revolution brings with it the promise of economic expansion into space that could benefit all of humanity. It also brings with it the possibility of extending war into space. Up to this point, the military use of space has been limited to the passive gathering and transmission of information which has promoted peace by allowing the great powers to monitor each other directly, removing uncertainty and generating stability. Satellites still do this, but today space is also a vital component of conventional military power, especially for the United States. This US dependence on space has not gone unnoticed, creating fear that space capable adversaries will turn space into an active warfighting domain and target US space assets in future conflicts. In response, the US is seeking methods of protecting itself from space capable adversaries. The result is a seemingly inevitable arms race as one nation's efforts to protect itself appear threatening, driving other nations to pursue arms in response even if they did not intend to. All sides then accuse the other of arming first, further fueling the perception of aggressive behavior, warranted or not. If an alternative to this cycle of arming is not found then a potentially unnecessary and costly race for militarily dominance in space is almost certain. Left unchecked this quest for space dominance will be as threatening to global stability as the nuclear arms race was during the darkest days of the Cold War.

At the dawn of the space age, few would have predicted that the current fragile peace in space would persevere for as long as it has. Military leaders envisioned a future where space was the ultimate high ground and space supremacy was a national imperative. General Bernard Schriever, one of the US Air Force's early space pioneers, called for national space supremacy, envisioning a future where "the important battles may not be sea or air battles, but space battles."[1] Gen. Schriever was not alone; his opinion was common in the military and civilian circles.[2] When the first simple satellites were being launched, the US was locked in competition with the USSR in what was viewed as a life or death ideological struggle for the future of humanity. It seemed naive to believe that the arms race between the two superpowers would not extend into space.

Despite these early fears of space war, the two great powers avoided an arms race in orbit. Instead, competition between them developed into a prestige contest with manned and scientific achievement in space serving as proxies for national power. Following the US success in landing on the Moon the intensity of this competition faded, though it briefly appeared that an arms race might replace it. Under President Reagan, the US launched an aggressive campaign to develop missile defenses that included a space component in direct violation of the Anti-Ballistic Missile Treaty. This space component promised to ignite a new arms race in space as the Soviet Union sought to match or counter proposed US capabilities. However, before the space arms race became more than a competition of ideas, the cost of sustained military competition combined with economic inefficiency and loss of faith in Soviet ideals forced the USSR to seek a peaceful resolution, and then suddenly collapse.[3]

After the fall of the Soviet Union, the US found itself with unquestioned dominance of the space domain. This control was not based on explicit military strength; rather it was based on presence and achievement. As the threat from the Soviet Union faded, the US was able to enjoy nearly two decades of complete freedom in space. That situation is rapidly changing, and today the world is entering a new era in space where the proliferation of space technology is challenging continued US dominance in an increasingly multi-polar world.

The US remains the preeminent space power, and as such it enjoys significant benefits from space. The economic, military, and scientific advantages that the US gains from space are greater than those enjoyed by any other nation and are the result of its substantial sustained investment in space capabilities. Exact figures vary from year to year, but on average the US government spends nearly $50 billion each year on space activities ranging from scientific research at National Aeronautics and Space Administration (NASA) to intelligence gathering activities at the National Reconnaissance Office (NRO).[4] No nation approaches US government spending on space. The nearest US competitor is China which lags significantly, spending between $6 and $11 billion annually.[5] US government expenditures on space have also provided a foundation for a robust US commercial space industry. The fruits of these US investments in space are clearly visible on orbit, with US-flagged satellites making up more than half of all active satellites.[6]

The rest of the world has also benefited from US investment in space. Scientific research, environmental monitoring, as well as accurate navigation and timing, are all benefits that the world freely enjoys as a side effect of US space presence. These capabilities enable many of the conveniences and necessities of modern life. For instance, the capabilities provided by the US Department of Defense (DOD) funded Global Positioning System (GPS) have revolutionized everything from global travel to banking. Most people are familiar with the navigation benefits of GPS; what they often fail to realize is that this is a side effect of the precise timing signal that these satellites produce. This timing signal regulates everything from the power grid to the banking system. Even tiny errors in this signal can cause disruptions in systems as diverse as cellular networks to credit card readers. Modern life could not function without the capabilities that uninterrupted access to space provides.

Enjoying the uninterrupted benefits provided from space capabilities is possible because space has long been a sanctuary from open conflict. Even at the height of the Cold War, the US and the USSR avoided any significant weaponizing of space. They did this despite agreeing to only a few formal treaties governing behavior in space and none that outlaw weaponizing space as long as nations refrain from placing weapons of mass destruction in orbit or on celestial bodies.[7] This lack of formal agreements governing behavior in space worked as long as the economic and military importance of space remained relatively small. In the period since the fall of the USSR these factors have changed, space is now an essential component of the US military's conventional warfighting capability as well as a vital economic asset. In an increasingly contested multi-polar world, the vulnerability of space systems to attack is becoming a significant national security issue.

National security concerns about space are the result of a recognition that the prolonged period of US dominance in space is being challenged by the rapid rise of China and by a seemingly resurgent Russia. US national leadership is concerned about the potential threat that these nation's actions pose to US space investment, and in response this has led to the re-establishment of a United States Space Command and the creation of a separate Space Force to deal with the threat.[8] This perception of the relative decline of US power in space is occurring at the same time that innovations from the US-based new space industry are disrupting every aspect of it.

New players in the US commercial space market such as SpaceX and Blue Origin are driving a renaissance in space. The cost of access to space is dropping in real terms for the first time, while at the same time technology miniaturization is allowing for the development of much smaller and more capable satellites. These developments are enabling the commercial space industry to realistically consider developing revolutionary new capabilities that were cost-prohibitive under the previous paradigm of launch and satellite development. Examples of these proposals range from mega-constellations of low-Earth orbiting (LEO) satellites providing global internet access to efforts to successfully mine asteroids. Whether these proposals will be a commercial success is not yet apparent, though they are indicative of the economic potential that space could provide. Military competition in orbit and the instability it generates could deter investors from exploring the economic potential of space, preventing these new capabilities from ever being realized.

Space is vast, but Earth orbit is not. It is in the surprisingly fragile environment of Earth orbit where the economic benefits of space provided by the collection and transmission of information are apparent. Each object placed in orbit remains in orbit until it is slowed enough by friction from the Earth's atmosphere to gradually re-enter and burn up. The higher the orbit, the slower this process is, and objects more than a few hundred kilometers above Earth's surface will never encounter enough friction to re-enter the atmosphere. These objects, ranging in size from a fleck of paint to the size of a passenger car, are moving at many times the speed of a bullet and will remain in orbit threatening anything that crosses their path. A Chinese anti-satellite (ASAT) test conducted in 2007 demonstrated the danger that space debris can pose. The Chinese ASAT struck a Chinese satellite and generated at least 3,312 pieces of debris, of which only 256 had reentered the atmosphere five years later.[9] This debris has not remained in its original orbit; orbital effects have caused it to spread out encompassing the globe posing a danger to hundreds of other satellites, potentially generating even more debris as it collides with other objects. The ultimate fear is that an orbital debris field could become self-sustaining and deny humanity access to space.

Even if a conflict in space can avoid creating a devastating cloud of debris, the threat of war in space still generates fear and uncertainty. Events in space are difficult to monitor under the best of conditions. Tracking objects in space relies heavily on the physics of objects remaining in predictable orbits with their orbital paths periodically updated by radar or optical systems at a handful of locations. Until an object passes over these sites, it generally remains unobserved. For those objects that are tracked, identifying them relies on the launching nation to self-identify their payloads. As satellites grow smaller, more numerous, and more capable, the possibility that some of them could be weapons rises. This fear is reinforced because the function of many of these government-controlled satellites is classified and cannot be verified through use, unlike commercial systems. This uncertainty generates fear that some of these satellites could constitute a threat to expensive and difficult to replace commercial and military satellites that are essential to terrestrial warfighting capability.

The linkage between space and terrestrial warfighting capabilities means that actions in space cannot be analyzed in isolation from other warfighting domains such as the air, sea, and land. Currently, the trend in military circles is to discuss warfighting in terms of cross-domain or multi-domain effects and operations. Since space remains a supporting domain, it is only relevant to the degree that it influences events on Earth. As a result, actions in space must always be considered in terms of their impact on other domains. Attempting to treat space as a domain independent of others can lead to flawed assumptions and analysis. Conflict and competition on Earth impact behavior in space, and it is now becoming possible that competition in space could lead to conflict on Earth. It is because of the intertwined and supporting nature of the space domain that this book does not remain confined to space and also takes into consideration the impact of actions and events outside of it.

The intentions, goals, and motivations of states influence how they interact across all domains. If one country views another as hostile to its interests and beliefs, it will be difficult for these two states to cooperate. Disagreements over territorial boundaries, trade disputes, or natural resources are examples of classic issues that can cause tension between states. That tension can then influence behavior in space. However, even when events are tense between states on Earth, they may choose to cooperate in space despite the mistrust that exists in other domains. The continued cooperation between the US and Russia's manned space programs in support of the International Space Station (ISS) is an example of this. Even so, cooperation in space is easier when competition between states is muted or if nations can establish a degree of trust between them that one state is not going to take advantage of another's willingness to cooperate. Encouraging cooperation, or at least peaceful competition, between the various competing space powers presents a real challenge as terrestrial tensions continue to expand into space.

The inability of states to fully trust each other is a function of the anarchic environment in which they exist. International organizations with the stated goal of regulating behavior among states do exist, yet these organizations lack the power to compel great powers forcibly. Even smaller states can only be forcibly compelled by international organizations with the cooperation of other states. Lacking any independent external enforcement mechanism, nations often find themselves living in a state of mutual distrust. Within this environment, states struggle to determine whether to adopt competitive or cooperative policies to ensure their security based on their assessment of their neighbors' motives and intentions.[10] In space all nations are neighbors, and given the increasing number of states with an active presence there, as well as the uncertainty surrounding their motivations, the temptation to adopt competitive security policies in space is increasing. Competitive security policies designed to ensure a state can protect itself from uncertain neighbors often leads these neighbors to pursue competitive policies of their own. This self-reinforcing cycle creates a security dilemma that is difficult to escape and can lead to the very conflict that both states were seeking to avoid. Does the borderless and anarchic nature of space mean competitive security policies are necessary and that conflict in space is inevitable? Is it possible that this is a future that could be avoided if the three largest space powers can find a way to build trust, cooperate, and establish new norms of behavior and then hold other nations to those norms?

The complex interaction of state power, motivations, and intentions that could answer these questions and determine whether competitive policies in space are necessary is difficult to untangle. Attempting to describe the motivations and behavior of states in an anarchic international system is the primary goal of international relations research. However, most international relations theories fail to capture many of the key variables that are influencing behavior in space or place excessive emphasis on just one, usually the relative power of states. Looking to the field of international relations for a sound theoretical basis for analyzing the possibilities for cooperation in space one theory stands out as uniquely applicable, Charles Glaser's rational theory of international politics.

Glaser's theory is a recent addition to the canon of international relations that provides a functioning framework that is well suited to space.[11] The rational theory that Glaser developed incorporates aspects of many different international relations theories into a single functioning framework. The difficulty with his theory is that in its effort to incorporate the best parts of many existing theories it becomes an exceedingly complex theory to apply. Other theories of international relations seem relatively simple in comparison, and it is their simplicity which makes them attractive and useful tools for analyzing and predicting state behavior. Also, despite their shortcomings, these older theories of international relations often still make accurate predictions of state behavior in many different circumstances. Glaser's theory attempts to take the best parts of many diverse theories and combine them in order to overcome the individual shortcomings of the component theories. In doing so, he creates a complex but useful tool for analyzing state behavior and determining how a state should behave that forms the functioning theory that underpins the research in this book.

Structure

This book is structured into three primary parts. First is a broad discussion of international relations theory's applicability to current conditions in space. Many books on international behavior in space have leaned on international relations theory to one extent or another. Some, such as Everett Dolman's *Astropolitik* framed their argument entirely within it much as this work does.[12] What differs is the starting assumptions and the structural nature of the theory used. Dolman's work framed his argument in terms of realist thought and started with the assumption that arming in space is inevitable, leading to the logical conclusion that one state, preferably the US as a democratic free society, should move to dominate Earth orbit while it can. This book also uses a primarily realist structure, but instead of assuming that arming is inevitable it argues that arming, while likely, is neither desirable nor inevitable. This leads to a search for alternatives and a structure to guide that search. The first portion of this work develops that structure which lays the foundation for the rest of the book.

The second portion of this book is the direct application of that theory to the space environment in order to determine the optimal strategic choice to preserve stability. This portion of the research also includes a brief analysis of existing space power theories, not for the purpose of creating a new theory, but rather it is to identify commonalities that can inform the application of strategic choice theory. This section ends once the theoretical basis has been firmly established and applied to develop several hypotheses on how to resolve the growing security dilemma in space. The last element of this research focuses on testing the hypotheses developed through theory application against reality.

Chapter 2 of this book begins with the development of Glaser's theory and its applicability to space. Here I explore the foundational concepts which underpin the rational theory of international politics and how they apply to space. I then develop the concept of the security dilemma which is fundamental to international relations theory and accurately describes the current space environment. The relevant elements of other international relations theories that contribute to Glaser's theory are also discussed before outlining rational strategic choice theory itself. With the fundamentals of rational strategic choice theory outlined, an analysis of the possible methods of achieving cooperation and reassurance in an environment dominated by the security dilemma follows. A brief discussion of the impact of the implications of the multistate environment of space on these reassurance strategies precedes a discussion on the mechanics of how to measure the security dilemma in space using the variables outlined in the chosen theory.

With the theory and the method outlined in Chapter 2, I can begin applying theory to the current situation in space. Chapter 3 begins with a brief history of US actions and rhetoric regarding space and how they have shifted over time from a state of unarmed strategic competition toward one typified by increasingly aggressive defensive postures. This transition has occurred gradually, accelerating in the last few years as Russia and China present new challenges to the US post-Cold War dominance of space. Before moving on from the US to analyze Chinese and Russian rhetoric and actions, I measure the power disparity between these nations in space to place the rest of the discussion in context. Separate analyses of Russia and China delve into these states' motivations and goals using the model developed in Chapter 2. This analysis provides the information necessary to estimate the motivations of each of the three states central to this analysis, including how they view themselves as well as each other. This chapter concludes with a determination that a security dilemma driven by misperception does exist in space and that it is driving a sub-optimal arms race.

Chapter 3 outlined most of the information necessary to determine the optimal reassurance strategy except for two key factors, the offense–defense balance and the degree of distinguishability within it. These two factors are the focus of Chapter 4, though determining them requires building a functioning theory of space power first. Since no accepted theory of space power exists and developing an entirely new theory is beyond the scope of this research, this chapter uses common factors found in multiple leading theories to develop a basic theoretical foundation. With this foundation in place, an in-depth analysis of the offense–defense balance and distinguishability can occur. Once that analysis is complete, all of the factors needed to determine an effective strategy for managing the security dilemma in space are present.

Two primary strategies for resolving the security dilemma in space are discussed in Chapter 4—deterrence and reassurance. A lack of mutual dependence and a neutral offense–defense balance lead to the conclusion that, for the US at least, competitive strategies built around deterrence are self-defeating and reassurance strategies that encourage cooperation are preferable. The challenge is finding a realistic reassurance strategy, or a combination of strategies, that can build enough trust to induce cooperation. Several different approaches are considered and rejected before settling on an acceptable method of reassurance. This approach—commercialization—and the reasoning behind it led to the development of three hypotheses that build upon each other and are the focus of the rest of this book.

The next three chapters analyze the policy and acquisition history associated with commercializing three of the major categories of national security satellites:

communications, remote sensing, and missile warning. The overarching purpose of these chapters is to determine the feasibility of a commercialization strategy. Each of these chapters generally follows a pattern. Each chapter begins with a brief history of the relevant policy and acquisition history for that category of space systems before transitioning to an analysis of case studies where available and then a discussion of commercialization options. These chapters conclude with a determination on the feasibility of each category in supporting a reassurance strategy based upon commercialization.

Chapter 5 is the first of these unit analysis chapters and focuses on the possibility of commercializing satellite communications (SATCOM). This chapter explores the shortcomings of military SATCOM and the troubled history surrounding the past acquisition of commercial capabilities. Despite this troubled history, the innovative approaches being pursued by the DOD to overcome its past issues with acquiring commercial communications are possible pathways to full commercialization. Using the existing policy structure and recent history as a guide this chapter looks at three primary approaches to commercialization and their implications. It then concludes by using the troubled past of the Iridium constellation as a case study for how a commercialization strategy might succeed or fail.

Chapter 6 looks toward the more difficult case of commercializing remote sensing. The commercial remote sensing industry is much smaller than commercial SATCOM and so presents a different set of challenges. It is an industry that is extremely sensitive to changes in US regulations and policy. It is also an industry where US government contracts have determined winners and losers and driven industry consolidation despite the accepted everyday use of satellite imagery by the average consumer in freely available tools such as Google Earth. This process of consolidation as the result of competition for government contracts is the subject of a case study that follows the growth and consolidation of the remote sensing industry. Thanks to the innovative application of small satellites, this industry is again experiencing growth and diversification. It is these capabilities which provide hope that replacing the existing national security remote sensing architecture with US-based and regulated commercial capabilities is feasible. Commercializing remote sensing may even lead to an advantageous shift of commercial providers back to the US, placing them under US regulatory control.

Chapter 7 tackles the greatest challenge to commercialization, missile warning. These satellites continuously monitor the globe in infrared bands looking for the telltale signature of missile launches. Originally designed to provide the earliest possible warning of nuclear strikes against the US, the missions and capabilities of these satellites have expanded over time. The expansion of these satellites into providing support for conventional missions threatens the protection these satellites traditionally enjoyed through their association with the nuclear deterrent.[13] With their protection from attack no longer assured by their exclusive association with the nuclear deterrent, these extraordinarily expensive satellites are fueling the security dilemma through the fear of loss they generate.

The extended and troubled acquisition process that the current generation of missile warning satellites underwent highlights the difficulty posed by attempting to commercialize them. Despite these difficulties, there is at least one example of the DOD successfully building and launching a hosted missile warning payload onboard a commercial satellite that serves as a case study for exploring a possible approach to commercialization.

The concluding chapter of this research analyzes the validity of a commercialization strategy. Three key questions are central to implementing this strategy. The first and most obvious is whether it is possible to follow a commercialization strategy. The second and more difficult question is whether such a strategy can provide enough reassurance to spark constructive dialog between the great powers in space that leads to lasting cooperation or at least peaceful competition. Finally, this chapter discusses what the signs are that the strategy is working and when should it be abandoned in favor of competitive strategies. Adopting a reassurance approach to escaping the security dilemma is built on an analysis of Chinese and Russian motives as well as on the difficulties associated with competitive strategies. Should ongoing Russian or Chinese behavior in space prove this analysis wrong, then abandoning reassurance and focusing on competitive strategies remains a possibility that cannot be disregarded.

This research is not without limitations that must be acknowledged. An analysis of the motivations of states and their attitude toward other states is always difficult given the complexity and competing interests that make up any large organization. The most common method of dealing with this complexity in international relations is to treat nations as unitary rational actors. This research makes an effort to move beyond the rational actor model and delve into the competition of ideas that lead to policy development within states; however, success in this area is limited. While the US is relatively open in allowing subordinate actors that influence policy within it to express their own opinions in public forums, Russia and China are not. Further, there is only a small amount of information available in the public domain reflecting Russian and Chinese positions on activities in space, at least in comparison to the US. For that reason, this research relies heavily on a fairly small sample size of data in reaching its conclusions on Chinese and Russian intentions.

There are also three key assumptions that underpin this research. The first is that states may choose to avoid or lessen their military competition in space while continuing to compete in other domains. This assumption has some historical supporting basis from the Cold War. The expense and dangers of competition in space as described in this book also support this assumption. A second related assumption is that the security dilemma in space can be at least partially separated from a larger security dilemma between the nations involved. Security dilemmas occur between states and are not usually limited to a specific warfighting domain. This research does not attempt to fully separate the two, but it does focus on behavior and actions within the space as drivers of a domain-centric security dilemma. At the very least, recent actions in space are indicative of the worrying expansion of security competition to the space domain. The final and perhaps most important assumption is that the complete weaponization of space is not inevitable, at least in the near term. Recent history in space provides some support for this assumption, but the broader scope of human history would argue that the domain will eventually become fully weaponized. One purpose of this book is to seek ways to delay that date for as long as possible in order to preserve the existing military advantages provided by space that promote peace and ensure the continued stability necessary to support humanity's continued economic exploitation of the domain.

As the dominant space power, the US is the pivot point for defusing the security dilemma. Someone must take the first step, and the US is best positioned for this. Selecting a viable reassurance strategy that is strong enough to overcome entrenched mistrust between the US, China, and Russia is a complex process. It must involve some risk or it cannot effectively overcome existing Chinese and Russian perceptions of US motivations. At the same time, it cannot place the US in a position of too much strategic risk that cannot be turned to its advantage should assumptions of Chinese and Russian motivations prove inaccurate. The goal of this book is to explore a possible method of keeping future competition in space peaceful, and should that fail, ensuring that the US remains in a position of continuing advantage.

Notes

1 Bernard Schriever, "ICBM-A Step Toward Space Conquest" (Astronautics Symposium, San Diego, CA, February 19, 1957).
2 Walter A. McDougall, *The Heavens and the Earth: A Political History of the Space Age* (Baltimore, MD: Johns Hopkins University Press, 1985), 141–150.
3 Andrew Krepinevich and Barry Watts, *The Last Warrior: Andrew Marshall and the Shaping of Modern American Defense Strategy* (New York, NY: Basic Books, 2015), 190.
4 "National Aeronautics and Space Administration FY 2018 Budget Estimates" (NASA, 2017), www.nasa.gov/sites/default/files/atoms/files/fy_2018_budget_estimates.pdf; "National Oceanic and Atmospheric Administration FY 2018 Budget Summary" (NOAA, 2017), www.corporateservices.noaa.gov/nbo/fy18_bluebook/FY18-BlueBook-508.pdf; Sandra Erwin, "Some Fresh Tidbits on the U.S. Military Space Budget," *Space News*, March 21, 2018, https://spacenews.com/some-fresh-tidbits-on-the-u-s-military-space-budget/; Wilson Andrews and Todd Lindeman, "The Black Budget," *Washington Post*, August 29, 2013, www.washingtonpost.com/wp-srv/special/national/black-budget/?noredirect=on.
5 See endnotes in Chapter 3 for a detailed discussion of estimates of Chinese space budgetary expenditures.
6 "Union of Concerned Scientists: Satellite Database" (Union of Concerned Scientists, May 1, 2018), www.ucsusa.org/nuclear-weapons/space-weapons/satellite-database#.W7eDtPZFxEZ.
7 "Treaty on Principles Governing the Activities of States in the Exploration and the Use of Outer Space, Including the Moon and Other Celestial Bodies" (Resolution Adopted by the UN General Assembly 2222 XXI, December 19, 1966).
8 Mike Pence, "Remarks by Vice President Pence on the Future of the U.S. Military in Space" (The Pentagon, August 9, 2018), www.whitehouse.gov/briefings-statements/remarks-vice-president-pence-future-u-s-military-space/.

 9 T.S. Kelso, "CelesTrak: Chinese ASAT Test," (CelesTrak, June 22, 2012), https:// celestrak.com/events/asat.php.
10 Charles L. Glaser, *Rational Theory of International Politics* (Princeton, NJ: Princeton University Press, 2010), 2.
11 Glaser's theory spent years in development and was finally codified into book form only in 2010.
12 Everett C. Dolman, *Astropolitik: Classical Geopolitics in the Space Age* (London: Routledge, 2001).
13 Lockheed Martin Corp., "SBIRS Fact Sheet" (Lockheed Martin Corp., 2017), www.lockheedmartin.com/content/dam/lockheed-martin/space/photo/sbirs/SBIRS_ Fact_Sheet_(Final).pdf.

2 The orbital security dilemma

Attempting to explain why states choose to cooperate or compete with each other is not a new effort. A myriad of theories exist that attempt to explain and predict state behavior. Even though space is a relatively new domain of human endeavor with many unique characteristics, many of these theories of state behavior remain applicable. At the core of these theories is the concept of the security dilemma. The security dilemma arises when a state's attempts to increase its security threaten other states, leading to unnecessary conflict or intensified security competition.[1] It is a relatively simple concept with complex outcomes. Since state behavior in space is beginning to resemble one of security seeking, the security dilemma can provide a framework for explaining and predicting state behavior. Most importantly, understanding the nature of the security dilemma may make it possible to determine a way to preserve the fragile peace within the space domain that best suits the desires of all concerned states.

The security dilemma is a term first used by John Herz more than 60 years ago to describe a situation that arises in an anarchic environment where one individual or group's quest for security through the accumulation of power creates insecurity in neighboring individuals or groups.[2] In an effort to ensure their security, these neighboring individuals or groups accumulate power in response. An action-reaction cycle then ensues with each party attempting to ensure their security by accumulating more power than their neighbor. In an anarchic world where individuals or groups are chiefly concerned with ensuring their own security, the security dilemma provides an explanation for competition and conflict. This chapter reviews the usage of the security dilemma within international relations theory and develops the basic theoretical foundation that forms the framework for this research.

There are three primary schools of thought which are prominent in international relations theory: realism, liberalism, and constructivism. Within each of these schools a myriad of variations exists, though generally these variations share a set of shared assumptions about the nature of states and human nature. Realism assumes states operate in an anarchic system and compete for power and survival. This is in stark contrast to liberalism which emphasizes the role of international organizations, economic cooperation, and democratic institutions in shaping state behavior. Constructivism varies from either of these in that it focuses on the idea that state behavior is guided by the structure of human society. Each of these theories has adherents who will argue vehemently in favor of one lens over another. However, this research will rely primarily on a realist structure. That realist structure, in turn, forms the basis of the rational theory of international politics that underpins large portions of this book. With realism's extensive application to arms races, security dilemmas, and great power competition it is the lens which seems best suited to studying the complex interaction of states in the space domain seemingly locked in all three of these situations.

The security dilemma and its description of competition and conflict is central to the realist school of thought. Realism has three core tenets: that states are the principal actors in international politics; that the behavior of states is influenced primarily by external and not internal factors; and finally, "that calculations about power dominate states' thinking."[3] Where approaches to realism begin to diverge is over the behavior of states in relation to power. Kenneth Waltz, for example, argues in *Theory of International Politics*, that "the first concern of states is not to maximize power but to maintain their positions within the system."[4] In contrast, John Mearsheimer contends that "the international system creates powerful incentives for states to look for opportunities to gain power at the expense of rivals, and to take advantage of those situations where the benefits outweigh the costs."[5] The key difference between these two viewpoints is that Waltz views states as seeking security by preserving the balance of power (status quo) while Mearsheimer views states as seeking security by aggressively trying to dominate other states and that states do not become "status quo powers until they completely dominate the system."[6] Separating these two different explanations for state motivations is vital to applying and understanding the security dilemma.

The motivation behind state behavior that drives the security dilemma must be understood in order to begin finding approaches to minimizing its effects. Mearsheimer cites Herz's original description of the security dilemma as support for his theory of offensive realism, concluding that the security dilemma shows that "the best defense is a good offense."[7] Herz's original formulation of the security dilemma was:

> Wherever such anarchic society has existed—and it has existed in most periods of known history on some level-there has arisen what may be called the "security dilemma" of men, or groups, or their leaders. Groups or individuals living in such a constellation must be, and usually are, concerned about their security from being attacked, subjected, dominated, or annihilated by other groups and individuals. Striving to attain security from such attack, they are driven to acquire more and more power in order to escape the impact of the power of others. This, in turn, renders the others more insecure and compels them to prepare for the worst. Since none can ever feel entirely secure in such a world of competing units,

power competition ensues, and the vicious circle of security and power accumulation is on.[8]

It is clear from Herz's original formulation of the security dilemma that it has a distinctly defensive character originating from fear and uncertainty over being attacked or dominated by others rather than Mearsheimer's offensive interpretation. Because under Mearsheimer's interpretation of realism only a hegemonic state will act to preserve the status quo, and as a security dilemma describes the interaction between at least two parties, then only one of the two states involved can be a status quo seeking power, the other must be revisionist.[9] In effect, within offensive realism a security dilemma does not really exist, rather it is a security competition since the dilemma arises only when both states are trapped in an undesirable situation. Therefore, when seeking to understand state behavior through the lens of the security dilemma, the pursuit of power cannot be the sole motivation of states.

Making the distinction between states' motivations for power seeking under the security dilemma allows for understanding how states can cope with it. If Mearsheimer's reasoning for why states seek power is correct then under his interpretation of the security dilemma states will have little reason to cooperate to reduce its severity. Since under the security dilemma as described by John Herz, states are seeking security in order to preserve the status quo then there are possibilities for cooperating to lessen its severity or at least minimize its effects.

In his landmark article, *Cooperation Under the Security Dilemma*, Robert Jervis laid out many of the challenges and conditions associated with achieving cooperation under the security dilemma.[10] Jervis framed the possibilities for cooperation in terms of the stag hunt and the prisoner's dilemma depending on specific conditions (see Figure 2.1 below). The prisoner's dilemma and stag hunt are models for social interaction in which there are varying rewards for

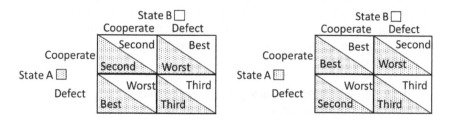

Figure 2.1 The classic prisoner's dilemma and stag hunt.

Source: author's original work.

Notes
The lower left value in each cell represents the relative reward or punishment for state A, while the upper right value in each cell represents the same for state B under the conditions for that cell. For example, within the prisoner's dilemma condition where state A defects and state B cooperates represented by the lower left cell, state A's defection is the best option, and so is represented by 'Best.' Cooperation is the worst option for state B and so is represented by 'Worst.'

cooperating and penalties for mutual noncooperation.[11] In the prisoner's dilemma the reward for cooperating exceeds the punishment for defecting, but the greatest reward is reserved for the side that chooses to defect when the other chooses to cooperate. The stag hunt is slightly different in that the rewards for mutual cooperation are the greatest, while unilateral defection by just one side is slightly less rewarding.

Using these models, the fear of being exploited in a scenario where one side chooses to cooperate and the other defects and chooses to compete is the primary driver of the security dilemma.[12] Defection and competition within this framework represent a state's decision to pursue arms or other forms of power that can threaten another state. The degree to which a state can afford to allow another state to defect from cooperation determines the severity of the dilemma. Under conditions where a state can afford for another state to defect, whether because it has the time or space to respond to defection, the security dilemma will be at its least severe. The converse is also true, as the amount of time or space to respond decreases the security dilemma becomes more intense. As a result, the degree of vulnerability of a state to defection helps determine a state's behavior. The stag hunt and the prisoner's dilemma differ in that the rewards for cooperation are greater under the stag hunt while the risk of mutual defection is the same.

While the primary driver of the security dilemma is the risk of defection when one party chooses to cooperate, Jervis also highlights the impacts on the security dilemma of mutual cooperation and mutual defection. In his words

> the main costs of a policy of reacting quickly and severely to an increase in the other's arms are not the price of one's own arms, but rather the sacrifice of the potential gains from cooperation (CC) and the increase in the dangers of needless arms races and wars (DD).[13]

The greater the cost associated with mutual defection, the more likely states are to seek mutual cooperation. This cost of mutual defection, risking it all on an unpredictable and potentially devastating war, explains why mutual defection is not more common under the security dilemma—however, in both the stag hunt and the prisoner's dilemma mutual defection is not the worst choice for both sides and remains an option.

If the costs of mutual defection are high enough then the situation begins to resemble a game of chicken (see Figure 2.2). In international relations, chicken is normally associated with crisis confrontations and military coercion rather than the security dilemma, but the possibilities associated with it must be addressed.[14] The essential nature of chicken is that one side "willfully creates a conflict by challenging the other and threatens to destroy an already enjoyed common interest if it does not get its way."[15] In the space domain, the game of chicken is most likely to manifest itself in one party threatening to use or construct weapons that deny all parties access to the common interest, meaning low-Earth orbit and beyond. Polluting low-Earth orbit easily could be accomplished

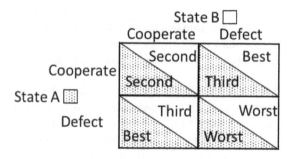

Figure 2.2 Chicken.

Source: author's original work.

through the use of nuclear weapons in orbit, overuse of ASATs generating significant amounts of debris, or simply launching payloads of sand. The challenge is that the threat must be seen as credible and it must be trying to achieve something.

Any power with the technical capability to reach orbit and do significant damage will of necessity have invested substantially in its technical base and must be willing to forgo any future benefit from that investment. Also, it must be willing to incur the combined wrath of all other states with a vested interest in space, which includes most of humanity, a proposition that no state is likely to survive. This situation could only occur under a scenario where a space capable nation is threatened with annihilation from space-based weapons. Short of that scenario, no nation is likely to create the fear necessary to drive a chicken scenario centered in the space domain. What this point demonstrates is that while establishing credibility in chicken involves creating fear, both the stag hunt and the prisoner's dilemma involve creating trust and are more applicable to the security dilemma and the fragile space environment.[16]

The central proposition of the security dilemma is that one state's efforts at seeking security through arms threatens its neighbors, driving them to arm in response. Jervis argues that there are two crucial drivers for this situation, the distinguishability of defensive from offensive weapons, and "whether the defense or the offense has the advantage."[17] If defensive weapons are easily distinguishable from offensive weapons, then a state can arm itself without threatening the security of its neighbors. In addition, when the "defense has the advantage over the offense, a large increase in one state's security only slightly decreases the security of [its neighbors]."[18] The result of this insight is that the balance between offense and defense is a key determinant of the severity of the security dilemma. For instance, if the offense has the advantage and states cannot distinguish between the nature of weapons, then the security dilemma is "doubly dangerous."[19] Alternatively, if the defense has a clear advantage and weapons types and uses are distinguishable, then the situation is stable and the security dilemma ceases to be an issue. This offense–defense balance can drive status quo powers to act aggressively if offense dominance exists or it can encourage cooperative behavior if defense dominates.[20]

In *Causes of War*, Stephen Van Evera even argues that the offense–defense balance can act as the centerpiece of a separate theory of international relations within the school of realism.[21] The core of his argument is that shifts in the offense–defense balance, real or perceived, have a large effect on the risk of war because these calculations drive "policymakers' estimates of relative power."[22] The result is that when conquest is easy war is far more likely. A perception of power imbalance, coupled with the ease of conquest in an offense-dominant environment, creates fear. This fear forces states to seek increased security through alliances, arms control agreements, or through the accumulation of arms.

While the offense–defense balance is a useful concept, relying on it as the key independent variable for a theory of state behavior has proved difficult. Attempts to apply the concept run into difficulties determining the exact balance. This gives the theory marginal utility as a stand-alone concept when substantial uncertainty existed in the single independent variable.[23] This is especially true as the perception of the balance prior to conflict is the defining factor in determining behavior as opposed to the reality, which could lead to *post hoc*, after the fact, justifications when analyzing historical case studies. A key goal of theories in international relations is to achieve parsimony, meaning economy of variables, without losing explanatory power and the offense–defense balance fails in this regard. While a significant influencing factor on the security dilemma and a component of this research, the offense–defense balance is not the sole influencing factor on state behavior and as a result is too parsimonious to stand as an independent theory.

Each of the realist theories of international relations discussed above have the concept of power and its impact on the security dilemma at the heart of their explanation for state behavior. Waltz's structural realism seeks to describe the behavior of states in the international system as stemming from power balancing and uses the relationship between the security dilemma and power to support his conclusion that bipolar international systems are more stable than multi-polar ones.[24] Offensive realism, as described by Mearsheimer, places the security dilemma at its heart even though when starting from the assumption that states are power maximizers it is less of a dilemma than an unavoidable competition. Defensive realists, which includes Van Evera, are more optimistic about international affairs believing that the defense is frequently prominent and consequently this reduces the severity of the dilemma. What each of these theories has in common is that they judge behavior based solely on material variables that contribute to power. Each theory assumes that uncertainty about other states' behavior prevents them from differentiating among potential adversaries so they must make assumptions about others' preferences, whether it be as security seekers bent on power balancing, preserving the status quo, or in maximizing power.

None of the theories mentioned above explicitly accounts for the uncertainty about other states' intentions that is at the dilemma's core.[25] These theories do not account for the information that a state has, or thinks it has, about an adversary state's motives as an independent variable. The core desires and behaviors of all states are established in advance for each theory as a starting point. From that starting point, deductions are made about state behavior. This lack of explicit consideration of the information available to states is a weakness in all these theories. Jervis described some of the issues associated with information available on state behavior type in *Perception and Misperception in International Politics* when discussing the psychological aspects of the spiral model in relation to the security dilemma, but did not provide a usable theory.[26] Charles Glaser later made a more successful effort to address this problem in *The Security Dilemma Revisited*, in which he argued that states communicate information about their motives through the number and type, offensive or defensive, of military forces that a state acquires.[27]

In his article, Glaser identified two variables in addition to power that are vital to understanding state behavior in a security dilemma, motive and information. However, it was not until recently in his book *Rational Theory of International Politics: The Logic of Competition and Cooperation* that he fully fleshed out this nascent realization and formally incorporated these two additional variables into his theory of rational strategic choice that guides this research.[28] The first new variable added is motive, which describes what the nature of a state is: a security seeker, a greedy state, or a hybrid of the two. The hybrid state has both greedy and security seeking motives or has neither and is unmotivated. This leads to four possible combinations of state motives as shown in Figure 2.3 below.[29]

Glaser also identified that a state's motives alone cannot adequately explain its behavior. Both a greedy state and a security seeking state can behave in a similar manner depending on the condition of the international environment that they find themselves in. A security seeking state may be extremely insecure and so willing to incur the large costs associated with arming and fighting in order to expand and ensure its security.[30] To an external observer this behavior would look identical to the behavior of a purely greedy state acting to upset the status quo. In these two situations states with different motives took the same actions. The difference between these two states is their intentions, or more accurately their observed behavior. The state with security seeking motives pursued expansionist policies to ensure its security within the status quo. Similarly, a greedy

		Greedy	
		Yes	No
Security	Yes	Greedy	Security Seeker
Seeking	No	Purely Greedy	Unmotivated

Figure 2.3 State motivations.

Source: Charles L. Glaser, *Rational Theory of International Politics* (Princeton: Princeton University Press, 2010), 37.

state may be deterred by its adversaries and so display status quo behavior. Intentions are important since states demonstrating the same external behavior would respond differently to strategies designed to appease them and encourage cooperation. An insecure security seeking state would respond positively to an adversary who pursues policies to reduce its insecurity by forgoing arming and fighting whereas a purely greedy state would not.[31] Including intentions in a theory accounts for states that display the same behavior, satisfied with the status quo or not (revisionist), despite having different motivations. This interaction of motives and intentions is shown in Figure 2.4 below.

According to Glaser's theory the other new independent variable necessary in determining behavior under the security dilemma is information. Specifically, "what the state knows about its adversary's motives and what it believes its adversary knows about its own motives."[32] This differs from the earlier structural theories discussed above which treated the uncertainty about states' motives as a static assumption, much like the rational actor assumption that Glaser retains. Instead, information about the level of uncertainty of states' motives becomes a variable for both parties. This does not mean that uncertainty can be eliminated, if that was so then the security dilemma would not exist. Using the information variable a state might believe with a high level of certainty that an adversary is a security seeking state and so influence it to pursue cooperative policies with only minor levels of hedging. In contrast, if a state was highly uncertain that an adversary was a security seeking state then it might decide that pursuing cooperative policies was too risky.

The other half of the information variable is what a state believes its adversary knows about its own motives. This reversal is necessary because it can lead to reaction and overreaction under the security dilemma. If state A believes that it is obvious to an adversary that it is a status quo security seeker then if the adversary, state B, continues to build up arms it must be a greedy revisionist state. However, the truth may be that state B does not see state A as a security seeker, or has a high level of uncertainty about state A's true intentions and so pursues a competitive policy to protect itself. This sequence of misperceptions was described by Jervis, but Glaser fully incorporates it into a functional theory.[33] The remaining independent variable in Glaser's rational strategic choice theory is a variation on the traditional power variable.

The quest for power to provide security from others' power is central to Herz's original formulation of the security dilemma.[34] Jervis recognized that a

| | | Intentions/Behavior | |
		Status Quo	Revisionist
Motives	Security	Secure or deterred	Insecure and not deterred
	Greed	Deterred	Not Deterred

Figure 2.4 Intention and motivation.

Source: modified from original in Charles L. Glaser, *Rational Theory of International Politics* (Princeton: Princeton University Press, 2010), 39.

way to describe power was in terms of the offense–defense balance and Van Evera took this thread to the extreme and tried to make it stand on its own as the independent variable in his own theory.[35] Glaser argues that the material variable's impact on the severity of the security dilemma is a function of the state's power, multiplied by the offense–defense balance.[36] Glaser defines power as the "ratio of states' resources that can be converted into military assets."[37] This definition can be understood as referring to military capability versus purely military assets as many normally non-military assets have military capability, such as commercial communications satellites. The offense–defense balance itself is defined as the ratio of the cost of offensive forces to defensive forces required to take territory.[38] In effect, in a situation where a known greedy state is significantly more powerful than a weaker security seeking adversary, yet the defense has the advantage, the weaker state will find it possible to balance the material power of the stronger state and the severity of the security dilemma will be moderate. If the offense has the advantage in the same situation then the weaker state is at a significant disadvantage and the security dilemma will be very severe.

The severity of the security dilemma is therefore determined by a combination of material, motive, and informational variables working together within the rational strategic choice framework. The explicit combination of these three variables explains why states sometimes pursue what would otherwise be seen as irrational policies under traditional realist structural theories. Since more than material factors impact the security dilemma in Glaser's theory as independent variables, a state might pursue cooperative policies when the material factors alone would point toward competition and vice versa.[39] When all of these variables are combined the severity of the security dilemma can be described by Figure 2.5 below.

For the purposes of this research, the value of determining the type and severity of the security dilemma that a state faces is to reach a decision on whether to pursue competition or cooperation in space. This choice is influenced by the three variables mentioned above and this makes it a complex and difficult decision that is not entirely confined within the space domain. Defaulting

		Information about Motives		
		Likely Greedy	Equally Likely Greedy or Security Seeker	Likely Security Seeker
Material Conditions	Offense Advantage	Very Severe	Severe	Moderate
	Defense Advantage	Moderate	Mild	Essentially Eliminated

Figure 2.5 Severity of the security dilemma.

Source: Charles L. Glaser, *Rational Theory of International Politics* (Princeton: Princeton University Press, 2010), 87.

toward cooperation seems to be the best option for escaping the security dilemma, but this is not always the case. Competing by pursuing arms can sometimes be the best, or optimal, choice for preventing war or at least for decreasing the probability of conflict. When facing a greedy state in an offense-dominant environment the optimal choice for a state is to pursue arms and seek to deter its adversary.[40] Of course war is still more likely when a security seeking state is faced with a greedy one, but choosing not to arm would only further increase the likelihood of conflict by encouraging the greedy state to take advantage of weakness and so the logic of deterrence becomes dominant. The pursuit of arms for security is then the optimal choice under these conditions because cooperation would be dangerous.

In contrast, a sub-optimal arms race can generate the insecurity that a state is attempting to avoid when cooperation would be a better option.[41] Sub-optimal arms races can create dangerous uncertainty and lead to conflict. At best sub-optimal arms races are a waste of resources that a state would be better off investing elsewhere, particularly in the space domain where changes in technology can rapidly offset the advantages gained through arms racing.[42] The overall logic of choosing to pursue arms or cooperation is shown in Figure 2.6 below. Both the upper left and lower right quadrant are optimal choices. In the upper left a state's best choice was to arm to deter a greedy adversary. In the lower right both states sought cooperation and correctly did so. In the upper right quadrant is the classic security dilemma where a state chose to arm when it did not need to with the attendant negative impacts on its security as other states responded. The other sub-optimal choice is when a state chose not to arm even when faced with a hostile adversary which leaves the cooperating state dangerously vulnerable. War is always possible with or without arms races though it is the uncertainty inherent within the security dilemma that drives these sub-optimal choices that "make war unnecessarily likely."[43]

Glaser's theory provides a significant increase in explanatory power over other realist theories by formally incorporating motive and information and by parsing the difference between state motive and intent, but it is not without shortcomings. While incorporating these additional variables provides additional explanatory power it leads to a lack of parsimony that makes this theory difficult to apply. Glaser is aware of this problem and argues that if a simpler theory with

| | | State Should Have Armed/Raced | |
		Yes	No
State Armed/Raced	Yes	Optimal Arming: Necessary Races	Suboptimal Arming: Dangerous Races
	No	Suboptimal Restraint: Dangerous Cooperation	Optimal Restraint: Desirable Cooperation

Figure 2.6 Quality of arming decisions.

Source: Charles L. Glaser, *Rational Theory of International Politics* (Princeton: Princeton University Press, 2010), 233.

fewer variables has equal explanatory power then the added complexity of his theory would be unnecessary.[44] For example, excluding the information variable greatly reduces the prospects for peaceful relations between great powers and so even if a variable is not always key to explaining behavior in one scenario it may be in another.[45]

A further issue with the theory is that it fails to account for irrational or sub-optimal decision making by states. While this is true, it does provide an optimal baseline from which deviations can be detected that can then be explained by a variety of other factors not accounted for in this theory, such as domestic politics or organizational dynamics. For the purposes of this research this shortcoming is less important since the goal is to determine what the rational decision in space should be, not to explain previous sub-optimal decisions though that may become apparent later.

Despite the complexity of rational strategic choice theory Adam Liff and John Ikenberry in *Racing Toward Tragedy* argue that it needs to include three further refinements.[46] First, they argue that not all arms build-ups are driven by security dilemmas, though they can exacerbate it. States may pursue military power as the result of domestic political interactions or for the sake of status and prestige. If internal factors are driving the accumulation of military power then there may be no mutually acceptable bargains to ensure state security that would prevent the accumulation of power.[47] Second, military power is not just a process of acquiring additional assets, it is also a process of increasing military capability. Qualitative increases in capability such as force modernization and enhancement efforts, changing force posture, and the formation of mutual defense pacts and other alliances could be perceived as power enhancing by other states and so impact the security dilemma. The third additional factor considers how leadership rhetoric and political statements generate insecurity when they are interpreted as offensive and threatening by other states.[48] Reflecting the logic of the information variable, these statements can be intended as displaying status quo security seeking motives and intentions while being interpreted as greedy and revisionist. The first of these three refinements reflects an effort to get at the internal domestic factors that Glaser explicitly excludes from his theory. The other two refinements reflect non-traditional ways to measure the material and information variables in Glaser's theory rather than in true adjustments to the core theory.

In answering the question posed of "what is the security dilemma?" a rich and deep body of explanatory theory is uncovered stemming from the seemingly simple observation, "that many of the means by which a state tries to increase its security decreases the security of others."[49] Glaser demonstrates that more than variations in power and the offense–defense balance influence the security dilemma and a state's optimal strategy. By including a state's information about other states' motives as well as its assessment of other states' information about its motives, cooperation rather than competition will be in a state's best interest under a much broader range of circumstances than material variables alone would dictate. While a security seeking state is still likely to compete, there are many circumstances where it might choose to cooperate when more than just the material variable is included. This leads to the question of how states can cooperate under the security dilemma and what the limitations on cooperation are in the anarchic environment that gives rise to it.

Cooperation and reassurance in a security dilemma

Cooperation between states is not an easy task as the world exists in an anarchic state lacking any common government with the power to enforce rules. International bodies with behavior-regulating goals do exist, the United Nations (UN) being an obvious example. These bodies simply lack the enforcement power to regulate behavior among great powers. Despite this lack of an external enforcement mechanism cooperation does occur. Once cooperation takes place, it can become sustainable as states are less willing to sacrifice the benefits of cooperation already achieved, at least until a state is convinced another state is willing to defect from the status quo arrangement. This willingness to cooperate is based on trust, whereas building security through competition is based on fear. The challenge is to determine how states signal a willingness to cooperate and build trust in a world driven by the uncertainties of the security dilemma. The following discussion outlines the advantages, disadvantages and challenges associated with achieving cooperation within the security dilemma.

The key to building trust lies in determining the motive and intent of an adversary state. A state that wants to cooperate rather than compete cannot be sure of its adversary's motives or the adversary's view of its motives. However, within the framework of the security dilemma a state can use the type and nature of the arms that it builds to signal its nature, whether as a greedy revisionist state or a security seeking status quo state and help resolve any misperception that other states have about its intentions. Though not all signals are equal; the greater the cost in security associated with sending the signal, the more likely it is to be interpreted as a signal of benign intent.[50] Efforts at signaling that involve little risk are likely to be dismissed as feints and will not achieve the intended goal of changing an adversary state's view of the signaling state's intentions.[51]

In descending order of cost, a security seeking state can indicate its benign intentions in three primary ways. First, through arms control agreements that limit offensive weapons, since greedy states are more likely to pursue offensive weapons than a security seeking state.[52] Second, a state may unilaterally adopt a defensive posture, though this is dangerous in an offense-dominant environment since a state will have to spend lavishly to ensure security.[53] Although this lavish defensive spending will have the added benefit of further reinforcing a state's signaling of its status as a security seeker. Finally, the most dangerous policy for a security seeking state is to exercise "unilateral restraint."[54] This means that a state can pursue a policy of unilateral disarmament by reducing its military capability to a point such that it is no longer capable of adequate deterrence or defense. This policy is dangerous because if a state has misinterpreted its adversary as a security seeker when it is, in fact, a greedy state, this policy can encourage that adversary to take advantage of real or perceived weakness and lack of resolve. While seemingly simple, each of these strategies can succeed or fail based on a complex series of actions and reactions.

Pursuing arms control agreements seems like a straightforward way of signaling cooperative intent while incurring a low cost to security since no disarmament need take place before a potential opponent also begins disarming. This low cost in pursuing arms control agreements comes at a price since it also makes them the weakest signal of intent, and therefore the least likely to be successful in transmitting intentions. There are two main categories of arms control that a state may pursue.[55] First, when offensive and defensive capabilities are easily distinguishable a state can seek agreements to limit offensive weapons. Examples of this type of arms control agreement include the 1972 Anti-Ballistic Missile (ABM) treaty, which under the inverted logic of nuclear deterrence was viewed as an offensive weapon. So, limiting an ostensibly defensive weapon actually ensured the continued viability of each state's defensive nuclear deterrent. Alternatively, a state can seek to limit the total size or capability of forces when offense and defense are not readily distinguishable. The Washington Naval Treaty of 1922 where several great powers agreed to limit the number of battleships and aircraft carriers while not limiting the number of cruisers, destroyers, or submarines since the former could be used for offensive or defensive purposes, while the latter were viewed as primarily defensive platforms.[56]

A significant challenge to successfully reaching arms control agreements is the risk of cheating. The primary method of cheating prevention is through inspection and verification. When the difficulty of inspection and verification is high and the advantages to be gained through cheating are large, arms control agreements are risky propositions.[57] Unless a state is very confident that the motives of the state proposing the arms control agreement under these conditions are security seeking, then it will be unlikely to accept the proposal. A state must also consider the behavior of third parties. If two security seeking states agree to arms control while a third greedy state does not, then those security seeking states have made themselves vulnerable. This multi-party behavior makes arms control agreements extremely difficult to successfully negotiate and helps explain why they are not more common in the international environment.

Not all states seek arms control agreements because they are benign; they may believe that arms control can provide them with some other advantage. A state might be uncertain about its ability to win an arms race and so seek to enter an arms control agreement that accepts the status quo rather than risk further upsetting the balance of military power.[58] Reasons for uncertainty in winning an arms race could depend on external or internal factors. A significant disparity in potential power could make it impossible for the weaker state to outcompete the stronger state. Domestically the economic burden of pursuing competitive policies may also be unaffordable; as economic potential is converted from growth to building arms, each additional increment of arms build-up becomes more difficult and costly.[59]

US and Soviet interaction following the launch of Sputnik is an example of the difficulties that arms control agreements face when initiated after a shift in the balance of power. Following the launch, US President Eisenhower reached out in a private letter to Soviet Premier Nikolai Bulganin where he proposed "that we agree that outer-space should be used only for peaceful purposes; we face a decisive moment in history in relation to this matter."[60] A week later, Soviet leader Nikita Khrushchev responded publicly in a speech in Minsk stating that these peace proposals "mean that they want to prohibit that which they do not possess."[61] This rejection by the Soviet leadership of a sincere peace proposal frustrated Eisenhower who was unable to identify the cause of the public rejection of his efforts to avoid a costly and dangerous arms race.[62] Though in the aftermath of Sputnik's impact on world perception of the power balance between the USSR and the US it is understandable that the Soviets would be unwilling to negotiate away a newly acquired technological advantage. This example highlights how the impact of technology shifts on the perceived balance of power can drive a previously dominant power to seek arms control. No matter how sincere the offer from the signaling state under these conditions, the receiving state is unlikely to interpret the effort as anything but weakness.

The offense–defense balance also can play a significant role in the eagerness of states to pursue arms control agreements. If new technology signals a shift toward the offense, security seeking states will no longer feel secure and so will be highly motivated to pursue arms control.[63] If arms control fails then two neighboring security seeking states that are uncertain about each other's motives may have no choice but to pursue arms races since they cannot agree to cooperate on arms limitations. Alternatively, in a defense-dominant environment, arms control agreements are less necessary as even security seeking states confronted by greedy ones are more secure, so states have little motivation to spend the time and effort pursuing them.[64]

The internal structure of a state could also influence trust in arms control agreements. States may have a historical record of keeping or breaking treaties or internal governance structures that make entering or leaving a treaty difficult. States also have varying degrees of internal transparency which could influence trust levels.[65] The US is relatively transparent which allows other states to observe and predict, or at least attempt to explain, national behavior. Meanwhile, China is notoriously secretive, with the internal politics in the politburo remaining opaque to both domestic and international observers. On balance, very low levels of domestic transparency can allow for accurate signaling of intentions almost as well as very high levels of transparency. In a somewhat transparent state, the noise of domestic politics can overwhelm diplomatic signals and confuse an adversary with too much information, so being at either end of the transparency spectrum allows for a more accurate interpretation of signals.[66] No matter the level of transparency within the observed state, any information received will be interpreted and filtered based on the observing state's existing mental frame of that state's values, culture, and political system.[67] These internal factors generally fall outside the scope of realist international relations theories.

Though when available the level of information available about other states' domestic decision-making processes can influence a state's interpretation of their motives and so impact the decision making involved in reaching cooperative agreements like arms control.

In addition to pursuing arms control to signal intent, a state may choose the costlier approach of unilaterally adopting a defensive posture. As noted earlier, the offense–defense balance can have a significant impact on the risk involved in adopting this strategy. For instance, in an offense-dominant environment, a state will have to spend lavishly on defense to ensure continued security.[68] An advantage of this approach to signaling intent is that a conciliatory gesture of this type is less likely than attempts at arms control to be interpreted as a cynical effort at manipulation to gain advantage through cheating. However, the effectiveness of the approach is dependent on the ability to distinguish offense from defense as well as on the offense–defense balance. If the level of distinguishability is low, then the effectiveness of adopting a defensive posture could be lost or open to different interpretations than the signaling state intends.[69] Also, if the offense–defense balance significantly favors the defense, then the risk of adopting this approach is low and so less costly, and less costly signals are typically less effective at reassuring an adversary that a state is trustworthy than ones that are costlier.[70] The need for sending a signal and reassuring an adversary is also low since when the defense dominates the security dilemma is at its least severe. Adopting a defensive approach when the offense has the advantage and is distinguishable from the defense is both expensive and costly and therefore sends the clearest signal of intent, but it carries extremely high risk, so a state is unlikely to adopt this approach. The remaining combination of variables is when the offense–defense balance does not greatly favor either offense or defense yet the two can be distinguished. Under these conditions a state has a choice of whether to adopt offensive or defensive weapons and that choice can signal intent, since a greedy state will be unlikely to choose defensive weapons.[71] The signal of intent that a state sends under these conditions is less costly than adopting a defensive posture when offense dominates, but it is also more reasonable as a strategic choice when a state is uncertain about its adversary's motives. In sum, the circumstances in which a state can pursue policies that involve unilaterally adopting a defensive posture to break out of an arms race spiral without creating an unacceptable level of vulnerability to attack are rare (see Figure 2.7).[72] Adopting a defensive posture when the offense–defense balance is neutral carries less risk, but still communicates intent and provides reassurance and so this option sits in the middle ground of signaling between pursuing arms control agreements and adopting a policy of unilateral disarmament.

The final, and costliest, method of signaling and reassuring an adversary is through unilateral disarmament. This action is fraught with danger so a state must be confident of both its immediate adversary's intentions and of all other potential adversaries' intentions before pursuing this course of action. Understandably actions along these lines are extremely rare. An alternative to completely disarming in the face of an uncertain adversary is to pursue incremental

Offense–Defense Differentiation

		Yes	No
	Defensive Advantage	Large reductions in defensive forces are necessary to reveal benign motives. Large concessions can still increase a state's vulnerability.	Signals that decrease a state's ability to attack also decrease its ability to defend. Large reductions necessary to reveal benign motives. Large concessions increase a state's vulnerability.
Offense– Defense Balance	Offense– Defense Balance Neutral	Benign states can reveal motives without increased vulnerability because: Differentiation allows states to choose clearly defensive forces. Defensive forces are as effective as offensive forces, so benign states are not at a disadvantage if they choose defense.	Signals that decrease a state's ability to attack also decrease its ability to defend. Moderate reductions in the number of forces will reveal benign motives. Moderate concessions will also increase a state's vulnerability.
	Offensive Advantage	Small Limits on offensive forces sufficient to reveal benign motives. Small Concessions increase a state's vulnerability.	Signals that decrease a state's ability to attack also decrease its ability to defend. Small reductions in the number of forces will be sufficient to reveal benign motives. Small concessions increase a state's vulnerability.

Figure 2.7 Offense–defense, reassurance, and vulnerability.

Source: Evan Braden Montgomery, "Breaking out of the Security Dilemma: Realism, Reassurance, and the Problem of Uncertainty," 169.

disarmament. Rather than completely disarm, a state may make a substantial reduction in its military capabilities and then observe its adversary's reaction to this conciliatory action. A complimentary reduction by its adversary to its arms in response would, in theory, result in a positive spiral of arms reductions. The example of Soviet Premier Nikita Khrushchev's troop reductions in the 1950s highlights the difficulties associated with incremental reductions.[73] Between 1955 and 1960 the Soviets reduced their Army in several publicly announced cuts by more than 3.3 million men, a reduction of more than half its total conventional force structure. Khrushchev felt that these actions had "convincingly proved our peace-making nature, and we will continue to prove it ... we will continue to reduce armed forces unilaterally."[74] While these actions did reduce the overall size of the Soviet military, the US did not interpret them as having any significant impact on Soviet military capability.[75] The absolute reduction in the size of the Soviet military was accompanied by efforts at force modernization, the launch of Sputnik, and the growth of the Soviet ICBM (inter-continental ballistic missile) force that greatly reduced the security cost and resulting effectiveness of Soviet signaling. In addition, the US national security community was still operating under the mental framework of NSC-68 which framed the US-Soviet relationship as an implacable ideological struggle.[76] This meant that any Soviet signals attempting to demonstrate benign motives and security seeking intent had little hope of success. The Soviet example demonstrates that the message sent by incremental disarmament can be lost or misinterpreted when accompanied by other actions that may signal a shift toward more economical weapons, especially when an adversary is already certain that the signaling state has hostile motives.

Signaling is about more than demonstrating the signaling state's nature. The other half of the equation is observing your adversary's response to your signal to determine their motives and intentions.[77] This adds additional complications and difficulties to the information variable in the security dilemma. If the signaling state is confident that it has signaled benign intent, yet the target of the signaling does not reciprocate, then the signaling state may conclude that its adversary is hostile. This uncertainty and misperception can be self-reinforcing when following failed overtures at signaling a benign state concludes its adversary has malign intent and pursues competitive policies. The adversary state, which may have completely missed or simply misinterpreted the cooperative signals, then only sees the competitive policies and the tragic spiral of misperception and competition that can lead to conflict ensues.

The three primary strategies for signaling benign intent and reassuring an adversary, pursuing arms control agreements, shifting toward a defensive posture, and unilateral disarmament all have challenges associated with them. The most significant challenge is that a state can never be certain of its adversary's intent. George Downs and David Rocke demonstrated in *Tacit Bargaining and Arms Control* that this uncertainty is key, as "a state will rarely be certain enough about an opponent's response to make a large cooperative gesture, and the opponent will rarely be trusting enough to respond enthusiastically to a small one."[78] The uncertainty that rests at the heart of the security dilemma makes it very difficult to escape. A state is always uncertain about other states' intentions to some degree, and when the power differential between two states is roughly comparable, states are rarely willing to undertake the costly signaling necessary for reassurance. Despite this uncertainty, cooperation is possible within the framework of the security dilemma since not every interaction between states leads to an adversarial relationship.

Measuring the security dilemma

The preceding discussion demonstrated that the three primary variables of material, information, and motive influence the security dilemma and determine how effectively cooperation is achieved within it. Accurately measuring the various aspects of these variables is therefore vitally important in determining what the optimal strategy should be for a state within the rational strategic choice framework. The challenge is that while some aspects of these variables can rely at least partially on objective measurements, primarily the material factor, it is through subjective interpretation of various pieces of information that states can best determine the nature and severity of the security dilemma. This section will explore in detail how each of the three primary variables can be measured when applied to the space domain, beginning with the relative certainty of the material variable and progressing to the much more difficult task of determining motive.

Within the rational strategic choice framework, the material factor is determined by a state's power as well as by offense–defense variables. Power lies at the heart of many realist theories of international relations, and so has a long history of attempts to define and measure it, though there is still no universally accepted approach. Within rational strategic choice theory power is defined as "the ratio of a state's resources that can be converted into military assets."[79] Using this definition, power is determined by a combination of a state's latent and explicit military power. Latent power is the various socio-economic factors that help build military power.[80] This can include variables such as population size as well as the strength and size of a nation's economy. Including population size in latent power calculations seems obvious since a large population is required to man a large military, however without a robust economy very little of a state's population can be effectively mobilized, armed, and equipped for modern war. Many nations today and throughout history have had large populations, but fallen far short of great power status. The conclusion is that while "a large population does not ensure great wealth … great wealth does require a large population."[81] Despite this conclusion, it is difficult to entirely separate population size from considerations of a state's latent power.

Since the best-available objective measurement of a nation's overall economic output is Gross Domestic Product (GDP) it is the primary factor for determining latent power. The fact that GDP measures the sum of consumption expenditures, business investment, and government spending less the value of imports means that population size still plays a substantial role in determining a nation's GDP.[82] An economy with greater levels of industrialization and productivity per worker has significantly more productive capability that is convertible from latent to military power in modern warfare than one that has a large percentage of its population engaged in subsistence farming.[83] Therefore, latent power must be more than just total economic output. It is relatable to measuring the available "slack" in an economy that has the potential to support military power, so utilizing GDP alone is not adequate for measuring latent power. Since modern warfare is a highly technical endeavor, especially outside of the land domain, the level of industrialization and technical capability must be considered in addition to GDP.

Two available measures that can be used to measure a state's level of industrialization and technical capability are the level of urbanization and the total manufacturing output of a state. The level of urbanization can serve as a proxy for how much of the population is directly or indirectly engaged in agricultural activities and so are unavailable for industrial activity. Urban populations also often have higher levels of education and technical skill that are required by modern militaries. Finally, manufacturing output is a useful addition as not all urbanites are engaged in industrial activities; many modern urban economies are service-centric and have little latent military value. These two factors taken together with GDP will serve as the primary measures of a nation's latent power, though some additional considerations must be taken into account when measuring latent power in the space domain.

Within the space domain, latent power includes more than just overall economic size; it also includes the size of a nation's space industry that can be converted to military purposes. The amount and ability of a nation's domestic engineering capacity to produce satellites and rockets, the number of annual space launches, and the size of its commercial space industry can all be considered latent power in the space domain. These modifiers mean that a state that under traditional economically based measures of latent power might be no match for another state may in fact be much closer to a near-peer in the space domain. This refinement of latent power regarding the space domain is necessary due to the substantial amount of time it takes to grow and mature a space industrial base capable of contributing to military power. This is time that great powers will not have in a world with nuclear weapons where conflicts are highly unlikely to last long enough for a state to alter its latent space power in any meaningful way.

The other half of the power variable is explicit military power which is not always directly proportional to latent economic power. While a nation's economy is the basis for its military strength, not all wealthy nations convert an equal share of their economy into military power. The reasons why a state chooses to avoid maximizing its military power vary. A state may not perceive that it has any powerful nearby external enemies that drive an arms build-up, or a state might face domestic political spending priorities that compete successfully against investing in additional security. Invariably, all government spending must be restrained to some degree to avoid harming the underlying economy that is the source of a state's wealth and military power.[84] How much to spend on military power is balanced against other priorities, though states should obtain enough of it to feel secure in the face of known threats. Finding this balance is extremely challenging as measuring military power can be surprisingly difficult, a major source of the insecurity that feeds security dilemmas.

Military power is typically defined as the size of a state's military forces, though size does not equate to capability.[85] Military capability can depend on the quality and suitability of doctrine, training, and equipment. The ability to project military power also matters. A large army is nearly meaningless in terms of offensive military capability if it is unable to reach an enemy that lies across a large body of water. Technical capability and training also can tip the balance of power between two equally sized forces significantly. In 1991, the Iraqi military fighting on its own soil was overmatched completely by a comparably sized force that possessed far better training and technical capability. Doctrine and method of employment cannot be ignored as important factors in military capability either. For example, an arguably more technically advanced French military suffered defeat in a matter of weeks by Hitler's Germany using new tactics and organizational structures that allowed the Wehrmacht to outmaneuver them. The impact of these factors is difficult to judge in advance, making it impossible to obtain an absolute measure of a nation's military power though an approximation is possible.

Two factors can be considered as a useful basis for measuring military power, expenditure on the military adjusted for purchasing power as well as the total percentage of GDP that a nation is devoting to military expenditure. These factors cannot account for inefficiencies in defense spending, but they can serve as a useful relative measurement of military power. Measuring a state's effective military space power also will rely mainly on cost data, but several other factors need to be considered to gain a more accurate measure of military space capability.

In the space domain military power is measured by the quantity and quality of a state's entire space infrastructure, from its launch capability to the number of satellites on orbit it can leverage to support military operations. It is more than a simple numerical comparison of explicitly flagged military satellites; it is a measure of how many space resources a nation leverages to support military operations. Since commercial capabilities in space also can be used for military purposes, more so than in any other domain, a degree of separation is necessary. Those space capabilities, explicitly military or nominally commercial, that a state uses in peacetime represent active military space power. Meanwhile, those non-military platforms that a nation could use for military purposes if it needed to represent readily available latent space power. With this differentiation in place, military space power can be measured based on total expenditures in space as well as on latent capabilities within a state's existing commercial architecture.

The remaining half of the material factor is the offense–defense variable. Discussed earlier, the offense–defense variable consists of two factors; the balance between offensive and defensive capabilities and the degree of differentiability that exists between them.[86] For the purposes of this space-centric research, the offense–defense balance is the ratio of the cost of offensive forces versus the cost of successfully defending against those forces without significant degradation of capability. This definition removes the troublesome reference to territory present in most definitions that is irrelevant in the space domain. Using this relative method of measurement is not without subjectivity as the cost of attacking versus the cost of defending must be categorized in subjective terms such as low, very low, or extremely high. Complicating this subjectivity, the process and methods of measuring the offense–defense balance are extremely controversial with some arguing that it cannot be done.[87] Despite this ambiguity, the perception of offensive or defensive advantage does play a significant role in determining states' arming choices and behaviors and so is included in rational strategic choice theory.

The primary factors that are used to determine the offense–defense balance are geography and technology.[88] Geography is usually the least controversial factor affecting the offense–defense balance between states.[89] If two states share a mountainous border that is difficult to cross or are separated by an ocean, then defense would have the advantage in any conflict between those states. In space, unlike on Earth, all states suffer from the same constraints imposed by orbital dynamics, so geography impacts all nations equally. Some might argue that access to launch sites near the equator, which allow larger masses to reach geosynchronous orbits for a given mass of fuel, represents a geographic limitation that may favor some states over others. However, the difference is small enough that it is not a driving factor in the offense–defense balance. For example, there is only a 20 percent gain in mass to geosynchronous orbit for a Soyuz launching from Baikonur, Russia (46 degrees North latitude), versus launching from Kourou, Guiana (5 degrees North latitude).[90] While this difference is economically significant, it is not enough to impact the balance of military power in space between great powers and so can be disregarded.

The second factor that impacts the offense–defense balance in space is technology, which is developing rapidly and so is the factor that will shift the balance over time the most. Within the security dilemma framework, a perception of offense dominance can have a large impact on its severity. Historically most perceptions of offense dominance turn out to be significantly overstated and are very uncommon.[91] For the moment, the balance in space is commonly perceived to favor the offense, though technology is potentially shifting this balance toward the defense. Despite this potential shift, the current perception of offense dominance remains firmly rooted and is driving significant and potentially unnecessary reactions. The shifting nature of the offense–defense balance in space is discussed in detail later.

Determining the degree of differentiation between offensive and defensive weapons is a further challenge that is becoming increasingly difficult in space. Inability to clearly differentiate weapons systems into categories of offensive and defensive has always presented problems, especially to attempts at arms control. Salvador de Madariaga, a Spanish diplomat, famously stated that "a weapon is either offensive or defensive according to which end of it you are looking at."[92] This statement highlights that the purpose of many weapons systems is dependent on how a state uses them and not on the intrinsic nature of the weapon. Even those that are explicitly defensive, such as fortifications, could be interpreted as supporting offensive purposes when they are used to free-up mobile forces for duty elsewhere. The space domain does not escape this confusion. Since space is primarily a domain for transmitting and gathering information, even a communications satellite could be construed as an offensive platform when used to support terrestrial offensive operations. To help alleviate this confusion of purpose only the role of platforms within the space domain will be considered. Those systems that do not explicitly harm space assets are considered defensive while those designed to harm or interfere with space assets are offensive. For example, an ASAT or a ground-based laser system is an offensive system, even if their use could be part of a defensive strategy, though this differentiation still does not entirely solve the problem of distinguishability.

The deployment of on-orbit repair and maintenance systems designed to service satellites or remove debris presents a dilemma. These systems are ostensibly designed for peaceful purposes, but a satellite with a repair arm or a net for catching debris could easily be used to damage or destroy a satellite. Unlike those explicitly offensive weapons categorized above, the purpose of these systems is dependent on how a state uses them. For the time being this challenge is mitigated by the fact that only a handful of systems on orbit fall into this category. In the future, as more of these systems are launched, they will become a more significant issue and represent a substantial barrier to attempts at arms control agreements in space.

The above discussion demonstrates that the difficulties associated with measuring the various components of the material variable are substantial, yet it is still the most suited of the three variables to objective, rather than subjective assessment. The remaining two variables, information and motive, are far harder to measure and rely entirely on subjective assessments. This is surprising given that information and motive are central to the security dilemma, which is itself a critical concept within the international relations field. When reviewing Glaser's theory, Jervis stated that

> the problem [with information and motive variables] is not that Glaser's analysis is not state of the art, but that the state of the art itself is weak; while his discussion of how he thinks information should be used has much to be said for it, there are severe inherent limits here.[93]

The inherent limit that Jervis refers to is the level of ambiguity in interpreting states' behavior.[94] Two analysts reviewing the same information about a state's behavior might reach opposing interpretations of that state's motives and intent.

How then can this confusion over motive and intent be resolved using the information available? One method is to rely on observing a state's military policy as a guide to interpreting its intentions.[95] A state that is increasing its investment in its military, reorganizing for a specific military purpose, or fielding disruptive technologies designed to defeat the capabilities of an adversary can all be indicators of motive and intent. Within the space domain the testing of offensive weapons platforms can send a variety of signals about motive and intent. If that testing is held openly it most likely signals that a state is not greedy since a greedy state would seek to prevent an adversary from gaining too much information about its offensive capabilities in advance of their use. In contrast, a state conducting open tests of offensive space weapons is most likely security seeking and attempting to create a deterrent effect. This is especially the case if the intimidation of any potential adversary is not a reasonable interpretation of their actions. If a security seeking state is pursuing offensive space weapons it is also demonstrating that it is either unsure of its adversary's motives and is assuming the worst, or believes its adversary has hostile intentions. Of course, it also may be difficult or impossible to hide the test which would create confusion over purpose if unaccompanied by clear messaging.

In contrast to clearly offensive systems, secret tests of defensive systems on orbit could represent rational behavior by a security seeking state seeking to pre-serve a fragile defensive advantage when confronted by a greedy state. Though difficulties arise if the system is not purely defensive, such as a co-orbital system that could be used to protect a nearby satellite or threaten other satellites, then the testing of that weapon could send different signals depending on the offense–defense balance. Under perceived offense dominance that system is more likely to be seen as threatening than when defense has a clear advantage. What a state builds and tests within the space domain serves as an indicator of intent and motive though these actions are still open to interpretation.

Completely relying on observation of a state's actions to determine motive and intent can create confusion and lead an observer to a false conclusion.[96] A state can have benign motives that manifest as hostile intent to another state. For example, a security seeking state might conclude that it is facing a greedy one when it is not. Its actions would then be targeted at building up its armed forces, forming alliances, and establishing military bases in strategically vital areas to increase its security. While the state performing these actions sees its own actions as strictly defensive in nature, the opposing state sees actions in line with greedy revisionist behavior and so an arms race is likely to ensue. The risks associated with signaling cooperative intent enumerated earlier might also prevent two states in this situation from effectively signaling their true purpose.[97] This could be true especially when the offense is perceived to have a strong advantage. The end result of this reaction and counter-action is a rational neg-ative political spiral that further reinforces each state's estimate that its opponent is greedy.[98] Finally, two states also can be motivated by security seeking yet have incompatible security needs. This could drive competition and conflict as one state attempts to decrease the relative power of another state by occupying key terrain or building weapons to offset its adversary's capabilities. These situ-ations where different motives and intent manifest themselves in similar actions can confuse an observing party so additional information to explain state behavior is necessary.

Additional information that can shape how actions by a state are perceived can come from statements and rhetoric from military and political leadership. These statements can provide valuable information about each state's motives and how it sees its adversary's motives.[99] The establishment of a new space focused arm of the military for the stated purpose of defense and security is less threatening than when accompanied by provocative language such as dominate

Independent Variable	Method of Measurement		
Material	Power	Latent	GDP; Level of urbanization; Manufacturing output; Domestic rocket production/capability; Number of annual launches; Size of commercial market; Number of commercial satellites
		Explicit	Total military expenditure adjusted for purchasing power; Percent of GDP devoted to defense spending; Total expenditure on military space; Number of satellites on orbit used for military purposes; Launch capability
	Offense–Defense Balance		Ratio of the cost of offensive forces versus the cost of successfully defending against those forces without significant degradation of capability
Information and Motive	Actions		Weapons testing/deployment in space; Military actions outside of the space domain; Pursuit of cooperative gestures such as arms control, disarmament, or defensive postures; Organizational and structure of military forces, Level of transparency
	Rhetoric		Statements by military and political leaders; Doctrine; Official documents

Figure 2.8 Approach to measuring the security dilemma.

Source: author's original work.

and defeat. Though the mixing of messages can create confusion and lead an adversary to simply assume the worst especially if it has a pre-existing narrative associated with that state.

Measuring the severity and nature of the security dilemma requires aggregating a mixture of material and information factors that determine motive and threat. The methods for measuring the variables outlined in the rational strategic choice theory are summarized in Figure 2.8. The material variable is measured through latent and explicit economic and military space power. This is further modified by the offense–defense balance and the level of distinguishability within the space domain. The information variable is open to more uncertainty, but is best determined by analyzing actions as well as rhetoric from political and military leaders. Using these variables within the strategic choice framework, we can attempt to answer the question of whether competition or cooperation is the optimal strategy within the space domain.

The multistate dilemma in space

A final problem that needs addressing is the bias in theories explaining behavior within a security dilemma toward analyzing interactions as being between just two states. This bias is evident in the language used as well as in the acceptance that the impact of geography has on making the interactions between any two states unique. Using this bipolar method of analysis is not without merit. A state will naturally be able to adopt different policies and accept higher levels of risk toward another state when they are separated by thousands of miles rather than when they share a common border. In recognition of this impact, geography and proximity are central to state behavior in realist theories. For example, this tyranny of geography is central to offensive realism where Mearsheimer argues that "the best outcome a great power can hope for is to be a regional hegemon and possibly control another region that is nearby and accessible over land."[100] Rational strategic choice theory suffers from the bipolar bias of geography as well by including it as "perhaps the least controversial of all the factors that affect the offense–defense balance."[101] The problem with existing theories of international relations is that they either implicitly or explicitly account for the impact of terrestrial geography; in Earth orbit, entirely different geographic limitations apply that make simple two-state solutions to strategic problems infeasible.

Orbital dynamics means that all states share a common border in space and so the actions of one state impacts all states. This interdependence is true even in geosynchronous orbit (GEO) where the static position of satellites relative to the Earth's surface might imply otherwise. Even in GEO satellites require station keeping and are not completely stationary relative to the Earth's surface. Just as with lower orbits, any debris generated in GEO can quickly pollute the entire orbit. The unique nature of the space domain means that the strategic choices that a state makes must be suitable for all the nations with a presence within the domain. In space, a state cannot signal benign intent toward one state by repositioning forces or pursuing defensive capabilities on a regional basis. In space, actions are global as are the risks.

Returning to the prisoner's dilemma and stag hunt models, a state pursuing a cooperative strategy must consider the risk of defection by any state in the international system, not just a single competitor (see Figure 2.1). While the benefits of all states cooperating in space are high, the difficulty of getting all space capable nations to cooperate is also high. These difficulties are only growing worse as more states gain the capability to reach space. Collective cooperation means that all states can realize the economic benefits of the space domain without fear of loss or paying the high cost of pursuing defensive capabilities. Orbital mechanics dictate that a conflict between two states in space will harm all states, not just those engaged in conflict. The debris and disabled satellites generated by a conflict in space pose a real and material threat to the shared space environment and could potentially limit future access to space. Therefore, environmental factors make cooperation in space the most desirable course of action. However, the difficulties of achieving collective cooperation mean that a state is likely to choose to compete even knowing that it is a worse option than cooperation. After all, mutual competition is not the worst option, and the dangers of even a single state pursuing space weapons and holding another nation's space assets hostage without the ability to respond in kind are very high. The result is that space can best be described as an environmentally driven stag hunt rather than a prisoner's dilemma.

The difficulty of selecting a strategy in space that signals benign intentions is complicated by the challenges associated with the geography of a shared domain. Despite environmental factors making cooperation the most attractive option, the challenges of achieving this goal encourages states to default to competitive policies. As the importance of a state's space assets to its economic and military power rises, it will be less likely to risk pursuing cooperative policies out of fear that another state will defect. The US and the USSR managed to achieve a level of cooperation in space because the domain was essentially bipolar and neither states' economic nor military power was as dependent on space as the US has become in recent decades. Space is now a multi-polar environment where the US has the most to lose if another state defects. The challenges associated with the geography of space exacerbate the security dilemma since a state must consider the risks posed by all states, not just those that it borders.

Summary

The seemingly simple concept of the security dilemma has generated a wealth of theories that use it to explain behavior and predict outcomes. The rational strategic choice theory captures the various implicit and explicit threads of these theories in a manner that allows for a structured analysis of state behavior that is neither too simplistic nor complex. Most importantly, it provides a framework for determining whether the optimal path for states within the space domain is cooperation or competition. The theory also provides an explanation for why cooperation is not always the rational choice and so incorporates the spiral model as well as deterrence theory, allowing for a richer milieu of strategic choices than would otherwise be possible.[102] The validity of the strategic choice that a state reaches is dependent on the material, informational, and motive variables within the theory as well as on the difficulties and uncertainties associated with choosing to cooperate when another state might instead defect and choose to compete.

Measuring the existence and severity of a security dilemma is a daunting task. There is no definitive method for concluding what the true motivations and intentions of states are. Even after the fact, when scholars have access to internal documents from both parties and the interaction in question has played itself out, there remains uncertainty and disagreement over what truly drove behavior. Even in seemingly clear-cut examples such as the Cold War, which had many of the characteristics of two security seeking states trapped in an arms race because of uncertainty over each other's motives, scholars cannot agree if it was a security dilemma or if a true conflict of interest existed.[103] Such examples, demonstrate yet again that there is no sure way of escaping the uncertainty that drives the security dilemma. Despite this uncertainty, states can still communicate motive and intent and achieve cooperation and avoid an unnecessary arms race. If two states have no real conflict of interest and are both seeking security then cooperation is possible. The ease of achieving cooperation is dependent on the level of uncertainty present about the nature of the security dilemma, which is in turn determined by the accuracy of how it is measured.

The interaction of the three variables within the rational strategic choice theory on a state's choice to arm in space or cooperate can be broken into three questions.[104] First, does choosing to pursue an arms race in space increase military capability? Second, does the increased military capability have a negative impact on an adversary's security? Third, are the benefits to security created by the increased military capability in space greater than the dangers generated by the increase in the level of insecurity to an adversary both in and out of the space domain? These questions are complex, and using the various aspects of strategic choice theory should make it possible to determine an answer on whether the optimal choice for states within the space domain is to cooperate or compete. In this chapter, we have explored the rationale as to why either of these paths might make strategic sense. To begin answering these last set of questions, the nature of the relationship between the three great competing space powers, the US, China, and Russia needs to be investigated using the measures developed in this chapter.

Notes

1   Evan Braden Montgomery, "Breaking out of the Security Dilemma: Realism, Reassurance, and the Problem of Uncertainty," *International Security* 31, no. 2 (2006): 151.
2   John H. Herz, "Idealist Internationalism and the Security Dilemma," *World Politics* 2, no. 2 (1950): 157.
3   John J. Mearsheimer, *The Tragedy of Great Power Politics* (New York, NY: W. W. Norton & Company, 2014), 17–18.
4   Kenneth Waltz, *Theory of International Politics* (Long Grove, IL: Waveland Press, 2010), 126.
5   Mearsheimer, *The Tragedy of Great Power Politics*, 21.
6   Ibid., 35.
7   Ibid., 36.
8   Herz, "Idealist Internationalism and the Security Dilemma," 157.
9   Mearsheimer, *The Tragedy of Great Power Politics*, 35.
10 Robert Jervis, "Cooperation Under the Security Dilemma," *World Politics* 30, no. 2 (1978): 167–214.
11 Glenn H. Snyder, " 'Prisoner's Dilemma' and 'Chicken' Models in International Politics," *International Studies Quarterly* 15, no. 1 (1971): 68.
12 Jervis, "Cooperation Under the Security Dilemma," 172.
13 Ibid., 176.
14 Snyder, " 'Prisoner's Dilemma' and 'Chicken' Models in International Politics," 82.
15 Ibid., 84.
16 Ibid.
17 Jervis, "Cooperation Under the Security Dilemma," 186–187.
18 Ibid., 187.
19 Ibid., 211.

20 Ibid., 188.
21 Stephen Van Evera, "Offense, Defense, and the Causes of War," *International Security* 22, no. 4 (1998): 5.
22 Stephen Van Evera, *Causes of War* (Ithaca, NY: Cornell University Press, 1999); Van Evera, "Offense, Defense, and the Causes of War," 33.
23 James W. Davis et al., "Taking Offense at Offense-Defense Theory," *International Security* 23, no. 3 (1998): 183.
24 Waltz, *Theory of International Politics*, 187.
25 R. Harrison Wagner, ed., "The Theory of International Politics," in *War and the State*, The Theory of International Politics (Ann Arbor, MI: University of Michigan Press, 2007), 33.
26 Robert Jervis, *Perception and Misperception in International Politics* (Princeton, NJ: Princeton University Press, 1976), 62–76.
27 Charles L. Glaser, "The Security Dilemma Revisited," *World Politics* 50, no. 1 (1997): 180.
28 Charles L. Glaser, *Rational Theory of International Politics* (Princeton, NJ: Princeton University Press, 2010).
29 Ibid., 37.
30 Ibid., 38.
31 Ibid.
32 Ibid., 3.
33 See Jervis, *Perception and Misperception in International Politics*, 62–86.
34 Herz, "Idealist Internationalism and the Security Dilemma," 157.
35 See Jervis, "Cooperation Under the Security Dilemma"; Van Evera, *Causes of War*.
36 Glaser, *Rational Theory of International Politics*, 78.
37 Ibid., 76.
38 Ibid., 78.
39 Ibid., 73.
40 Ibid., 236.
41 Ibid., 231–232.
42 Ibid., 232.
43 Ibid., 229.
44 Ibid., 91–92.
45 Ibid., 206.
46 Adam P. Liff and G. John Ikenberry, "Racing toward Tragedy?," *International Security* 39, no. 2 (2014): 60–61.
47 Ibid., 60.
48 Ibid., 61.
49 Jervis, "Cooperation Under the Security Dilemma," 169.
50 Andrew Kydd, "Trust, Reassurance, and Cooperation," *International Organization* 54, no. 2 (2000): 326.
51 Ibid.
52 Glaser, "The Security Dilemma Revisited," 181.
53 Ibid.
54 Ibid.
55 Evan Braden Montgomery, "Breaking out of the Security Dilemma: Realism, Reassurance, and the Problem of Uncertainty," 157.
56 Jervis, "Cooperation Under the Security Dilemma," 196.
57 Ibid., 173.

58 Charles L. Glaser, "Realists as Optimists: Cooperation as Self-Help," *International Security* 19, no. 3 (1994): 59.

59 James D. Morrow, "Arms Versus Allies: Trade-Offs in the Search for Security," *International Organization* 47, no. 2 (1993): 214.

60 Dwight D. Eisenhower, "Letter from President Eisenhower to Soviet Premier Bulganin," January 12, 1958, White House Office of the Special Assistant for Disarmament (Harold Stassen): Records, 1955–1958, Box 7–8 Eisenhower-Bulganin Letters, Dwight D. Eisenhower Library, Abilene, Kansas.

61 Dwight D. Eisenhower, "Letter from President Eisenhower to Soviet Premier Bulganin," February 15, 1958, White House Office of the Special Assistant for Disarmament (Harold Stassen): Records, 1955–1958, Box 7–8 Eisenhower–Bulganin Letters, Dwight D. Eisenhower Library, Abilene, Kansas.

62 Ibid.

63 Glaser, "Realists as Optimists: Cooperation as Self-Help," 64–65.

64 Jervis, "Cooperation Under the Security Dilemma," 188.

65 Bernard I. Finel and Kristin M. Lord, "The Surprising Logic of Transparency," *International Studies Quarterly* 43, no. 2 (1999): 315–339.

66 Ibid., 335.

67 Andrew Kydd, "Game Theory and the Spiral Model," *World Politics* 49, no. 3 (1997): 372.

68 Glaser, "The Security Dilemma Revisited," 181.

69 Evan Braden Montgomery, "Breaking out of the Security Dilemma: Realism, Reassurance, and the Problem of Uncertainty," 154.

70 Kydd, "Trust, Reassurance, and Cooperation," 326.

71 Evan Braden Montgomery, "Breaking out of the Security Dilemma: Realism, Reassurance, and the Problem of Uncertainty," 154.

72 Ibid., 153.

73 See ibid., 174–177 for a detailed discussion of the interplay of actions and reactions that made Khrushchev's signaling attempts inadequate, including the simultaneous growth in the size of the Soviet Strategic Rocket Forces.

74 Quoted in ibid., 174.

75 Ibid., 176.

76 "National Security Council Report, NSC-68, 'United States Objectives and Programs for National Security,'" April 12, 1950, Truman Library, www.trumanlibrary.org/whistlestop/study_collections/coldwar/documents/pdf/10-1.pdf.

77 Glaser, *Rational Theory of International Politics*, 83.

78 George W. Downs and David M. Rocke, "Tacit Bargaining and Arms Control," *World Politics* 39, no. 3 (1987): 322.

79 Glaser, *Rational Theory of International Politics*, 41.

80 Mearsheimer, *The Tragedy of Great Power Politics*, 55.

81 Ibid., 62.

82 International Monetary Fund (IMF), "Components of the Gross Domestic Product (GDP) Data Series in International Financial Statistics (IFS)," accessed August 21, 2018, http://datahelp.imf.org/knowledgebase/articles/498480-what-are-the-components-of-the-gross-domestic-prod.

83 See Mearsheimer, 63–74 for a detailed breakdown of the challenges and issues associated with using GNP and related measurements.

84 Mearsheimer, *The Tragedy of Great Power Politics*, 78.

85 Ibid., 56.

86 Robert Jervis, "Cooperation Under the Security Dilemma," *World Politics* 30, no. 2 (1978): 186–187.
87 James W. Davis et al., "Taking Offense at Offense-Defense Theory," *International Security* 23, no. 3 (1998): 186–187.
88 Jervis, "Cooperation Under the Security Dilemma," 194.
89 Charles L. Glaser and Chaim Kaufmann, "What Is the Offense-Defense Balance and Can We Measure It?," *International Security* 22, no. 4 (1998): 64.
90 "Soyuz Data Sheet," Space Launch Report, accessed August 23, 2018, www.space launchreport.com/soyuz.html.
91 Stephen Van Evera, "Offense, Defense, and the Causes of War," *International Security* 22, no. 4 (1998): 42.
92 Quoted in Jervis, "Cooperation Under the Security Dilemma," 201.
93 Robert Jervis, "Dilemmas About Security Dilemmas," *Security Studies* 20, no. 3 (July 2011): 420.
94 Ibid.
95 Charles L. Glaser, "When Are Arms Races Dangerous? Rational versus Suboptimal Arming," *International Security* 28, no. 4 (2004): 45.
96 Robert Jervis, "Was the Cold War a Security Dilemma?," *Journal of Cold War Studies* 3, no. 1 (2001): 37–38.
97 Charles L. Glaser, *Rational Theory of International Politics* (Princeton, NJ: Princeton University Press, 2010), 145.
98 Ibid.
99 Adam P. Liff and G. John Ikenberry, "Racing toward Tragedy?," *International Security* 39, no. 2 (2014): 61.
100 Mearsheimer, *The Tragedy of Great Power Politics*, 41.
101 Glaser and Kaufmann, "What Is the Offense–Defense Balance and Can We Measure It?," 64.
102 Glaser, *Rational Theory of International Politics*, 85.
103 Jervis, "Was the Cold War a Security Dilemma?"
104 These questions are modifications of questions found in Glaser, *Rational Theory of International Politics*, 235.

3 Power, motive, intent, and information

Determining whether to compete or to pursue cooperation in space is entirely dependent on the nature of the relationship between the major powers, both in and outside the space domain. The shared nature of the space domain and the lack of differing geographic limitations means that any strategy selected by a nation must be suitable for all its potential adversaries. Reassuring a potential competitor by unilaterally disarming might be effective in signaling benign intent and creating positive cooperation, but it could also prove disastrous if another geopolitical competitor with greedy motives chooses to compete. From the US perspective, the major security threats that an appropriate space strategy must be capable of addressing are China and Russia.[1]

These two nations represent security challenges to the US in space for differing reasons and to differing degrees. China is a rising power with a rapidly growing space sector while Russia is a former great power struggling to maintain and revitalize its existing space industrial base. China is a relatively new space power, and so has only a limited history of interacting with other great powers in the space domain. This relatively short history and rapid rise make Chinese behavior in space unpredictable, leading to additional uncertainty and fear. Meanwhile Russia, through its predecessor the USSR, and the US have a long history of interaction, and more recently active cooperation in the space domain. Though this history of interaction and active cooperation on some projects does not eliminate competition, it does significantly decrease uncertainty about technical capability, and as a result some of the fear associated with the unknown. Russia and China each present different challenges, and the US must develop a strategy that addresses the issues presented by both a potentially growing threat and another resurgent one.

The security challenges between the US, Russia, and China are not confined to the space domain. The reality is that the various security challenges created by issues outside of the space domain are the main drivers of competition between these three states. However, this does not make the space domain irrelevant. The recognized importance of space to US power projection places it at the center of any security competition. Any sub-optimal space policy, whether competitive or cooperative, could further exacerbate an overall security dilemma between the states concerned. To avoid choosing a sub-optimal policy and creating a condition where space could act as a trigger for future conflict the power, motives, and intentions of the competing space powers require accurate assessment. If there is a possibility that both US competitors are trapped in security dilemma rather than a pure security competition then it is worth attempting to defuse the dilemma. At the very least the attempt to defuse an orbital security dilemma will remove any doubt about the motivations and intentions of Russia and China in the space domain and have grave implications for the possibilities of cooperation outside of it.

A brief history of US actions and rhetoric in space

The material balance of power in space continues to strongly favor the US. This should encourage the US to act as a secure status quo power within the domain, yet it does not. US rhetoric concerning space warfare and associated actions shows a growing level of insecurity and fear about its status as the dominant power. This behavior by the US is a crucial driver of the information that other states have about US motives, and it also shows what the US thinks about the motives of other states. This section will show that the US attitude toward its control of space has evolved from one of cautious coexistence in the Cold War-era through a period of largely complacent dominance to a state of severe insecurity today.

The importance of space to geopolitics and strategy began when the USSR launched Sputnik into orbit in 1957. This act shocked the US public out of a deep sense of complacency about US technological superiority and led to demands for action.[2] The Eisenhower Administration was surprised by the level of public outcry in response to Sputnik's launch as well as the impact it had on global opinion and the perceived balance of power. The Administration accurately assessed that there were four primary impacts on US security as a result of Sputnik:

1   Soviet claims of scientific and technological superiority over the West and especially the US have won greatly widened acceptance.
2   Public opinion in friendly countries shows decided concern over the possibility that the balance of military power has shifted or may soon shift in favor of the USSR.
3   The general credibility of Soviet propaganda has been greatly enhanced.
4   American prestige is viewed as having sustained a severe blow, and the American reaction, so sharply marked by concern, discomfiture and intense interest, has itself increased the disquiet of friendly countries and increased the impact of the satellite.[3]

Sputnik's launch surprised the US public and placed the nation in a position of perceived vulnerability. Overnight the perception of American technological and military dominance that had persisted since the end of WW2 disappeared. Space moved overnight from the realm of scientific curiosity to one of prime national importance.

These reactions and impacts should not have come as a shock to the Administration. The US had its own plans to launch a satellite in conjunction with the International Geophysical Year (IGY) year that lasted from 1957–1958 and was fully aware of Soviet plans announced at IGY forums.[4] Also, nearly three years before Sputnik's launch NSC 5520 recommended a US satellite launch in order to reap some of the same benefits the Soviets did with Sputnik. Among the three primary justifications used by NSC 5520 for the satellite project was that "the prestige and psychological benefits resulting from the demonstration of advanced technology by the US and the impact of such a demonstration on the political determination of free world countries to resist Communism."[5] Despite the intellectual awareness by some within the Eisenhower Administration of the importance of being first in space, the Administration had missed the opportunity and so placed the US in a position of perceived vulnerability.

The fear that the sudden Soviet dominance of space created led the US down dual paths of competition and cooperation. While the US and the USSR entered into a fierce competition for national prestige that culminated with the US Moon-landing, the US also cooperated with the USSR to avoid the active weaponization of space. Taking the lead on early efforts by the UN to regulate activities in space the US actively supported the peaceful use of space while at the same time shaping the definition of peaceful to include "non-aggressive military activities" in order to preserve the advantages gained by space-based reconnaissance.[6] By taking this position and matching actions to rhetoric, the US was able to ensure that competition in space between the USSR and the US was limited primarily to the prestige projects of manned and unmanned space exploration. Early efforts by the Department of Defense to upset this delicate balance and develop dedicated anti-satellite (ASAT) weapons were successfully opposed, and the US and USSR settled into a relationship where neither country pursued space weapons in any serious manner.[7] Protecting vulnerable US reconnaissance satellites from attack or interference resulted in a de facto US policy of legitimizing those activities in space that it considered beneficial, and opposing those that were not.[8]

The US and the USSR avoided an outright arms race in space following the launch of Sputnik and over the next two decades signed several treaties that further reduced the possibility of space weaponization. The most significant of these treaties was the Outer Space Treaty which limited the deployment of weapons of mass destruction on orbit and the militarization of celestial bodies.[9] This treaty did not stop both the US and USSR from pursuing limited ground-based ASAT weapons research and development though it did help prevent an arms race in space. Following the signing of the Outer Space Treaty in 1967, the US and USSR agreed to the Anti-Ballistic Missile (ABM) treaty in 1972 which banned space-based interceptor missiles and limited each nation to two ABM sites.[10] The understandings and treaties reached during this period appeared to settle the issue of space weaponization in favor of passive and peaceful uses, and following the Moon-landings space retreated into the background of national security issues. This decade of calm between the two great space powers ended suddenly with the announcement by President Reagan of the Strategic Defense Initiative (SDI).

In a March 23, 1983, address to the nation, President Reagan cast the USSR as a greedy revisionist power building an "offensive military force" designed to "permit sudden, surprise offensives."[11] To counteract the growing threat posed by the USSR, President Reagan proposed building a defensive system capable of negating the Soviet missile threat. This speech did away with both the entire strategic doctrine of Mutually Assured Destruction (MAD) upon which peace ultimately rested, as well as the ABM treaty. Despite President Reagan's efforts to cast SDI as a defensive system, its announcement was more disruptive than any offensive capability could be. With the ABM treaty in place, both the US and USSR possessed an assured nuclear deterrent that limited competition for more effective nuclear delivery systems. An attempt to build a defensive system like SDI decreased the effectiveness of the Soviet nuclear deterrent and so had a destabilizing effect not normally attributed to defensive weapons. Though President Reagan did not mention space explicitly in the speech, it quickly became apparent that an effective missile shield would require the deployment of space-based defenses and so SDI earned the moniker "Star Wars." Formally initiating SDI with National Security Decision Directive 85 (NSDD 85), the president further confused matters by directing that research on SDI be "carried out in a manner consistent with our obligations under the ABM Treaty" thereby avoiding formally withdrawing from it.[12]

Though SDI was ultimately technologically infeasible and unaffordable, it did energize the long-dormant military space community. Signaling a renewed interest space, the DOD established a unified US Space Command in 1985 to oversee the US Air Force Space Command established three years prior.[13] The Army and Navy also established their own space focused commands within US Space Command, and serious work on the proactive employment of space assets in support of tactical military operations started.[14] This new thinking contrasted sharply with the strategic focus centered around national reconnaissance and secure nuclear communications that had previously dominated the US use of space. Just a few years later, during the 1991 Gulf War, space-enabled conventional warfare would make its formal debut in what journalists dubbed the first "Space War."[15]

Desert Storm demonstrated to the world the advantages that space can provide to conventional military power. Space-based navigation, precision targeting, and satellite communications enabled the speed of the US-led coalition's success and its dominance on the battlefield. Though nascent in its employment and with many shortfalls, the way in which the US leveraged its space assets demonstrated the need for further integration of space to aid the warfighter.[16] This success did have one significant downside. Since space now served as a vital enabling capability for conventional forces, the door was open for potential rivals to look hard at ways to limit this US advantage and for the US Military to begin looking at ways to protect its space assets.

Growing US military dependence on space accelerated in the decade following the Gulf War as broader efforts at transformation toward a more lethal and networked force required greater integration of space capabilities. To address deficiencies in the organization and management of military space, the National Defense Authorization Act for Fiscal Year 2000 directed the establishment of a Space Commission to study these issues.[17] Led by former Secretary of Defense Donald Rumsfeld—who again assumed that post shortly before the study's publication—the commission consisted of an eminent group of former military leaders, politicians, and space experts which promised to ensure the report had a significant impact on the US military space community. Recognizing the linkages between successful adoption of technology and organizational innovation and flexibility, the commission developed a series of unanimous recommendations for reorganizing the national security space architecture. These reorganization efforts hinged on the recognition that "the relative dependence of the US on space makes its space systems potentially attractive targets."[18] Raising the specter of a "Space Pearl Harbor," the commission highlighted how a surprise attack on US space assets was possible and could have significant repercussions for US military capabilities.[19] Believing that war in space was inevitable, the report argued that the US must develop the means to "deter and defend against hostile acts in and from space."[20] The report's emphasis on the inevitability of war in space signaled a new era of US insecurity in its ability to freely operate in space and a shift in strategic thinking about the space domain.

The commission's chairman, Defense Secretary Donald Rumsfeld, was in the ideal position to implement the report's bold recommendations. However, the events of 9/11 overwhelmed the national security community and efforts at transforming the US military stance in space largely disappeared from the Bush Administration's agenda.[21] Despite this shift in priorities, the Administration did make some progress in implementing some of the less significant recommendations in the report and made one other significant change to US space posture. In late 2001, President George W. Bush leveraged the events of 9/11 to formally withdraw from the ABM treaty arguing that "I have concluded that the ABM treaty hinders our government's ways to protect our people from future terrorist or rogue state missile attacks."[22] Russia and China strenuously opposed the withdrawal, with the latter also considering an increase to the size of its nuclear arsenal in response.[23] While it was not immediately obvious, the US withdrawal from the ABM treaty would set in motion the sequence of events that lead to the current arms race in space.

In another significant change to space policy, the Bush Administration released a new US National Space Policy (NSP) in 2006 which departed sharply from the previous 1996 Clinton Administration policy. The 1996 space policy would have been largely recognizable to an Eisenhower era official with its language promoting the peaceful use of space and asserting the right of space systems from any nation to have freedom of passage without interference.[24] While including much of the legacy language on US support for the peaceful use of space in a *pro forma* manner, the 2006 NSP took on a much more aggressive character. The most significant departure from past policy was a change in the key principles of the policy which stated that:

Consistent with this policy, the United States will: preserve its rights, capabilities, and freedom of action in space; dissuade or deter others from
either impeding those rights or developing capabilities intended to do so;
take those actions necessary to protect its space capabilities; respond to
interference; and deny, if necessary, adversaries the use of space capabilities
hostile to U.S. national interests.[25]

The addition of the word "deny" was a significant addition to the terms "defend
and deter" used in the 2001 Space Commission report and the 1996 space
policy. It represented a public and dramatic departure from long-established US
policy toward space activities. The 2006 policy also took a distinctly hostile attitude toward arms control treaties, stating that the US would "oppose the
development of new legal regimes or other restrictions that seek to prohibit or
limit US access to or use of space."[26] In sum, the 2006 policy was a significant
departure from previous US space policy with an aggressive tone that went even
further than the 2001 Space Commission report in its assertion of US rights in
space.

The Obama Administration took a softer tone in its 2010 National Space
Policy, but the core of the Bush-era stance on space remained. Replacing the
term "deny" with "defeat," the 2010 policy expanded the language of defense to
include allies and other responsible parties in space.[27] It also stepped back from
the hostile language in the 2006 policy toward arms control, and instead encouraged the development of confidence-building measures and increased multi-
lateral transparency in the pursuit of the peaceful use of space.[28] The new
national space policy was accompanied by the release of the first-ever National
Security Space Strategy (NSSS). The NSSS nested within the 2010 NSP emphasizing a shift toward promoting the "responsible, peaceful, and safe use
of space," while preserving the language of "deter and defeat."[29] Overall, the
2010 space policy and NSSS took a softer and less nationalistic tone, yet continued the Bush-era policy of treating space as an environment with significant
evolving threats.

The Obama-era return to a more cooperative stance on space, at least in publicly available documentation, took a sharp turn with the election of President
Trump. In 2018, the Trump Administration released a new national space
strategy. This strategy supersedes the 2011 NSSS but does not technically
supersede the 2010 National Space Policy, though it represents such a drastic
departure from the 2010 document that the apparent discontinuities between it
and the Obama-era NSP are difficult to reconcile. The full strategy is not publicly available, but the fact sheet released by the Administration bluntly states in
its title that it is an "America first strategy."[30] Dropping any pretense about promoting the peaceful use of space, the strategy mentions peace only twice within
the phrase "peace through strength." Also, discarding the softened language in
the 2010 Policy which broadened defense of space assets to include allies and
responsible parties, the new strategy openly blames unnamed competitors and
adversaries for turning space into a warfighting domain. The most dramatic idea

included in the fact sheet is the statement that "the strategy affirms that any harmful interference with or attack upon critical components of our space architecture that directly affects this vital interest will be met with a deliberate response at a time, place, manner, and domain of our choosing."[31] The key word in that statement is "domain," by affirming that the US is willing to respond to attacks in space outside of the space domain, the new strategy turns space into an area ripe for escalation. For example, implementation of this policy means that a ground-based jammer interfering with a US satellite could be attacked by US military assets in response, representing a violation of national sovereignty much more visible to the world than the interference that incited it. These changes mean that actions in space can now act as a trigger for a broader military response both in and outside the space domain. The 2018 NSSS created the conditions for conflict escalation and open warfare in space while blaming the need for this stance on unnamed competitors.

A single fact sheet for a strategy document representing a dramatic departure from the previously established policy is significant. However, by itself, it might not force competitors to re-evaluate US motives, though rhetoric and actions reinforcing that new space strategy cannot help but force other nations to see US behavior as insecure and aggressive. The Trump Administration has provided the rhetorical fuel for this viewpoint. A week prior to the release of the new strategy President Trump stated in a speech that "space is a warfighting domain, just like the land, air, and sea."[32] This reference to space as a warfighting domain was nothing new in military circles, but it was the first time a US president had referred to space in that way. He also mused on the possibility of creating a separate Space Force, though with a president famous for throwing out ideas, the remark was not taken especially seriously even by the president himself. These remarks received substantial media attention that eventually died down without renewed emphasis provided by the president.

On June 18, 2018, President Trump re-energized the military space debate when he gave a speech at a National Space Council meeting for the signing of Space Policy Directive-3 (SPD-3), an innocuous document on space traffic management, where he called for US dominance in space.[33] In the very next breath, he directed the Chairman of the Joint Chiefs, General Joseph Dunford, to establish a separate US Space Force. Linking the establishment of a new space focused military force with American dominance in space cannot help but signal to competitors that the US is no longer satisfied with the status quo. It also shapes the perceptions of any future Space Force as a disruptive organization designed to dominate the space domain, not preserve peace.

President Trump is famous for inadvertent remarks, so his statements do not carry the same gravitas that previous presidential speeches and remarks do unless others in the Administration further reinforce them. In an August 2018 speech at the Pentagon, Vice President Pence again referred to space as a warfighting domain and as a future battlefield in need of a separate branch of the military.[34] He also repeated the president's statements calling for "American dominance in space."[35] Unlike the Administration's space strategy, the vice president explicitly named the countries that the US considers hostile adversaries in space: Russia, China, North Korea, and Iran.[36] Blaming these nations for transforming space into a warfighting domain and framing US actions as simply a response to preserve peace, the vice president could not help but reinforce to China and Russia that the US views them as hostile adversaries.

Administrations change and with them US policy. So, while the current Administration's policy toward China and Russia in space is clear, long-term trends are better represented by the attitudes of the organizations which implement policy, particularly the DOD. The attitudes the members of these organizations have toward space will persist and influence the decisions of future Administrations. Former Secretary of Defense, prior Marine General James Mattis, represents a blend of the Administration and the organization in which he spent most of his life. Secretary Mattis opposed earlier congressional efforts to create a Space Corps within the Air Force. In a 2017 letter to Congress, he stated that "I do not wish to add a separate service that would likely present a narrower and even parochial approach to space operations."[37] When the president later directed the creation of the Space Force, Secretary Mattis moderated his earlier comments by stating that "what I was against was rushing to do [Space Force] before we could define the problem."[38] Then Deputy Defense Secretary Pat Shanahan further clarified that Secretary Mattis' former resistance to a separate Space Force was only due to budgetary concerns.[39] Secretary Mattis' prior comments on the need for a Space Force indicate that President Trump's desire for a Space Force tasked to dominate the space domain faced significant organizational opposition due to cost and concern over potential future service parochialism and very little opposition to militarizing the domain.

Concern over organizational competition rather than the further militarization of the domain is echoed in comments made by senior US military officers. The Chief of Staff of the Air force, Gen. Goldfein, when questioned by the Senate on the need for a separate Space Corps, replied that "I do not support it at this time in our history based on where we are in this transition from a benign environment to a warfighting domain."[40] He went on to argue that the large organizational change required to create a separate Space Corps would "actually slow us down right now."[41] These comments were echoed in many ways by the US military's senior space officer, Gen. John Hyten, in testimony to the Senate Armed Services Committee nearly a year later. Following President Trump's March speech musing on the possibility of a Space Force, Gen. Hyten stated during testimony that he did "not think the time is right for [Space Force] right now, but I loved the fact that the president talked about space as a warfighting domain."[42] Gen. Hyten did express a more measured attitude toward China and Russia than the Administration. While concerned about Chinese and Russian military modernization efforts, Gen. Hyten stated in his testimony that "we remain committed to strategic stability with China and Russia."[43] Gen. Hyten's and Gen. Goldfein's comments and testimony show that US Air Force efforts were centered on gaining recognition for space as a warfighting domain for the purposes of winning funding for efforts to protect

US space infrastructure. Their support is more measured with their statements and verbiage more reserved than the aggressive language of "dominance" used by the president and vice president. Rather, their language more broadly echoed Bush-era verbiage that focused on defense through deterrence rather than dominance. Ultimately, the resistance of Air Force leadership to establishing a separate space focused force was overcome with the re-establishment of US Space Command followed closely by the creation of the United States Space Force (USSF) in December 2019.

Two new organizations, one focused on protecting and defending US space assets and the other on organizing, training, and equipping US space forces introduced an uncertain new organizational dynamic to US actions in space. Created as the result of political horsetrading the USSF officially established a new organization dedicated to "protecting US and allied interests in space."[44] Despite this seemingly benign mission statement the new force will inevitably begin drifting toward offensive doctrines for several reasons.[45] First, as a new service with a history as a supporting domain subservient to the needs of other military branches, the USSF will pursue offensive doctrines to enhance its relative autonomy and importance. Second, offensive strategies are preferable because they reduce institutional uncertainty in a domain with little historical baseline for what warfare in the new domain entails. Finally, offensive approaches to warfare require more technical complexity and are more demanding on an organization, driving a need for increases in organizational size and wealth. These tendencies toward offensive behavior and inflated threats will be reinforced as the smallest service with a correspondingly tiny budget competes with the other more established services for money and manpower.

Internal resistance within the national security community to the creation of an independent Space Force demonstrated a disconnect with the desires of the Trump Administration. The comments by two senior Generals and the former Secretary of the Air Force taken together show a reluctance to pursue a separate Space Force, combined with a desire to ensure military preparedness to protect space assets. Their language is defensive and not dominance oriented, though their organizational actions ultimately reflected Administration priorities and the subsequent organizational needs of the USSF will trend towards offensive behavior. This differentiation shows a disconnect between the Trump Administration and the attitudes of the US national security community. How competitor nations interpret this initial reluctance to the creation of the USSF will determine the information they will use to form their idea of how the US sees their motives and how they see US motives.

Looking at US attitudes toward space, Russia and China would see Administration verbiage that points to a greedy and revisionist US policy while institutional verbiage signals insecurity and security seeking. The Administration also sees Russia and China as hostile to US interests with prior Administrations and the US military being more measured in their interpretation. Combining these viewpoints to achieve a middle ground paints a picture of the US firmly casting aside the concept of space as a peaceful sanctuary and developing an aggressive attitude toward defense in space that is taking on an increasingly offensive cast. The basis for this aggressive defensive posture is the perceived threats from other space powers, especially China and Russia, against the space architecture on which US military power projection is dependent. This architecture, by far the largest of any nation in space, is vulnerable though US fears may be grounded as much in the seemingly sudden appearance of any threat in a previously benign environment rather than the development of a true existential threat. A comparison of material measures of space power still puts the US substantially ahead of any nation in latent and explicit space power as the next section will show.

Reaching for space dominance

The US remains the dominant world power in space, for the moment. After decades of competition during the Cold War, the US emerged as the leader in space technology and development. In the decade following the fall of the Soviet Union, no other state came close to matching US achievements in space or the size of the US space industrial base that supported them. Since the turn of the century, the degree of dominance that the US enjoys has decreased as space technology has proliferated, though this relative decline has been countered by an absolute increase in US space assets and capability. With more than 59 nations now having a presence in space it is no longer just the domain of great nations, rather it is a global commons increasingly vital to military and economic power.[46] This decline in space power relative to other nations was inevitable, though it is causing concern within the US. As a result of this relative decline, the US is becoming increasingly strident and aggressive in its language and behavior toward space. This section describes the degree of US power in space and demonstrates that while the US remains the dominant space power it is increasingly challenged by a rising China.

The most significant measure of modern US space dominance is how much it spends on space, far more than any other state. There are three major categories of US space spending: defense, civil, and intelligence. The National Aeronautics and Space Administration (NASA) represents the majority of the civil funds and the single largest beneficiary with a budget of $19.6 billion.[47] A distant second in the civil space realm is the National Oceanic and Atmospheric Administration (NOAA) which receives between $2.3 and $1.8 billion a year to support its network of environmental monitoring satellites.[48] Following civil space is defense spending on space which was $11.4 billion in 2018 and will rise to $12.5 billion in 2019.[49] Though not officially available, the portion of the US intelligence budget for 2013 devoted to space was $15.2 billion.[50] The National Reconnaissance Office (NRO) with a budget of $10.3 billion represents most of this spending, though the National Geospatial-Intelligence Agency (NGA) received a respectable $4.9 billion. With nearly $50 billion devoted to it, the annual US space budget is greater than total defense spending in all but five other nations and far ahead of the $6–11 billion China spends on space or the $3–5.2 billion Russian space budget.[51,52]

This massive space budget relative to other nations is manifested in US space capability. The US government and military have 334 active satellites on orbit, nearly twice China's 183 and three times Russia's 114.[53] The numbers do not necessarily equate to capability as US systems also represent the majority of the most technically advanced platforms available to any government. For instance, with the Global Positioning System (GPS) it has the world's premier space-based navigation system while Russia's equivalent GLONASS system has significant issues.[54] It also possesses the most accurate and distributed space tracking network available to any nation with 18 globally distributed ground tracking stations that can track most small objects in LEO.[55] Though even this system has significant weaknesses with minimal coverage in the southern hemisphere and the ability to only update orbital data on objects when they pass over a ground site. Despite these weaknesses, the US government possesses by far the most capable on-orbit infrastructure of any nation with more explicit space power than any other nation by far.

In the commercial market, the US possesses an even more impressive lead than it does with US government on-orbit infrastructure. In 2017 there were 505 US-flagged commercial satellites in orbit compared to 60 for China and only 26 for Russia.[56] The number of US commercial satellites in orbit is more than the total number of active satellites from all countries combined as recently as ten years ago and growing rapidly. This massive growth in the US commercial market is unprecedented and a recent phenomenon. Driven by decreasing launch prices and an increasing ability to place greater levels of capability in smaller satellite platforms, the growth in on-orbit satellite numbers is only just beginning. Small satellite constellations delivering high-speed global communications will soon appear with realistic plans for constellations as large as 1,980 platforms.[57] If successful, these constellations will revolutionize global communications for commercial and military users. As the home for nearly all new space companies, the long-term US investment in its space industrial base is becoming self-sustaining. With its rapidly growing commercial satellite industry, the US possesses far more latent space power in this area that could be mobilized for military purposes than any other nation.

The situation does not favor the US quite as much in the launch market. The launch market is far more competitive for the US than its lead in on-orbit infrastructure would imply. With 29 successful launch attempts in 2017, compared to 19 for Russia and 17 for China, the US seemed for a moment to possess a significant lead. However, 2017 was the first year that the US led the world in the total number of successful launches since 2003 and in 2018 China staged a significant comeback.[58] In 2018 China launched a post-Cold War record of 39 rockets exceeding the US total of 31 and far surpassing its 2017 total.[59] In addition, US launch growth is almost entirely driven by SpaceX, a new space company that accounted for 18 of the US launches in 2017. That same year, the established launch provider for the US government, United Launch Alliance (ULA), which operates two different vehicles, only accounted for eight total launches. While China is currently the leader in space launches, it is a fragile lead, and there is stiff competition between the US and China, with Russia coming in as an increasingly distant third.

The launch situation does favor the US when the variety and capability of launch vehicles are considered. The US currently possess four different private launch companies that compete for government and private contracts: SpaceX, Blue Origin, Orbital ATK, and ULA. SpaceX and ULA have three different launch vehicles between them capable of reaching GEO, and Blue Origin may soon join them with the New Glenn Rocket. Both Blue Origin and SpaceX have independently developed and demonstrated partial reusability on their rockets, a primary contributor to the global decrease in launch costs as these new players are creating rapid disruption in the space launch industry for the first time. These technological innovations and the vibrancy of the domestic commercial launch market currently provide the US government with faster, better, and cheaper access to space than any other nation.

The variety of launch vehicles and providers in the US is in stark contrast with Russia and China's reliance on state-owned companies. China relies almost entirely on a state-owned company, China Aerospace Science and Technology Corporation (CASC) and its family of Long March (LM) rockets for access to space.[60] This family of rockets is entering its fifth generation with a single booster capable of either LEO or GEO. Though with a payload capacity to GEO of only 14 metric tons, the latest generation of Chinese rocket, the Long March-5 (LM-5), significantly lags emerging US rocket capability.[61] While legacy US rocket systems like the ULA Delta Four Heavy have a comparable payload to GEO transfer orbit to the Chinese LM-5, newer US commercial systems such as the Falcon Heavy rocket produced by SpaceX have a far greater GEO capability of nearly 27 metric tons.[62] Even with its shortcoming in technology and lift capability, however, CASC has managed to remain cost competitive with US providers.[63]

The situation is not as good for China in the private sector as it is in the US, though that situation is changing rapidly. A dedicated program of technology sharing through public–private partnerships with various Chinese state agencies has resulted in several successful semi-private launches.[64] While the rapid success of these partnerships is remarkable, it is unclear if these are truly viable private companies capable of innovating and competing separate from their parent agency. Even with these privatization successes without significant new investment CASC will have difficulty remaining competitive in the commercial marketplace as rapidly evolving US competition deprives it of a significant source of external income. Despite these challenges, the Chinese launch industry is in a much stronger position than its increasingly outdated Russian counterpart.

The Russian space launch industry suffers from significant problems and is rapidly falling behind both the US and Chinese space launch industry. After a string of high-profile launch failures, the Russian government abolished its Federal Space Agency in 2015 and merged it with the state-run corporation

United Rocket and Space Corporation (USRC) to form a new state-run corporation, Roscosmos.[65] This effort to address the collapse of the Russian launch industry has only been partially successful. As recently as 2013, USRC controlled almost half the commercial launch market generating more than $2 billion annually.[66] With the emergence of new space competitors in the US, Roscosmos has been unable to compete and fallen to a minuscule share of the market in just a few years. The loss of market share is partially due to the failure to replace Soviet-era Soyuz and Proton rockets since modernized replacements remain on the drawing board. In 2018, Russian Deputy Prime Minister Dmitry Rogozin announced that it was not worth the effort to recapture market share and "elbow Musk [SpaceX] and China aside."[67] With this announcement, Russia essentially abandoned the commercial space market and shifted its focus toward preserving national capability.

Comparing a recent snapshot of the US with Russia and China seems to show that by any material measure the US is the dominant space power and should have little to fear. Its two largest competitors, Russia and China, are struggling to keep up with the emerging US-based new space industry which is rapidly reshaping the space domain. China is investing in new technology and nursing a nascent commercial industry though it continues to lag the US in terms of technical capability. Meanwhile, Russia is struggling to keep its space industry relevant and is still reliant on Soviet-era technology for its access to space. With all these material power advantages, it seems that the US should have little to feel insecure about in space and no reason to disrupt the status quo. Even so, US actions and rhetoric demonstrate a level of insecurity that at first blush seems both alarming and unwarranted.

Moving beyond both a snapshot in time and an exclusive view of just the space domain the reasons for US insecurity become more apparent. In a generation China has progressed from an insignificant space power hardly capable of reaching orbit with just a single successful space launch in 2001 to a nation that consistently competes with or supersedes the US (see Figure 3.1). The rapid pace at which China has closed the capability gap is a primary driver of US insecurity, even as the US remains the leader by most material measures. It is not difficult to imagine that if China continues to invest in its space industry and progress at pace it will surpass the US far sooner than expected. Further, the ability to generate raw space power does not necessarily equate directly to the ability to deny it to others. Even Russia with its waning space power can leverage its proven technical base to present a substantial threat to a dominant space power. While the situation in space gives some cause for alarm, explanations for US insecurity become even more apparent when material measures of power outside of the space domain are included.

China

The rise of China is perhaps the most significant success story of the last 30 years. When the Cold War was ending and the US was demonstrating its

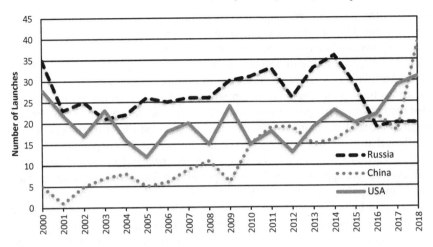

Figure 3.1 Number of space launches per year.

military prowess over Iraq in 1991, the ratio of American per capita GDP to Chinese per capita GDP was 67 : 1 which as of 2017 had fallen to 6.8 : 1.[68] In less than 30 years China has risen from an impoverished country to a middle-income nation and manufacturing giant with the second largest aggregate GDP. This rapid rise and potential to surpass the US has created significant concern about potential conflict between the US, whose power is in relative decline, and China whose power is increasing rapidly. This section discusses the strategic, material, informational, and motive factors that shape US and Chinese behavior outside of the space domain yet have a significant impact on it. Once the broader strategic picture is established, it is possible to analyze the limited information on Chinese actions and rhetoric in space to determine China's likely motives and intentions.

The outcome of China's rise and US relative decline is a focus of intense debate on whether this rise will be peaceful or result in conflict. Using Kenneth Waltz's standard structural realism argument, China's rise will result in a power bloc that will balance against US power in Asia.[69] War will not necessarily result from the balancing of US power in Asia, though intense competition is likely. Alternatively, offensive realism predicts that China will maximize its power and attempt to achieve regional hegemony over Eastern Asia. The US will attempt to prevent that rise and intense competition resulting in a high likelihood of conflict will result.[70] In contrast to these two theories, Glaser's rational theory includes more variables than just relative measures of power, so a more nuanced prediction of behavior is possible; a prediction that is much less pessimistic about the possibility of war between the US and China.[71]

China and the US have several varied sources of potential conflict in Eastern Asia, the most important of which is disagreement over the future of Taiwan. Following the thaw in relations with mainland China in the 1970s, President

Carter withdrew from the long-standing mutual defense treaty between the US and Taiwan designed to resist communist aggression. This withdrawal, made with only minimal congressional consultation, created significant concern within Congress over the future of Taiwan, resulting in the passage of the Taiwan Relations Act (TRA). Attempting to ensure the future safety of Taiwan while not upsetting the developing relationship with mainland China, the act made clear that the US would not permit communist China to interfere with Taiwan's sovereignty or social structure.[72] The act carefully avoided directly challenging the position that there is only one China, instead pledging the support of the US to ensure Taiwan's future "will be determined by peaceful means."[73] The TRA achieves an unhappy balance between preserving Taiwan's independence and the Chinese policy that "there is only one China in the world; both the mainland and Taiwan belong to one China; the sovereignty and territory of China cannot be split."[74] Continued US interference in what China views as a domestic matter makes Taiwan's future the primary source of contention between the US and China, but it is far from the only issue.

In addition to disagreements over the future of Taiwan, China has significant territorial disputes and strategic vulnerabilities in the South China Sea that involve the US. China has established bases on artificial islands in the region and is opposing US ship and aircraft movement near them. These actions are in support of Chinese claims that it "has indisputable sovereignty over the islands of the South China Sea and the adjacent waters."[75] These claims are disputed by the other nations that border the South China Sea, particularly Vietnam and the Philippines. Vietnam has responded by seeking closer military ties with India and the US in an effort to balance Chinese power in the region. These efforts have been marked by a lifting of the ban on selling weapons to Vietnam by President Obama in 2016 and in March 2018 the first visit of a US Aircraft Carrier to the country since the end of the Vietnam War.[76] Vietnam's efforts at seeking closer military ties with the US are in contrast to the Philippines which is trying to maintain its relationships with both China and the US.

The Philippines previously pursued and won a Court of Arbitration case against China over disputed reefs and islands in the South China Sea in 2016. China, perhaps seeing its case as weak, refused to participate in the tribunal claiming that the court had no jurisdiction.[77] The tribunal found no basis for Chinese assertions and invalidated essentially all of China's territorial claims in the region. Calling the tribunal a farce in official news outlets and that "its true purpose is to violate China's territorial sovereignty," China refuses to recognize the ruling and it remains a sore point between the two countries that the Philippines cannot resolve through force.[78] When asked two years after the ruling about efforts to enforce the ruling President Rodrigo Duterte remarked "I can send my Marines there, I can send every police there [sic], but what will happen? They will all be massacred."[79] A clear recognition of Chinese power relative to the Philippines has not driven them into a closer relationship with the US, though joint military exercises have continued. Balancing its relationship with China and the US, Philippine President Rodrigo Duterte avoids criticizing

Chinese actions in the South China Sea where possible. Instead, he is pursuing a dual track of closer economic ties with China while continuing limited military cooperation with the US.[80] Both the Philippine and Vietnamese approaches to handling Chinese power in the region involve the US as the primary balancer and serve as yet another point of tension between the two countries.

Beyond territorial disputes specific to Taiwan and the South China Sea, China faces significant strategic vulnerabilities at sea. Chinese economic well-being is highly dependent on its ability to import raw materials and export finished goods. Since China lacks a powerful blue-water Navy to secure its sea lines of communication they are vulnerable to disruption. The Strait of Malacca that marks the western boundary of the South China Sea is the most critical area of Chinese concern. Just 1.7 miles wide at its narrowest point, the strait supports 90 percent of China's oil shipments and 64 percent of its global maritime trade.[81] This strait is vital to Chinese energy security and trade so any disruption would harm economic development and therefore social stability.[82] Recognizing this vulnerability, China is very sensitive to perceived threats to the security of the strait. Former Chinese president, Hu Jintao, even accused "certain major powers" of trying to control the strait and called for efforts to mitigate the vulnerability it presents.[83] It is not hard to infer that the Chinese president was referencing US involvement in the area and the threat that could represent to Chinese interests.

China has legitimate security concerns in the South China Sea and an unresolved conflict with Taiwan that it is increasingly able to address. As recently as 1996, the US was able to sail a carrier battle group through the Taiwan Strait as a deterrent against China.[84] Growing Chinese military power supported by its rapid economic growth has made this militarily impossible today. China, recognizing that any conflict with the US will require the US to project power into the region in a military campaign that will be air and sea-centric has developed an Anti-Access/Area Denial (A2/AD) strategy. This strategy focuses on making it too dangerous for US naval or air assets to operate close to China. The centerpiece of this strategy is a modern missile, air, and submarine force that makes operations near the Chinese mainland extremely risky.[85] The effectiveness of these forces is dependent on their ability to identify, track, and target US assets across the vast Pacific region. Outside of sight of the Chinese mainland, these enabling capabilities can be provided only by space-based assets.[86]

The growth of China's space program and satellite infrastructure has enabled it to pursue its A2/AD strategy in a way that was not possible in 1996 when it had fewer than ten satellites in orbit.[87] At that time Chinese missiles were prevented from accurately striking targets by what the Chinese military suspected was deliberate US interference with the commercial grade US GPS systems those missiles relied on.[88] According to a former PLA officer, this episode was "a great shame for the Peoples Liberation Army ... an unforgettable humiliation, that's how we made up our mind to develop our own global navigation and positioning system, no matter how huge the cost."[89] Supporting the cost of building the satellite infrastructure and systems required to support an effective A2/AD strategy and secure Chinese interests is possible only because of the rapid growth of the Chinese economy.

Chinese latent power in the form of its manufacturing and economic growth has increased dramatically since the 1996 incident over Taiwan. Displacing the US at the top of relative GDP charts has greatly contributed to the "rise of China" narrative. News stories supporting this narrative are the most consistently read about topic in the media this century.[90] The raw numbers support the idea that the US is in relative decline. Adjusted for purchasing power parity, Chinese GDP surpassed US GDP in 2013 and it continues to outpace US GDP growth significantly (see Figure 3.2). While Chinese GDP in absolute dollar terms has yet to surpass the US, it is on track to do so within the next decade despite slightly lower levels of predicted annual growth. Also, manufacturing capacity surpassed the US a few years earlier in 2010 (see Figure 3.3).

Chinese military expenditures have kept pace with its economic growth. These expenditures now approach the total level of US military expenditure during the 1996 crisis, though the substantial growth in US military expenditure since 2001 has kept China from approaching US spending levels (see Figure 3.4). Total expenditure as a function of the percentage of GDP has remained fairly constant for China in the last 20 years at about 1.9 percent of GDP. The US, in contrast, has varied between 3 and 4.6 percent of GDP spent on its military.[91] While increasing in real terms, Chinese military expenditure has remained steady in relation to its economic growth. This rise is difficult to keep in perspective. In 2005, US Defense Secretary Rumsfeld saw the absolute increase in spending as threatening, arguing that "since no nation threatens China, one must wonder: Why this growing investment? Why these continuing large and expanding arms purchases?"[92]

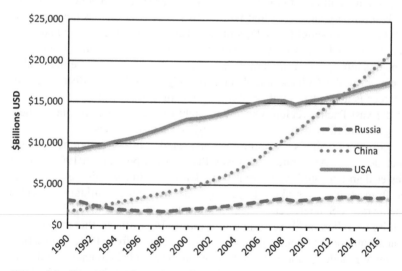

Figure 3.2 Gross domestic product adjusted for purchasing power parity in constant 2011 US$.

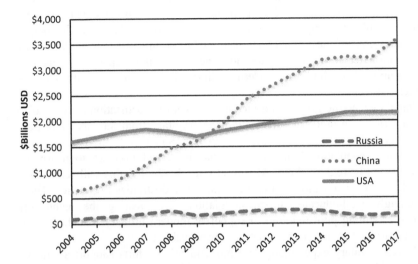

Figure 3.3 Manufacturing value added.

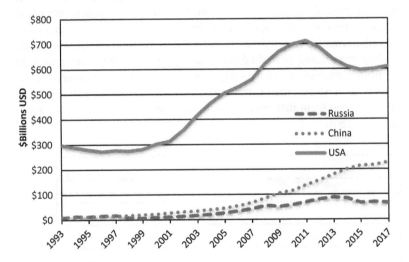

Figure 3.4 Military expenditure in billions of US$.

China's known military budget is three times larger today than it was then, which can further feed fears of Chinese expansionism despite the fact that it has kept its public budget constant relative to its GDP. Something that China is keen to highlight as a sign of its peaceful intentions though its messaging suffers due to suspicions that it is not revealing all of its military expenditures. The expenditures that it does reveal are not focused on growth in the absolute size of the military but rather in modernization efforts. The Chinese military cut 1.7 million personnel from its force structure from 1985 to 2005. These cuts have continued with a further 300,000 in reductions occurring between 2015 and 2017 resulting in an active duty force size of approximately 1.57 million personnel across all branches, which is comparable to the 1.28 million the US has on active duty.[93] The stated goal of these reductions in the Chinese military is to build a more "informationized military" and develop "new and high technology weaponry and equipment to build a modern military force structure with Chinese characteristics."[94]

China's economic growth and increasing level of urbanization have enabled the transition to a modern "informationized" military from the largely peasant armies that China employed in the Korean war. Since 1990, the percentage of the Chinese population living in cities has increased from 25 to nearly 60 percent (see Figure 3.5 below). At current rates, by 2030 China will achieve the same level of urbanization as the most modern industrialized countries. These increasingly well-educated urban populations are creating the technically capable recruiting pool required to support a modern military that the subsistence agrarian economy of the recent past was unable to.

Chinese military expenditures and modernization efforts support a strategy that is shifting from a focus on local defense and toward a more expansive view of strategic defense which entails a degree of global power projection in support of overseas interests. Clues to the motivations behind Chinese actions are found in its national military strategies, which like US strategies, serve as formal messaging platforms for national intent. Released in the form of variously titled defense white papers and published on an irregular basis, these statements of Chinese military strategy are remarkably consistent. While the primary themes and messages in these statements vary little, they do have subtle yet significant differences between them that provide insight into what China views as emerging security concerns.

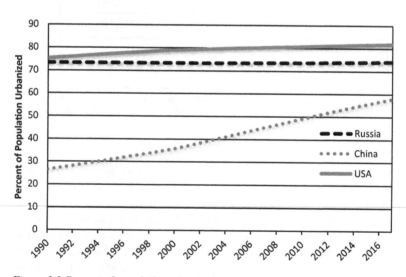

Figure 3.5 Percent of population urbanized.

China has emphasized several key themes in the four defense white papers published since 2010. First, is that China's military strategy is centered on active defense which in practice means a defensive posture at the strategic level coupled with an offensive posture at the tactical and operational level. A combination that accurately describes China's A2/AD approach to warfare. Second, all recent white papers have very specifically noted that the role of its armed forces is safeguarding national sovereignty and that it would "not attack unless we are attacked; but we will surely counterattack if attacked."[95] Third, the importance of territorial integrity, particularly in regards to "separatist" Taiwan is always an area of emphasis. Finally, these white papers highlight Chinese public opposition to hegemony, power politics, and expansionism in all forms.

The strategic concept that China will only attack when attacked seems reasonable and in accordance with the stated goal of protecting national sovereignty. That is until consideration is made for the gray area of defining what exactly constitutes an attack, especially within the space domain. Recent white papers have increasingly highlighted the importance that China places on protecting its national rights in space. Beginning in its 2010 document China accused "some powers [of having] worked out strategies for outer space."[96] This document then focused on Chinese efforts at arms control in space with no mention of the domain as an area that it was prepared to compete in militarily. By 2015 this had changed; space was now a "commanding height in international strategic competition" with China promising to "deal with security threats and challenges in that domain."[97] It is not entirely clear how closely China links protecting its national sovereignty with security threats in space, but Chinese doctrine does view threats against space systems as a way to "generate an impact upon a nation's policymakers and associated strategic decision making."[98] In light of this, China would likely see foreign military space assets passing overhead during times of heightened tension as both a threat and an opportunity for messaging. This combination would likely make them among the first systems targeted by China's growing arsenal of electronic warfare and directed energy anti-satellite weapons.[99]

Different national thresholds for security threats and responses create the potential for conflict, especially in space. The 2018 US National Space Strategy clearly defined harmful interference as an aggressive act requiring a response without defining precisely what form the response would take. If China does not view interfering on a temporary basis with US satellites transiting above it as an attack, merely part of a compellence centric messaging strategy, while the US does consider it an attack, then a US response could quickly lead to escalation. The US policy articulated in the national space strategy of reserving the right to respond outside of the space domain would only increase the potential for escalation. While there might be some confusion about what constitutes harm in space, China will surely consider a military response outside of the space domain as an attack. Absent other factors causing increased tensions, actions confined to space are unlikely to lead to conflict.[100] However, if these tensions are present, then confusion over what constitutes an attack could quickly escalate with the two contrasting policies.

These white papers also provide a rare insight into China's broader national strategic thinking. Casting itself as opposed to "any form of hegemonism" portrays China as a nation focused on protecting its territorial sovereignty and ensuring that it is free to pursue "peaceful development."[101] Chinese President Xi Jinping reiterated these sentiments recently and provided insight into what constitutes national sovereignty and territorial integrity. He repeated the commitment to peace while stating that China cannot give up "even one inch of territory that the country's ancestors left behind."[102] This statement shows that protecting Chinese territorial integrity is not limited to its current national boundaries. It is not clear precisely what China considers ancestral territory, though the South China Sea and Taiwan are clearly included. If the historical boundaries of China are represented by the Qing dynasty that ruled China just prior to the arrival of Europeans, then China is at or very near its territorial goals. Ongoing uncertainty associated with its territorial goals and disagreements over what it has already claimed creates the single most significant potential flashpoint for conflict with China.

Finally, these military strategies provide a useful insight into Chinese views of US behavior. Frequently in their description of global threats and trends, these papers cite the behavior of "some country [that] has strengthened its Asia-Pacific military alliances, expanded its military presence in the region, and frequently makes the situation there tenser" or some similar text.[103] Oddly, with the exception of its 2015 military strategy, China carefully avoids explicitly naming the United States, but there is no other nation to which this statement could refer. These statements and others demonstrate that to China, US actions are inflammatory and aggressive, not defensive.

This position that US actions are aggressive is not unique to Chinese strategy statements and is generally undisputed in Chinese academic and government literature. A Chinese professor at the Chinese National University of Defense Technology analyzing US behavior called the current Trump Administration policy toward China "borderline aggressive, even risky" and described it as casting China as "a competitor in a zero-sum environment."[104] This analysis is echoed in a 2018 Chinese government white paper responding to US trade tariffs that described US behavior as having "brazenly preached unilateralism, protectionism, and economic hegemony."[105] The white paper assured readers that Chinese policy is focused on peace, prosperity, and stability and declared that "cooperation is the only correct option for China and the United States" and that China's "position is clear, consistent and firm."[106]

Fears of US hegemonic behavior extend into the space domain. In 2014, PLA researchers analyzing US space policy under the Obama Administration determined that the shift toward more cooperative language was not a sign of changed US attitudes, noting that "the US has not abandoned its goal of space hegemony."[107] Certainly, the dominance centric rhetoric of the Trump Administration has not done anything to lessen concerns about US hegemonic goals in space. US rhetoric and actions within the space domain can only serve to reinforce the broader sense of the US as a hegemonic power opposed to China's peaceful rise. This perception of the US as an aggressive hegemon contrasts sharply with China's self-professed view of its own behavior as a rising, yet peaceful power focused only on protecting its national sovereignty.

The viewpoint that China has peaceful intentions is reinforced, at least publicly, by its support for a ban on weapons in space. In cooperation with Russia, China has long pursued an international treaty banning the deployment of weapons in outer space. In 2008, China and Russia submitted a draft "Treaty on Prevention of the Placement of Weapons in Outer Space and of the Threat or Use of Force Against Outer Space Objects" (PPWT) to the UN Committee on the Peaceful Use of Outer Space (COPUOS). China and Russia have updated the PPWT several times in an effort to win more support, but the US has consistently opposed these efforts. Calling the original version of the treaty and subsequent updates as "fundamentally flawed," US officials argue that the treaty lacks an adequate verification mechanism as well as any restrictions on ground-based anti-satellite weapons.[108] According to the US representative to the Geneva-based Conference on Disarmament, these shortcomings in the PPWT mean that a nation "could develop a readily deployable space-based weapons break-out capability" shortly after withdrawing from the treaty.[109]

Further work on this treaty and efforts to prevent an arms race in space have continued with strong Russian and Chinese support. In 2017, the UN general assembly voted to establish a group of diverse governmental experts to study practical measures for the Prevention of an Arms Race in Outer Space (PAROS). The UN established a similar committee in 1985 in reaction to the brief period of elevated tensions related to the Reagan-era SDI program. Until 1994, this committee met annually to examine issues relating to PAROS until it was disestablished in the face of US resistance to further space legal regimes and fading Cold War tensions. Prior to the first session of the new committee, China, in cooperation with Russia, hosted an international workshop for the committee members in Beijing. The Chair of the committee was effusive in his praise of the Chinese and Russian led workshop, crediting it with being instrumental to the group's work.[110] At least at the surface level, continued Chinese and Russian support for arms control in the face of US resistance portrays them as responsible security seeking nations in space.

While US resistance to a binding treaty has not stopped China from continuing to support a treaty designed to prevent an arms race in space its motivations may be less than benign. A public push for a treaty fits neatly within Chinese doctrine of seeking psychological advantage by "presenting oneself as the aggrieved party [while] holding the moral and legal high ground."[111] A long and public push for a legal regime in space combines aspects of psychological and legal warfare that work to China's advantage. A treaty that at the surface level portrays peaceful intent despite being difficult to enforce gives China public legitimacy as an aggrieved state reacting to US aggression in space. At the same time, the treaty offers the possibility of creating a legal regime that constrains the US. The shortcomings highlighted by the US in the PPWT only reinforce this view. Further support for the idea that Chinese peace efforts are hypocritical or possibly even outright malicious are found in China's actions in space, notably its anti-satellite weapons (ASAT) tests.

China has conducted one very public test of an ASAT and is suspected of carrying out several others. The first successful test, conducted in 2007, destroyed a non-functioning weather satellite named Fengyun 1-C operating in LEO.[112] This test created international outcry due to the debris field that it generated in a relatively high orbit which is expected to remain in orbit for decades. It followed three previous failed attempts and was the first successful test of an ASAT since the end of the Cold War.[113] The negative global response forced China to announce that it planned no further tests. Despite this statement, in 2010 and twice in 2013 China likely carried out other much more sophisticated tests that attempted to escape notice by being non-destructive. US sources claimed the May 2013 test was for an ASAT system that could reach much higher orbits than the 2007 test, possibly demonstrating the ability to strike US GPS satellites in Medium Earth Orbit (MEO) or even satellites in GEO.[114] China never acknowledged any of these tests, claiming in the case of the May 2013 event that it had only launched a science experiment with "high-energetic particle detectors, magnetometers, and a barium-powder release experiment."[115] When asked directly by the media about the test, Chinese Foreign Ministry spokesman Hong Lei said he was not aware of any test and repeated the official position that "China has consistently advocated the peaceful use of outer space and is opposed to militarizing and conducting an arms race in outer space."[116]

Following this suspected test of a GEO capable ASAT, China tested yet another ASAT in 2014 which then US Air Force Lt. Gen. Jay Raymond confirmed was "successful" in public comments.[117] In response to this test, the US State Department called on China to "refrain from destabilizing actions—such as the continued development and testing of destructive anti-satellite systems."[118] China characterized the test as nothing more than a "land-based anti-missile technology experiment" still refusing to acknowledge the existence of an ASAT program.[119] Testing still has not stopped with the most recent test conducted in July 2017 according to US military officials.[120]

Chinese behavior in space has several possible explanations. First, China must know that testing ground-based ASATs is unlikely to escape notice by the US intelligence community. Launches into space are virtually impossible to hide from either an observer on the ground or assets in orbit. Even the aging US Defense Support Program (DSP) missile warning satellites can detect small theater ballistic missiles which would not be nearly as easy to spot as a space launch, and the DSP satellites have since been replaced by the more effective Space Based Infrared System (SBIRS) constellation.[121] Once detected, US ground tracking stations can observe on-orbit behavior and an effective ASATs trajectory is much different from that of a newly launched satellite.[122] Given this information, the most likely explanation for conducting these tests in this manner is to create a deterrent effect against the US while maintaining deniability so that some credibility is still attached to its peace efforts. Behavior along these lines would most likely signal that China is a security seeking state with benign intent in space.

An alternate explanation is that ASAT testing is not the limit of Chinese space weaponization and its pursuit of an arms control agreement is merely an attempt to paint the US as the aggressor. While an ASAT has enough unique characteristics to make testing hard to hide, a weapon disguised as a typical satellite might escape any special notice. In its annual report to Congress on China, the US DOD described the PLAs ability "to use space-based systems— and deny them to adversaries—as central to modern warfare."[123] In pursuit of this strategy, the US report accused the PLA of developing "kinetic-kill missiles, ground-based lasers, and orbiting space robots."[124] If this accusation is true, then explanations for Chinese behavior based on its actions in space become more complex and are certainly not entirely benign. If ground-based lasers are designed to destroy or damage satellites passing overhead rather than create temporary and reversible effects, then it could signal a more malign intent, or it may still be acting as a deterrent. Orbiting space robots could be part of a broader effort to advance the technology of orbital repair and refueling as at least one commercial US company is attempting to do.[125] If this is the case, the US is assigning malicious intent to a dual-use capability simply because it exists. Alternatively, this could hint that China is pursuing a much broader active space weapons program beyond just ASATs and poses a serious threat to US space assets.

A third explanation is that the US military is magnifying the threat posed by China's space program. Earlier analysis showed that while China does have a robust space program, it is still small relative to the US. It is also true that the US military and national security community possess an information monopoly in space more so than anywhere else. As Stephen Van Evera argued, "militaries are prone to inflate threats, states will overspend groundlessly when militaries have an information monopoly that lets them alone assess the threat."[126] With a wealth of resources to analyze threats that no other state or private entity can come close to matching, it is impossible to verify US claims of Chinese actions outside of the most obvious ASAT tests. It is interesting though that in a 2017 interview US Gen. John Hyten, the senior US space officer, stated that he was more concerned about Chinese than Russian space capabilities even though "the threats we face are actually very small."[127] This single statement by Gen. Hyten does not invalidate his other testimony, but it can serve as a sign that while real threats do exist they are not yet on the scale that budget-driven congressional testimony would lead an observer to believe.

A final explanation for Chinese behavior in space is tied to the generally low level of transparency it provides into its military capabilities. China is a rising power by any measure of the latent material variables discussed above, but it is most likely still significantly weaker than the US in terms of explicit military power. Oriana Mastro theorizes that China is attempting to hide its military weak-ness from the US behind a façade of high technology to create uncertainty about

China's true deterrent capabilities while it addresses its vulnerabilities.[128] The most obvious support for this theory is that more is known about China's most advanced weapons than is known about the most fundamental aspects of the PLA such as its military budget, training, or readiness levels. She cites the fact that China actively advertises its high-tech capabilities, such as the J-20 fighter, in addition to making no real effort to hide its ASAT tests and other counterspace capabilities from the US.[129] Essentially creating uncertainty about where it is in its military development by exaggerating its deterrent capability is advantageous to China when faced with an existing hegemonic power that might be tempted to take pre-emptive action to prevent China's rise. Mastro believes that in order to compensate for the low level of capability transparency and the uncertainty this creates, China is instead maintaining a high level of transparency and consistency in messaging its peaceful intent.[130]

Of the four explanations for Chinese space behavior presented above no single explanation is likely complete. The verifiable facts are that China does have an ASAT program, it is pursuing arms control agreements in space, and it does have a substantially smaller budget for space programs. Its demonstrated ASAT capability stokes fear in the US national security space community; a community that is already primed to magnify threats given the vulnerability of the individual satellites it possesses. The US national security community also controls access to its information sources, so its conclusions on additional weapons testing are difficult or impossible to challenge. Combine this uncertainty with the fact of China's growing latent power, high-profile military capabilities in other areas, and aggressiveness in the South China Sea, and it is easy to see why US policymakers would attribute malign intent to Chinese space activities.

On the other hand, China's military budget has remained steady relative to its GDP, and it has invested a relatively small amount in its space program. Its A2/AD strategy is enabled by its growing space capabilities just as US power projection is, and China probably suffers from the same fears of losing its growing investment in its space assets that the US does. In addition, direct ascent kinetic-kill ASATs are a messy weapon as demonstrated by the debris generated during its 2007 test; a characteristic that limits their usefulness. The fact that China is also pursuing an arms control agreement albeit with significant flaws from the US perspective is also a significant data point. Combine these facts with China's limited historical and security-driven territorial goals discussed above and China looks very much like a security seeking state whose limited expansion is targeted at ensuring its future economic and territorial security. Under this framework, Chinese behavior is that of a state that has little desire to enter into a space arms race but is hedging its bets with capabilities designed to either deter or compel the US because it actively fears US intentions and capabilities.

From the Chinese perspective, American rhetoric and actions in space are not doing anything to relieve fears that the US is striving for space hegemony. The continued use of nationalist and aggressive language directed at China, as well as active opposition to a new space legal regime, paint the US as at best a dangerously insecure status quo state and at worst as the domineering hegemon that China fears. Also, while the US accuses China of a lack of transparency in space, US actions there are cloaked in just as much, if not more, secrecy. The occasional announcement of previously unknown space capabilities like the Geosynchronous Space Situational Awareness Program (GSAP) only reinforces the secrecy that other US space programs operate under.[131] With a massive military budget and a space budget larger than all but a handful of other nations' entire defense budgets, China cannot help but see itself as under threat.

The interaction between the US and China has all the hallmarks of a security dilemma that is increasing in severity. Both nations publicly state that they want peace, yet both are also pursuing arms. China is more consistent in its peaceful language though its actions demonstrate that it is not entirely satisfied with the security provided by the status quo. Meanwhile, US language with regard to space is rapidly moving away from the traditional view of space as a sanctuary and instead emphasizing a strategy of dominance and warfighting designed to protect its space assets. The evidence presented shows that both states see themselves as security seekers addressing insecurities. Despite being defensive in intent, the actions that are taken by both nations to address their insecurities outwardly manifest themselves as revisionist behavior which reinforces existing suspicions that the other state has malign intent and is greedy. Under these conditions, a severe security dilemma is present where dangerous and sub-optimal arms racing is likely to occur.

Russia

Russia is in many ways the opposite of China. It has not enjoyed the economic success that China has despite predictions of high future economic growth by Goldman Sachs' analysts in 2001.[132] Formerly a superpower at the heart of the USSR, today it is much diminished in territory and prestige. Russian power and influence are a shadow of what the Soviet Union once wielded. Despite this, it remains a nuclear power with significant space capability, who in US eyes has regained the status of a great power competitor through its aggressive actions toward its neighbors and involvement as a powerbroker in US conflicts in the Middle East. This section will look closely at the material and strategic factors creating tension between the US and Russia. With the strategic friction points established an analysis of Russian actions and rhetoric toward space follows.

After the end of the Cold War, Russian and US relations improved dramatically even as the Russian economy collapsed and a state of relative anarchy existed in the country. Economic reforms designed to privatize state-owned assets benefited only a relative few, and in 1996 and again in 1998 the Russian Federation needed humiliating multi-billion-dollar bailouts from the International Monetary Fund and international donors to stave off financial collapse.[133] Russia needed other less obvious bailouts as well. The US kept Russia's formerly world-leading space industry on life support throughout the 1990s ensuring that it received more than $1.2 billion in support from NASA in the form of contracts for the International Space Station and other joint space ventures.[134] These contracts awarded by NASA to the Russian Space Agency were the space equivalent of the US Department of Energy's (DOE) Initiatives for Proliferation Prevention (IPP) program which provided jobs to Russian weapons scientists to prevent them from selling their knowledge elsewhere.[135]

A drastic turn around in Russia's economic deterioration coincided with Russian President Vladimir Putin's rise to power in 2000. Driven by rising gas and oil prices, between 1999 and 2007 Russian GDP recovered to levels last seen in the Soviet-era (see Figure 3.2).[136] This period of economic growth allowed Russia to once again reassert itself on the global stage and address grievances built up since the end of the Cold War.

The origin of modern disputes between the US and Russia lies in the eastward expansion of North Atlantic Treaty Organization (NATO) after the fall of the Soviet Union. In 1990, during the final months of the Soviet Union, Soviet President Mikhail Gorbachev and US Secretary of State James Baker reached an informal agreement to not expand the US-led NATO eastward if the Soviets agreed to German reunification.[137] Secretary Baker publicized this agreement at a press conference stating that there "should be no extension of NATO forces eastward in order to assuage the security concerns of those [to] the East of Germany."[138] On the basis of this understanding, negotiations for German reunification and Soviet retrenchment in Eastern Europe commenced and in little more than a year the Soviet Union itself collapsed.

The failure to obtain written assurances against NATO expansion proved to be a mistake for Russia, as the absence of a formal agreement has since become a major point of contention. Russia argues that NATO expansion threatens Russian security and is in violation of the agreement reached between Baker and Gorbachev. NATO's counter-argument is that "no such pledge was made, and no evidence to back up Russia's claims has ever been produced."[139] In the context of the Cold War, where informal agreements between the US and USSR were a vital part of interstate relations, NATO's claims while technically true, are disingenuous.

NATO expanded significantly in the years following the breakup of the Soviet Union. In the period up to 2008, NATO added ten states that were formerly within the Soviet Union or counted as its close allies. These new additions included Poland in 1999 and the Baltic states in 2004 which share a border with Russia. Each move toward expansion of NATO eastward met resistance from the Russian government. The inclusion of the Baltic states was especially contentious, prompting the Russian Parliament to adopt a formal resolution denouncing it and called for a reconsideration of Russian defense strategy.[140] Russia's foreign minister, Sergei Lavrov, stated that the addition of the Baltic states was a mistake and that "the presence of American soldiers on our border has created a kind of paranoia in Russia."[141] Disregarding Russian complaints, the Lithuanian defense minister typified NATO's dismissive attitude when he attributed Russian concerns to those that "are sad to lose the territory of the old empire."[142]

In 2004, Russia was still too weak to oppose US and NATO encroachment eastward with anything but words. Its economic recovery was in progress, but its military was hollowed out by underinvestment. In 1989 the USSR spent $225 billion on defense. In 1992, the Russian Federation's first full year of existence, it spent only $41 billion, this number decreased further to $23 billion in 1997, and following the 1998 financial crisis, reached a nadir of just $13.6 billion.[143] This represented just 3.2 percent of the US defense budget for the same year and matched what the US spent on NASA's budget alone.[144] The Russian defense budget had only just returned to 1997 levels in 2004, and it was far from prepared to provoke the US which had successfully invaded Iraq the year before and seemed to be enjoying unquestioned military success.

The situation had changed dramatically by 2008 when US President George W. Bush proposed expanding NATO again and including both the Ukraine and Georgia. US military campaigns in Iraq and Afghanistan had turned into bloody insurgencies, and the US military and its NATO allies were preoccupied in dealing with the situation. Meanwhile, Russian military investment had been increasing rapidly and had returned to its 1992 peak. Under these conditions, President Bush proposed putting both the Ukraine and Georgia on the path to full NATO membership. Surprisingly, the other NATO members rejected the US proposal for Georgian and Ukrainian membership though they agreed to reconsider it later that year and issued a statement declaring that "these countries will become members of NATO."[145] The temporary rejection of President Bush's proposal was likely due to the fierce resistance from President Putin who was attending the NATO conference for the first time. He argued that "NATO cannot guarantee its security at the expense of other countries' security."[146]

The inability to get a firm commitment to halt NATO expansion into the Ukraine and Georgia was a failure for President Putin that could not remain unresolved. Just two weeks after leaving the NATO meeting Putin issued a decree essentially recognizing the two breakaway Georgian provinces of Abkhazia and South Ossetia.[147] Following this action, a rapid series of provocations occurred between Georgia and Russia that ultimately resulted in a brief war between the two countries. Russian forces defeated their Georgian opponents, though not without some difficulties that exposed significant military weaknesses, and Russian forces took control of the two nominally independent breakaway provinces.[148]

The sequence of events that resulted in the Russian invasion of Georgia should have made it clear to the US that Russia would not tolerate further encroachment along its borders. Despite this action, NATO never withdrew its promise of future NATO membership for the Ukraine or Georgia. So, when the pro-Russian Ukrainian government was overthrown in a revolution after rejecting a proposal for a partnership with the European Union, Russia could not help but see a western plot.[149] Once the pro-Russian Ukrainian government fell at the end of February 2014, Putin moved quickly to secure Russian interests by seizing Crimea under the pretense of an independence vote less than a month later.[150] This action was strategically necessary for Russia due to the presence of its only warm-water naval base at Sevastopol which would be at risk under a pro-western regime intent on joining the EU and NATO. Maintaining access to the base, even under a nominally pro-Russian regime, had caused enough problems. Just a year prior, Russia negotiated a lease extension for the base that had required using various economic threats such as gas price hikes to convince the Ukrainian government to accept a continued Russian military presence.[151] Conveniently, this large Russian military presence enabled the quick seizure of the peninsula, though Russia made every effort to disguise its involvement and give itself a degree of deniability. These "little green men" as the Russian troops came to be known, made a weak attempt to disguise their origin by removing unit patches and other identifying insignia but were otherwise clearly Russian. Despite the obvious nature of Russian involvement, President Putin insisted that these troops were members of "self-defense groups" unaffiliated with the Russian government.[152]

This pattern of subversive political action supported by thinly disguised military action has since become the Russian strategy of choice, particularly in Eastern Ukraine. Western military analysts refer to this mixture of military and non-military means as Russia's strategy of "hybrid warfare," and it has caused significant concern in the US and other NATO countries due to its success in destabilizing Ukraine and the potential for use in the Baltic states.[153] The hybrid warfare approach of relying on proxy forces, leveraging internal political dissent, and using covert military forces to destabilize and overthrow a government presents a special challenge to NATO because it clouds the issue of just who is behind the attack. This confusion provides just enough deniability for Russian involvement that it could avoid triggering NATO's mutual defense clause when used against a member state. In mid-2018 NATO decided to clarify the issue of its response to Russia's hybrid threat strategy by announcing that "in cases of hybrid warfare, the Council could decide to invoke Article 5 of the Washington treaty, as in the case of armed attack."[154] NATO's action does not entirely remove the threat posed by Russian hybrid warfare, but it does mean that using it in the future will require further increasing the degree of deniability beyond the flimsy excuses used in Ukraine.

Hybrid warfare is a strategy of weakness, and Russian reliance on it is telling. The key characteristic of hybrid warfare is that it economizes the use of force by avoiding overt actions that might trigger a response.[155] By choosing this strategy, Russia is signaling not only that it is relatively weak militarily compared to its enemies, but also that it knows it. In order to prevent further western encroachment on its borders, Russia has been forced to rely on deception and disruption rather than on conventional capabilities. This reliance on deception and disruption has significant implications for its future military actions. Russia is unlikely to pursue a strategy that relies primarily on conventional forces knowing that it does not have the strength to prevail. Instead, when facing a conventionally superior opponent, its past actions imply that it would rely on asymmetric methods of warfare that have a high degree of deniability to slow or delay any response until it could present an opponent with a *fait accompli*. Cyber-attacks, sabotage, and political subversion meet this requirement, as does interfering with space assets vital for communications, reconnaissance, and navigation.

Knowing that the US military, and by proxy NATO, is heavily dependent on space-enabled capabilities, Russia might choose to pursue a hybrid strategy that includes attacks on space systems. Among the most significant problems with space warfare is establishing attribution, meaning establishing which nation is the source of the attack. The difficulty of establishing attribution in space makes it an attractive option for attacks when relying on a strategy of confusing and delaying an opponent while masking involvement. The issue is that unlike cyber-attacks or political subversion, building genuinely effective counterspace capabilities that can avoid easy attribution can become extremely expensive. For a nation with minimal economic growth and a struggling space program, this should be a deterrent to space weapons development. However, there are fears that Russia has decided to pursue this course of action despite the costs, reviving Soviet-era capabilities.

Russia has substantial historical experience with ASATs and other space weapons that could make developing updated systems economically feasible. From 1963 until 1982 the USSR fielded a co-orbital ASAT system that conducted more than 20 on-orbit tests.[156] This system, appropriately named Istrabitel Sputnikov (IS), "satellite destroyer" in Russian, had significant problems in testing and was limited to targets within a narrow range of low-Earth orbits. These limitations meant that IS did not cause as much excitement as an equivalent system would today, especially since at the time the US military was not heavily dependent on space for conventional capabilities and most vital national security satellites were in higher orbits beyond the Soviet system's reach.[157] Even so, the presence of a functioning ASAT system did put the USSR far ahead of the US in space weapons development, at least until the Reagan Administration. In 1983, following the announcement of SDI by President Reagan and a renewed focus on developing space warfare capabilities, then Soviet leader Yuri Andropov proposed a moratorium on further ASAT testing if the US would reciprocate.[158] Nothing came of this proposal as there were substantial issues with verification and the timing of the proposal raised suspicions that it was designed to hamstring the US from developing equivalent capabilities to the Soviets.[159]

The continued pursuit of SDI and the failure of Andropov's proposed space weapons ban spurred the Soviets into further research and development of space weapons systems. By 1985 the Soviets were working on a variety of designs including a ground-based kinetic-kill vehicle that could target satellites up to GEO, a constellation of space mines to be deployed near potential targets, and a system that would use a laser to disable electronic systems on satellites from orbit.[160] Most of these programs were just as prohibitively expensive and technically infeasible as the SDI program they were designed to counter. With the exception of efforts to improve the existing IS ASAT system, the various weapons programs were all canceled within a few years of inception. The one remaining major program, the Naryad-V ASAT, had only just begun initial testing when the Soviet Union disintegrated in 1991.[161]

Soviet-era research on ASAT technology was not completely wasted. In the spirit of new-found capitalism, Russian rocket designers quickly realized that a market existed for the commercial use of some of its former ASAT technology. The upper stage of the Naryad-V ASAT system, called "Briz," which was designed to sit atop a Soviet SS-19 ICBM was a prime candidate for commercialization.[162] Disguising the original origins and purpose of Briz, the Russian design firm KB Salyut teamed with Daimler-Benz Aerospace (DASA) to offer its services as a commercial launch vehicle with Briz as the upper stage and a convenient supply of former Soviet SS-19 ICBMs as the rocket body.[163] The combined venture between KB Salyut and DASA was named Eurockot and is currently an ongoing joint venture of ArianeGroup and the Russian Khrunichev Space Center.[164] When President Putin visited the Khrunichev Space Center in 2002 the company's leaders mentioned that some elements of the Naryad-V program existed within its current systems and raised the possibility that the program could be resumed.[165] While the successful commercialization of Soviet ASAT technology has likely preserved some knowledge of Soviet-era weapons programs, it is unlikely that it has done much beyond preserving the Russian space industrial base.

Russian development of systems designed to counter US space capabilities are not all unfounded fears of the rebirth of Soviet-era ASAT technology. For example, Russia is an active user and developer of GPS jamming systems with some Russian made GPS jamming devices observed in use by Iraq during the 2003 invasion.[166] Russian development of these systems has only progressed since then with widespread installation of sophisticated GPS jamming devices across Russia starting in 2015.[167] A national-level installation of GPS jamming systems within its borders is a prudent defensive move for Russia since many US weapon systems rely on GPS for targeting, though other Russian actions undercut the defensive nature of this deployment. In 2018, the US National PNT Advisory board accused Russia of active jamming and spoofing of GPS along its borders.[168] Spoofing is especially dangerous as it gives users of GPS false position data which could lead to serious navigation errors for planes and ships. The reason for these jamming and spoofing actions is not apparent, but it does have the effect of undermining confidence in the US GPS constellation and encouraging the use of the Russian GLONASS PNT system in regions bordering Russia. The interference could also be the result of careless testing of defensive systems in its border regions since relatively small GPS jammers can have regional effects.[169]

GPS jammers are not the probable limit of Russian space weapons research; there are also signs that Russia is developing new ground-based kinetic ASAT systems. One of these potential systems is the latest Russian air-defense system, the S-500 which according to President Putin is "capable of operating at ultra-high altitudes including near space."[170] Western experts disagree over whether this system is just a missile and air-defense system or whether it also has an ASAT mission, but it is probably the Russian equivalent to the US THAAD theater missile defense system.[171] In 2017, the Russian state news site Pravda did report that the S-500 was capable of destroying low-flying satellites.[172] This could be a case of dual-purpose, as any system capable of destroying ballistic missiles during mid-course can also be adapted to targeting satellites flying within its missile engagement area. Another more capable Russian air-defense system, the PL19 Nudol, is more likely to have an ASAT mission in addition to its primary ballistic missile defense mission. With a reported altitude envelope of 1,000 km, it can reach most satellites in LEO but falls far short of threatening the large number of satellites in orbits beyond that.[173] The PL19 is nominally a ballistic missile defense system equivalent to the US Ground-Based Midcourse Defense (GMD) system based in California and Alaska, so its role as an ASAT is open to interpretation.

That Russia and China can develop these systems is entirely the result of the US withdrawal from the ABM treaty. Once the US withdrew from the ABM treaty, it opened the door for both countries to pursue similar systems. President Putin argued in 2015 that it was the unilateral US withdrawal from the ABM treaty more than any other factor that "pushes us to a new round of the arms race, because it changes the global security system."[174] The dual-use nature of ABM technology means that Russia cannot avoid developing an ASAT system at the same time it develops an ABM system. The US demonstrated this in 2008 during Operation Burnt Frost when it used the ship-based SM-3 missile to shoot down a malfunctioning NRO satellite. The SM-3 was designed to intercept incoming ballistic missiles yet in a period of just a few weeks was repurposed to destroy a satellite at an altitude of 246 km.[175] US actions during Burnt Frost make its condemnation of other countries' ASAT and missile defense tests seem hypocritical. When confronted with this comparison in 2008, then USSTRATCOM Commander, Gen. Kevin Chilton remarked that "there's no comparison" between China's ASAT test the previous year and US actions.[176] While China's tests were clearly targeted at satellites, Russia's system remains a missile defense system, at least publicly. This does not stop the US from grouping the two together as US Director of National Intelligence Dan Coats did when he stated that "both Russia and China continue to pursue anti-satellite weapons as a means to reduce US and allied military effectiveness."[177] Accusations that Russia is pursuing a dedicated ASAT system may be the result of exaggerated US fears, but that does not mean that Russia is incapable of developing sophisticated strategic weapons.

In a highly publicized speech to the Russian Federal Assembly in March 2018 President Putin referenced a new laser weapons system, a hypersonic cruise missile, and other strategic capabilities. Pointing out that these new capabilities are not "something left over from the Soviet Union," Putin reassured the assembly that despite its relatively small military budget Russia is still capable of developing complex state-of-the-art strategic weapon systems.[178] The announcement of these sophisticated weapons systems taken at face value is a threatening statement of strength reminiscent of the Cold War which is how it was received by many.[179] However, looking beyond the descriptions of sophisticated weapons technology featured in his speech it is not hard to see that Putin is looking for recognition of Russia's security concerns. He took special care to again highlight that these new weapons were developed in direct response to the US withdrawal from the ABM treaty and development of missile defenses.[180] In addition, he warned against bringing US missile defenses and other "NATO infrastructure closer to the Russian border." The speech also included the reasoning behind revealing these weapons systems. Expressing exasperation with the repeated failure of arms control talks, he finished with "nobody wanted to listen to us, so listen now."[181] Rather than a return to Cold War levels of mutual hostilities, this speech shows that Putin is seeking a return to Cold War levels of respect for Russian security concerns and the weapons announcement was a way of gaining US attention.

The technical sophistication and continued development of Russian space capabilities alluded to in Putin's speech is getting the attention he sought, just not in the manner he might have desired. At the Geneva-based Conference on Disarmament, the US State Department accused Russia of testing a space weapon in 2017 under the guise of a "space apparatus inspector" as part of a space weapons program that has been in existence since at least 2009.[182] The US representative also pointed to Putin's March 1 speech as proof that Russian "space troops have received a mobile laser system."[183] As further proof of Russian space weapons programs the US representative pointed to statements by Russia's Space Force Commander that "assimilating new prototype weapons [into] Space Forces' military units" is the "main task facing the Aerospace Forces Space Troops."[184] The representative then contrasted the evidence of Russian weapons programs with Russian support for the PPWT space treaty and its public commitment to not being the first to deploy space weapons, pointing out Russian duplicity and lack of transparency. The US representative summarized the litany of complaints against Russia as "yet further proof that Russian actions do not match their words."[185] This complaint, as well as other statements by US officials, leave little doubt that the US will interpret almost any Russian action in space as hostile to US interests.

Despite US concerns, Russian actions and behavior seem remarkably consistent and transparent. Since coming to power almost two decades ago, President Putin has repeatedly stated that he opposes further NATO expansion and will do everything possible to prevent it. From his perspective NATO and the US seem deaf to his clearly stated desires leaving him no option but to oppose further expansion with increasingly aggressive behavior. In a 2018 diplomatic conference in Moscow he again reiterated precisely how he would respond to further NATO expansion:

> The key to providing security and safety in Europe is in expanding cooperation and restoring trust, and not in deploying new NATO bases and military infrastructure near Russia's borders, which is what is taking place now. We will respond appropriately to such aggressive steps, which pose a direct threat to Russia. Our colleagues, who are trying to aggravate the situation, seeking to include, among others, Ukraine and Georgia in the orbit of

the alliance, should think about the possible consequences of such an irresponsible policy. We need a new, positive agenda aimed at collaboration and attempts to find common ground.[186]

In light of statements like this backed by actions that are entirely consistent with its stated policy, it is hard to see how Russian leadership could be any more transparent. President Putin is acutely aware of the economic shortcomings of his country and the challenges associated with an arms race with the US, but he is also unwilling to compromise on what he interprets as Russian security needs which creates the conditions for competitive policies toward the US.

The relationship between the US and Russia is one that shows all the signs of a severe security dilemma. While Russia sees itself as a security seeker desperately trying to preserve or restore the status quo, its actions in support of that effort are seen as greedy, competitive, and revisionist by the western world. The cause of this misunderstanding is that the US sees itself and NATO as benign security seeking entities and armed resistance to its expansion must, therefore, signal malign intent. After all, from the US perspective why would any nation fear a defensive alliance like NATO? From the Russian perspective, NATO is a US proxy that is, and always will be, a hostile military alliance formed to threaten Russian security. Combine NATO's creeping expansion with the US's unilateral withdrawal from the ABM treaty and active resistance to serious discussion on arms control in space, and from the Russian viewpoint the US is a hostile and greedy power with malign intent.

The one moderating feature of the security dilemma between the two states is that there is a significant and growing disparity in material power between the two countries. This forces Russia to pursue strategies designed to achieve maximum results with minimum investment while pursuing whatever degree of deniability is necessary to prevent anything more than a token response. With the Russian legacy of Soviet space-flight technology, the material disparity is smaller in the space domain than in almost any other. Russia has every incentive to try and blunt US advantages by competing in space where it is operating at less of a disadvantage and where attribution is difficult. Especially since US withdrawal from the ABM treaty, actions in space, and resistance to an admittedly flawed space arms control treaty place the US in a compromised moral position. The result is that the security dilemma between the US and Russia is severe, and this is likely to result in sub-optimal competition within the space domain.

Summary

The purpose of this chapter was to examine the material, motive, and informational variables between the US and China and the US and Russia to determine the existence and severity of security dilemmas within the two rivalries. The evidence shows that a security dilemma does exist and that the space domain is subject to sub-optimal arms racing as a result. Each state sees itself as a security seeker and its competitor as most likely a greedy state (see Figure 3.6 below). This misperception and sub-optimal arms racing is driving a negative security spiral as each nation's defensive actions further decrease the security of its adversary. There are different drivers behind the security dilemma between the US and each adversary, though given the nature of the space domain any strategy for reassurance must be valid for all states. This creates difficulty as China and Russia have different motivations and intentions.

While China is likely a security seeker, it does have what appears to be some greedy tendencies from the US perspective, the same is true for China's view of US behavior. The US has status quo motives but is also increasingly insecure in its global status, which is creating revisionist tendencies, most obviously in space. Meanwhile, China is not entirely satisfied with the status quo which was established when it was much weaker. With the gap in both latent and explicit power between the US and China decreasing each year, China is seeking to exert greater control over regional security at the expense of the US. This behavior manifests through assertive claims to the South China Sea as well as the development and testing of A2/AD and ASAT capabilities designed to deter US military action. Essentially the US and China are both security seekers whose information about the other leads them to conclude that the other state could be either an insecure security seeker or have outright greedy motives. Doubt, mistrust, and a lack of transparency lead each nation to treat the other as a greedy state, leading to actions and reactions that only reinforce this viewpoint.

Unlike China, the evidence shows that Russia is not a revisionist state but rather an extremely insecure security seeker trying to preserve the status quo. At first glance, Russian actions since the invasion of Georgia in 2008 seem to support the position that it is highly revisionist. Looking closer, it becomes evident that Russia is in fact trying and failing to preserve the status quo that existed prior to US withdrawal from the ABM treaty and the inclusion of

Figure 3.6 State motives and information.

numerous Eastern European countries into NATO. That Russia is actively resisting further changes to the status quo does not mean that Russia is a greedy state, the opposite is actually true. From the Russian perspective, it is the US that appears as the greedy revisionist state by constantly seeking NATO expansion along Russia's borders, canceling or avoiding treaties, and hypocritically denouncing the Russian development of missile defenses.

This chapter established that a security dilemma does exist in space driven by misperception of each state's security needs leading to a negative security spiral. The question then is how can this spiral be broken? What specific strategy can the US pursue that would avoid a sub-optimal arms race that only harms the security of all states involved? In the previous chapter, we discussed the three primary strategies for signaling benign intent and reassuring an adversary: pursuing arms control agreements, shifting toward a defensive posture, and unilateral disarmament. The likely success of these strategies requires information on one more material variable that remains to be discussed, the offense–defense balance. The next chapter will investigate the nature of the offense–defense balance in space and based on this, and the information about motives from this chapter, examine the feasibility of various approaches to reassurance that can end the space arms race that is only just beginning.

Notes

1 "Global Trends: Paradox of Progress" (National Intelligence Council, 2018), www.dni.gov/index.php/the-next-five-years/space.
2 Yanek Mieczkowski, ed., *Eisenhower's Sputnik Moment, The Race for Space and World Prestige*, The Race for Space and World Prestige (Ithaca, NY: Cornell University Press, 2013), 20.
3 "Reaction to the Soviet Satellite—A Preliminary Evaluation" (White House Office of the Staff Research Group, October 16, 1957), 1, Box 35, Special Projects: Sputnik, Missiles and Related Matters, Eisenhower Library.
4 "Proposed News Release from the National Academy of Sciences Regarding Soviet Plans to Launch an Earth Satellite as Part of the International Geophysical Year Program" (National Academy of Sciences—S.D. Cornell, June 10, 1957), Box 625, OF 146-F-2 Outer Space, Earth-Circling Satellites, Eisenhower Library.
5 "Progress Report on the US Scientific Satellite Program (NSC 5520)" (White House Office, National Security Council Staff, October 3, 1956), 1, Box 38 Outer Space (1)-(6), Eisenhower Library.
6 "Draft Position Paper for UN Ad hoc Committee on Peaceful Uses of Outer Space: Legal Problems Which May Arise in the Exploration of Space" (White House Office of the Staff Secretary: Records, 1952–1961, April 22, 1959), 8, Box 24, Space Council (7), Eisenhower Library.
7 James R. Killian, "Minutes from the Meeting of the President's Science Advisory Committee: Public Information Handling of Reconnaissance Satellites" (President's Science Advisory Committee, January 19, 1960), Box 15, White House Office of the Special Assistant for Science and Technology (James R. Killian and George B. Kistiakowsky): Records, 1957–1961 Pre-Accession and Accession 76–16, Eisenhower Library.

8   Paul B. Stares, *The Militarization of Space: U.S. Policy, 1945–1984* (Ithaca, NY: Cornell University Press, 1985), 54.

9   "Treaty on Principles Governing the Activities of States in the Exploration and the Use of Outer Space, Including the Moon and Other Celestial Bodies" (Resolution Adopted by the UN General Assembly 2222 XXI, December 19, 1966).

10 "Treaty Between the United States of America and the Union Of Soviet Socialist Republics On The Limitation Of Anti-Ballistic Missile Systems," May 26, 1972, www.state.gov/t/avc/trty/101888.htm#text.

11 Ronald Reagan, "Address to the Nation on Defense and National Security," www.atomicarchive.com/Docs/Missile/Starwars.shtml.

12 "National Security Decision Directive 85: Eliminating the Threat from Ballistic Missiles" (White House, March 25, 1983), www.hsdl.org/?view&did=463005.

13 Roger Handberg, *Seeking New World Vistas: The Militarization of Space* (West Port, CT: Praeger Publishers, 2000), 82.

14 Ibid., 91.

15 Steven Lambakis, "Space Control in Desert Storm and Beyond," *Orbis* 39, no. 3 (Summer 1995).

16 Handberg, *Seeking New World Vistas: The Militarization of Space*, 105.

17 "Commission to Assess United States National Security Space Management and Organization" (US government, January 11, 2001), 1.

18 Ibid., 8.

19 Ibid.

20 Ibid., 10.

21 Jeremy Singer, "War on Terror Supersedes 2001 Space Commission Vision," *Space News*, June 29, 2004, https://spacenews.com/war-terror-supersedes-2001-space-commission-vision/.

22 Terrence Neilan, "Bush Pulls Out of ABM Treaty; Putin Calls Move a Mistake," *New York Times*, December 13, 2001, www.nytimes.com/2001/12/13/international/bush-pulls-out-of-abm-treaty-putin-calls-move-a-mistake.html.

23 Ibid.

24 "1996 US National Space Policy" (The White House National Science and Technology Council, September 19, 1996), 1.

25 "2006 US National Space Policy" (The White House National Science and Technology Council, August 31, 2006), 1–2.

26 Ibid., 2.

27 "2010 National Space Policy of the United States of America," June 28, 2010, 3.

28 Ibid., 7.

29 "2010 US National Security Space Strategy, Unclassified Summary" (Department of Defense and the Director of National Intelligence, January 2011), 5.

30 "President Donald J. Trump Is Unveiling an America First National Space Strategy," The White House, March 23, 2018, www.whitehouse.gov/briefings-statements/president-donald-j-trump-unveiling-america-first-national-space-strategy/.

31 Ibid.

32 Christina Wilkie, "Trump Floats the Idea of Creating a 'Space Force' to Fight Wars in Space," *CNBC*, March 13, 2018, www.cnbc.com/2018/03/13/trump-floats-the-idea-of-creating-a-space-force-to-fight-wars-in-space.html.

33 Donald J. Trump, "Remarks by President Trump at a Meeting with the National Space Council and Signing of Space Policy Directive-3" (White House East Room, June 18, 2018), www.whitehouse.gov/briefings-statements/remarks-president-trump-meeting-national-space-council-signing-space-policy-directive-3/.

34 Mike Pence, "Remarks by Vice President Pence on the Future of the U.S. Military in Space" (The Pentagon, August 9, 2018), www.whitehouse.gov/briefings-statements/remarks-vice-president-pence-future-u-s-military-space/.

35 Ibid.

36 Ibid.

37 Robert Burns, "Pentagon Chief Defends His Reversal on Space Force, Says It's the Right Thing to Do," *Military Times*, August 13, 2018, www.militarytimes.com/news/your-military/2018/08/13/pentagon-chief-defends-his-reversal-on-space-force-says-its-the-right-thing-to-do/.

38 Ibid.

39 Travis J. Tritten, "Jim Mattis Originally Opposed Space Force Over Sequester, Deputy Says," *Washington Examiner*, August 10, 2018, www.washingtonexaminer.com/policy/defense-national-security/jim-mattis-originally-opposed-space-force-over-sequester-deputy-says.

40 David L. Goldfien, "Military Space Organization, Policy, and Programs," § Subcommittee on Strategic Forces (2017), 22, www.armed-services.senate.gov/imo/media/doc/17-46_05-17-17.pdf.

41 Ibid.

42 John E. Hyten, "Hearing to Receive Testimony on United States Strategic Command in Review of the Defense Authorization Request for Fiscal Year 2019 and the Future Years Defense Program," § Committee on Armed Services (2018), 25.

43 John E. Hyten, "Statement of John E. Hyten, Commander United States Strategic Command, Before the Senate Committee on Armed Services," § Senate Committee on Armed Services (2018), 3.

44 "Space Force Fact Sheet," accessed February 8, 2020, www.spaceforce.mil/About-Us/Fact-Sheet.

45 Barry Posen, *The Sources of Military Doctrine: France, Britain, and Germany Between the World Wars* (Ithaca, NY and London: Cornell University Press, 1986), 47–50.

46 "Union of Concerned Scientists: Satellite Database" (Union of Concerned Scientists, May 1, 2018), www.ucsusa.org/nuclear-weapons/space-weapons/satellite-database#.W7eDtPZFxEZ.

47 "National Aeronautics and Space Administration FY 2018 Budget Estimates" (NASA, 2017), BUD-1, www.nasa.gov/sites/default/files/atoms/files/fy_2018_budget_estimates.pdf.

48 "National Oceanic and Atmospheric Administration FY 2018 Budget Summary," Budget Summary (NOAA, 2017), 42, www.corporateservices.noaa.gov/nbo/fy18_bluebook/FY18-BlueBook-508.pdf.

49 Sandra Erwin, "Some Fresh Tidbits on the U.S. Military Space Budget," *Space News*, March 21, 2018, https://spacenews.com/some-fresh-tidbits-on-the-u-s-military-space-budget/.

50 Wilson Andrews and Todd Lindeman, "The Black Budget," *Washington Post*, August 29, 2013, www.washingtonpost.com/wp-srv/special/national/black-budget/?noredirect=on.

51 "SIPRI Military Expenditure Database," Database (Stockholm International Peace Research Institute), accessed September 12, 2018, www.sipri.org/databases/milex.

52 Exact numbers for China's space budget are not available, though estimates in various sources range from as low as $1.3 billion to as high as $20 billion. The lower end estimates are certainly too low since China charges $70 million for a space

launch to commercial customers under significant price pressure from SpaceX. With 17 launches in 2017, launch costs alone were at least $1.1 billion even before the cost of maintaining its own broader space program is considered. The truth is most likely between $6 and $11 billion. The OECD estimate was $6.1 billion for 2014 Chinese space expenditures and $5.2 billion for Russia. As a comparison, their estimate for US spending was $39 billion for 2014 which is low compared to the calculation discussed above. Russian spending for its human space-flight program is public and was cut by two-thirds from a projected $6.4 billion annually to $2.2 billion annually in 2014 due to budgetary issues. Similar budget cuts likely impacted the secret military portion of the Russian space budget given its close ties with Roscosmos which means a total budget of $5 billion is probably accurate. For the OECD reference see: OECD, "Space budgets of selected OECD and non-OECD countries in current USD, 2013," in *Readiness Factors: Inputs to the Space Economy* (Paris: OECD Publishing, 2014).

53   "Union of Concerned Scientists: Satellite Database" Note: Iridium is considered a US government satellite constellation by the UCS due to its heavy reliance on US military contracts.

54   Gerhard Beutler et al., "The System: GLONASS in April, What Went Wrong," *GPS World*, June 24, 2014, http://gpsworld.com/the-system-glonass-in-april-what-went-wrong/.

55   Brian Weeden, "Space Situational Awareness Fact Sheet" (Secure World Foundation, May 2017), https://swfound.org/media/205874/swf_ssa_fact_sheet.pdf.

56   "Union of Concerned Scientists: Satellite Database."

57   Caleb Henry, "OneWeb Asks FCC to Authorize 1,200 More Satellites," *Space News*, March 20, 2018, https://spacenews.com/oneweb-asks-fcc-to-authorize-1200-more-satellites/.

58   Eric Berger, "In 2017, the US Led the World in Launches for the First Time since 2003: With 18 Orbital Flights, SpaceX Drove the Surge in US Missions Last Year," *ArsTechnica*, January 3, 2018, https://arstechnica.com/science/2018/01/in-2017-the-us-led-the-world-in-launches-for-the-first-time-since-2003/.

59   Ibid.

60   Andrew Jones, "Chinese Space Launch Vehicle Maker Provides Updates on Long March Launch Plans, New Rockets," *GB Times*, March 2, 2018, https://gbtimes.com/chinese-space-launch-vehicle-maker-provides-updates-on-long-march-launch-plans-new-rockets.

61   "Long March-5 (LM-5)" (China Aerospace Science and Technology Corporation), accessed September 12, 2018, http://english.spacechina.com/n16421/n17215/n17269/n19031/c125250/content.html.

62   "The World's Most Powerful Rocket: Falcon Heavy" (SpaceX), accessed September 12, 2018, www.spacex.com/falcon-heavy.

63   Sutirtho Patranobis, "China Plans to Reduce Satellite Launch Prices, ISRO Says We Can Do That Too," *Hindustan Times*, November 14, 2017, www.hindustantimes.com/world-news/india-china-in-race-to-reduce-rocket-launch-prices/story-mF7X9RwS5ai1rCjUDYb2zH.html.

64   Andrew Jones, "Chinese Commercial Rocket Smart Dragon-1 Reaches Orbit with FirstLaunch,"*SpaceNews*,August 19, 2019, https://spacenews.com/chinese-commercial-rocket-smart-dragon-1-reaches-orbit-with-first-launch/.

65   "Federal Space Agency and United Rocket and Space Corporation Was Mergered in the New State Corporation—Roscosmos" (Roscosmos, January 23, 2015), http://en.roscosmos.ru/20365/.

66 Eric Berger, "Russia Appears to Have Surrendered to SpaceX in the Global Launch Market," *ArsTechnica*, April 18, 2018, https://arstechnica.com/science/2018/04/russia-appears-to-have-surrendered-to-spacex-in-the-global-launch-market/.

67 "Russian Deputy PM Sees No Reason for Competing With Musk on Launch Vehicles Market," *TASS*, April 17, 2018, sec. Science and Space, http://tass.com/science/1000229.

68 Joshua R. Itzkowitz Shifrinson and Michael Beckley, "Debating China's Rise and U.S. Decline," *International Security* 37, no. 3 (December 13, 2012): 173, https://doi.org/10.1162/ISEC_c_00111.

69 Kenneth Waltz, *Theory of International Politics* (Long Grove, IL: Waveland Press, 2010), 123–128.

70 John J. Mearsheimer, *The Tragedy of Great Power Politics* (New York, NY: W. W. Norton & Company, 2014), 40–42.

71 Charles L. Glaser, "A U.S.–China Grand Bargain?," *International Security* 39, no. 4 (Spring 2015): 53.

72 Jacob K. Javits, "Congress and Foreign Relations: The Taiwan Relations Act," *Foreign Affairs* 60, no. 1 (1981): 59, https://doi.org/10.2307/20040989.

73 "Taiwan Relations Act," Pub. L. No. 96–98, 3301 22 (1979), sec. 2b.

74 Chinese Vice Premier Qian Qichen quoted in Chen-yuan Tung, "An Assessment of China's Taiwan Policy under the Third Generation Leadership," *Asian Survey* 45, no. 3 (2005): 346.

75 Chinese Foreign Minister Yang Jiechi quoted in David Scott, "Conflict Irresolution in the South China Sea," *Asian Survey* 52, no. 6 (2012): 1029–1030.

76 Derek Grossman, "Why March 2018 Was an Active Month in Vietnam's Balancing Against China in the South China Sea," *The Diplomat*, March 23, 2018, https://thediplomat.com/2018/03/why-march-2018-was-an-active-month-in-vietnams-balancing-against-china-in-the-south-china-sea/.

77 "The South China Sea Arbitration (The Republic of the Philippines v. The People's Republic of China)," *The International Journal of Marine and Coastal Law* 31, no. 4 (November 22, 2016): 1.

78 Chengliang Wu, "People's Daily Slams South China Sea Arbitration Tribunal for Being Political Tool," *Peoples Daily*, July 14, 2016, http://en.people.cn/n3/2016/0714/c90000-9085753.html.

79 Chad de Guzman, "Duterte Asserts Arbitral Ruling on South China Sea Not during His Term," *CNN*, May 21, 2018, http://cnnphilippines.com/news/2018/05/21/duterte-arbitral-ruling-south-china-sea.html.

80 Jim Gomez, "Duterte: China Should Temper Its Behavior in Disputed Waters," *Associated Press*, August 15, 2018, www.apnews.com/21030e69e14345bb9ce6b8bd91ce308a.

81 "Strait of Malacca Key Chokepoint for Oil Trade," *The Maritime Executive*, August 27, 2018, www.maritime-executive.com/article/strait-of-malacca-key-chokepoint-for-oil-trade.

82 ZhongXiang Zhang, "Why Are the Stakes So High?," in *Rebalancing and Sustaining Growth in China*, vol. 2012 (ANU Press, 2012), 335.

83 Ibid., 336.

84 B. Gellman, "U.S. and China Nearly Came to Blows in '96; Tension over Taiwan Prompted Repair of Ties," *Washington Post*, June 21, 1998, sec. 1 of 2.

85 Terrence K. Kelly et al., eds., "Land-Based Anti-Ship Missiles in the Western Pacific," in *Employing Land-Based Anti-Ship Missiles in the Western Pacific* (RAND Corporation, 2013), 5.

86 Anthony H. Cordesman and Joseph Kendall, "How China Plans to Utilize Space for A2/AD in the Pacific," *The National Interest*, August 17, 2016, https://nationalinterest.org/blog/the-buzz/how-china-plans-utilize-space-a2-ad-the-pacific-17383.

87 Ibid.

88 Minnie Chan, " 'Unforgettable Humiliation' Led to Development of GPS Equivalent," *South China Morning Post*, November 13, 2009, www.scmp.com/article/698161/unforgettable-humiliation-led-development-gps-equivalent.

89 Ibid.

90 Michael Beckley, "China's Century? Why America's Edge Will Endure," *International Security* 36, no. 3 (2011): 41.

91 "Data Bank" (World Bank), accessed August 15, 2018, http://databank.worldbank.org/data/home.aspx.

92 Associated Press, "Rumsfeld: China Buildup a Threat to Asia," *NBC News*, June 4, 2005, www.nbcnews.com/id/8091198/ns/world_news-asia_pacific/t/rumsfeld-china-buildup-threat-asia/#.W7edKfZFxEZ.

93 "China Armed Forces Estimate" (Janes IHS, 2018) Data combined with latest US DOD report on Chinese Army force numbers which are significantly lower than Janes estimates (1.34 million vs 915,000). Interestingly US estimates for the PLAA are even lower than stated Chinese numbers which place their Army at 850,000.

94 "The Diversified Employment of China's Armed Forces" (Information Office of the State Council of the People's Republic of China, April 16, 2013), www.china.org.cn/government/whitepaper/node_7181425.htm.

95 Ibid.

96 "China's National Defense in 2010" (Information Office of the State Council of the People's Republic of China, March 2011), 5.

97 "China's Military Strategy 2015" (State Council Information Office of the People's Republic of China, May 2015), 16.

98 Cited in Dean Cheng, *Cyber Dragon* (Denver, CO: Praeger Publishers, 2017), 166.

99 "Challenges to Security in Space" (Defense Intelligence Agency, January 2019), 20.

100 This is due primarily to the lack of casualties among military personnel or visceral imagery of destruction capable of generating the political will necessary to support conflict outside the domain. This idea is explored more thoroughly in Chapter 4.

101 "The Diversified Employment of China's Armed Forces."

102 Phil Stewart and Ben Blanchard, "Xi Tells Mattis China Won't Give up 'even One Inch' of Territory," *Reuters*, June 26, 2018, https://uk.reuters.com/article/uk-china-usa-defence/xi-tells-mattis-china-wont-give-up-one-inch-of-territory-idUKKBN1JN06O.

103 "The Diversified Employment of China's Armed Forces."

104 Ge Hanwen, " 'Say No to Decline' and 'American Fortification': Trump's Grand Strategy," *Journal of International Security Studies,* 国防科技大学国际关系学院国际安全研究中心 *(International Studies College, National University of Defense Technology)*, no. 3 (March 11, 2018): 3.

105 Mingming Du, "China Releases White Paper on Facts and Its Position on Trade Friction With U.S.," *Peoples Daily*, September 24, 2018, http://en.people.cn/n3/2018/0924/c90000-9503026.html.

106 Ibid.

107 Yangyu Gao and Long Ke, "New Trends in US Space Cooperation Policy," *International Research Reference, People's Liberation Army Foreign Languages Institute*, no. 6 (2014): 5.

108 Jeff Foust, "U.S. Dismisses Space Weapons Treaty Proposal As 'Fundamentally Flawed,'"*SpaceNews*, September 11, 2014, https://spacenews.com/41842us-dismisses-space-weapons-treaty-proposal-as-fundamentally-flawed/.

109 Ibid.

110 "Report by the Chair of the Group of Governmental Experts on Further Practical Measures for the Prevention of an Arms Race in Outer Space" (United Nations office for Disarmament Affairs, January 31, 2019), https://s3.amazonaws.com/unoda-web/wp-content/uploads/2019/02/oral-report-chair-gge-paros-2019-01-31.pdf.

111 Cheng, *Cyber Dragon*, 44.

112 Mike Gruss, "U.S. Official: China Turned to Debris-Free ASAT Tests Following 2007 Outcry," *Space News*, January 11, 2016, https://spacenews.com/u-s-official-china-turned-to-debris-free-asat-tests-following-2007-outcry/.

113 Carin Zissis, "China's Anti-Satellite Test" (Council on Foreign Relations, February 22, 2007), www.cfr.org/backgrounder/chinas-anti-satellite-test.

114 Brian Weeden, "Through a Glass, Darkly: Chinese, American, and Russian Anti-Satellite Testing in Space" (Secure World Foundation, March 17, 2014), 4–19.

115 Bill Gertz, "China Conducts Test of New Anti-Satellite Missile," *The Washington Free Beacon*, May 14, 2013, https://freebeacon.com/national-security/china-conducts-test-of-new-anti-satellite-missile/.

116 Ibid.

117 Colin Clark, "Chinese ASAT Test Was 'Successful:' Lt. Gen. Raymond," *Breaking Defense*, April 14, 2015, https://breakingdefense.com/2015/04/chinese-asat-test-was-successful-lt-gen-raymond/.

118 Zachary Keck, "China Conducted Anti-Satellite Missile Test," *The Diplomat*, July 29, 2014, https://thediplomat.com/2014/07/china-conducted-anti-satellite-missile-test/.

119 Zachary Keck, "China Conducts Third Anti-Missile Test," *The Diplomat*, July 24, 2014, https://thediplomat.com/2014/07/china-conducts-third-anti-missile-test/.

120 Bill Gertz, "China Carries Out Flight Test of Anti-Satellite Missile," *The Washington Free Beacon*, August 2, 2017, https://freebeacon.com/national-security/china-carries-flight-test-anti-satellite-missile/.

121 "Defense Support Program Satellites" (US Air Force Space Command Public Affairs Office, November 23, 2015), www.af.mil/About-Us/Fact-Sheets/Display/Article/104611/defense-support-program-satellites/.

122 Weeden, "Through a Glass, Darkly: Chinese, American, and Russian Anti-Satellite Testing in Space," 12–14.

123 "Annual Report to Congress: Military and Security Developments Involving the People's Republic of China" (Office of the Secretary of Defense, May 16, 2018), 61.

124 Ibid.

125 See "Orbit Fab: Fuel Supply for Satellites" (Orbit Fab), accessed September 22, 2018, www.orbitfab.space/for an example of a firm commercializing on-orbit refueling.

126 Stephen Van Evera, "Offense, Defense, and the Causes of War," *International Security* 22, no. 4 (1998): 13–14.

127 Gertz, "China Carries Out Flight Test of Anti-Satellite Missile."

128 Oriana Skylar Mastro, "The Vulnerability of Rising Powers: The Logic Behind China's Low Military Transparency," *Asian Security* 12, no. 2 (May 3, 2016): 71–72.

129 Ibid., 72.

130 Ibid., 72–73.

131 Amy Butler, "USAF Space Chief Outs Classified Spy Sat Program," *Aviation Week*, February 21, 2014, http://aviationweek.com/defense/usaf-space-chief-outs-classified-spy-sat-program.

132 Jim O'Neill et al., "Goldman Sachs Economic Research Group," no. 66 (2001): 3.

133 Michael R. Gordon and David E. Sanger, "The Bailout of the Kremlin: How the US Pressed the IMF," *New York Times*, July 17, 1998, www.nytimes.com/1998/07/17/world/rescuing-russia-special-report-bailout-kremlin-us-pressed-imf.html.

134 Kathy Sawyer, "NASA Wants to Bail out the Russian Space Agency," *Washington Post*, September 21, 1998, www.washingtonpost.com/wp-srv/national/longterm/station/stories/russia.htm.

135 Phillip H. Hemberger, "The Initiatives for Proliferation and Prevention Program: Goals, Projects, and Opportunities" (Los Alamos National Laboratory, 2001), 1, www.osti.gov/servlets/purl/788224.

136 William H. Cooper, "Russia's Economic Performance and Policies and Their Implications for the United States" (Congressional Research Service, June 29, 2009), 1, https://fas.org/sgp/crs/row/RL34512.pdf.

137 Joshua R. Itzkowitz Shifrinson, "Deal or No Deal? The End of the Cold War and the U.S. Offer to Limit NATO Expansion," *International Security* 40, no. 4 (2016): 16.

138 Quoted in ibid.

139 "Russia's Accusations-Setting the Record Straight," Fact Sheet (North Atlantic Treaty Organization, April 2014), 2, www.nato.int/nato_static/assets/pdf/pdf_2014/20140411_140411-factsheet_russia_en.pdf.

140 Steven Lee Myers, "As NATO Finally Arrives on Its Border, Russia Grumbles," *New York Times*, April 3, 2004.

141 Ibid.

142 Ibid.

143 "SIPRI Military Expenditure Database."

144 NASA, "National Aeronautics and Space Administration Fiscal Year 1998 Estimates," n.d.

145 NATO, "Bucharest Summit Declaration," April 3, 2008, www.nato.int/cps/en/natolive/official_texts_8443.htm.

146 S. Erlanger, "NATO Leaders Hear Putin on Russia Security Worries," *Pittsburgh Post–Gazette*, April 5, 2008.

147 Vladimir Socor, "Russia Moves toward Open Annexation of Abkhazia, South Ossetia," *Eurasia Daily Monitor*, April 18, 2008, https://jamestown.org/program/russia-moves-toward-open-annexation-of-abkhazia-south-ossetia/.

148 Christian Lowe, "Georgia War Shows Russian Army Strong but Flawed," *Reuters*, August 20, 2008, www.reuters.com/article/us-georgia-ossetia-military/georgia-war-shows-russian-army-strong-but-flawed-idUSLK23804020080821.

149 N. Obermueller, "Protesters Expect Russia Will Intervene," *USA Today*, February 7, 2014.

150 "Crimea Declares Independence, Seeks UN Recognition," *RT News*, March 17, 2014, www.rt.com/news/crimea-referendum-results-official-250/.

151 "Ukraine Crisis: Why Russia Sees Crimea as Its Naval Stronghold," *Guardian*, March 7, 2014, www.theguardian.com/world/2014/mar/07/ukraine-russia-crimea-naval-base-tatars-explainer.

152 Vitaly Shevchenko, " 'Little Green Men' or 'Russian Invaders'?," *BBC*, March 11, 2014, www.bbc.com/news/world-europe-26532154.

153 Christopher S. Chivvis, "Understanding Russian 'Hybrid Warfare,' And What Can Be Done About It," § House Armed Services Committee (2017), 1, www.rand.org/content/dam/rand/pubs/testimonies/CT400/CT468/RAND_CT468.pdf.

154 "Brussels Summit Declaration: Issued by the Heads of State and Governments Participating in the Meeting of the North Atlantic Council in Brussels 11–12 July 2018" (North Atlantic Treaty Organization, July 11, 2018), www.nato.int/cps/en/natohq/official_texts_156624.htm.

155 Chivvis, "Understanding Russian 'Hybrid Warfare,' And What Can Be Done About It," 2.

156 Todd Harrison, Kaitlyn Johnson, and Thomas G. Roberts, "Space Threat Assessment 2018," A Report of the CSIS Aerospace Security Project (Center for Strategic and International Studies, April 2018), 13.

157 Fred Kaplan, "The Pentagon A War in the Stars?; Reagan Team Presses Outer Space Arms Plan," *Boston Globe*, October 16, 1983.

158 S.R. Reed, "Soviets Offer to Pull Arms from Space," *Philadelphia Inquirer*, August 19, 1983.

159 Ibid.

160 Bart Hendrickx, "Naryad-V and the Soviet Anti-Satellite Fleet," *Space Chronicle* 69 (2016): 5.

161 Ibid., 9–10.

162 Ibid., 15.

163 Ibid.

164 Eurockot, "Eurockot Launch Services," accessed October 2, 2018, www.eurockot.com/launch-services/.

165 Pavel Podvig, "Did Star Wars Help End the Cold War? Soviet Response to the SDI Program," *Science & Global Security* 25, no. 1 (2017): 19.

166 Anne Marie Squeo, "The Assault on Iraq: U.S. Bombs Iraqi GPS-Jamming Sites," *Wall Street Journal*, March 26, 2003.

167 US National PNT Advisory Board, "Russia Undermining Confidence in GPS," 2, www.gps.gov/governance/advisory/meetings/2018-05/goward.pdf.

168 Ibid., 7.

169 Kashmir Hill, "Jamming GPS Signals Is Illegal, Dangerous, Cheap, and Easy," *Gizmodo*, July 24, 2017, https://gizmodo.com/jamming-gps-signals-is-illegal-dangerous-cheap-and-e-1796778955.

170 Vladimir Putin, "Meeting with Defense Ministry Senior Officials: The President Met with Defense Ministry Senior Officials in the First of a Series of Meetings on the Development of the Armed Force." (Office of the President of Russia, May 15, 2018), http://en.kremlin.ru/events/president/transcripts/57477.

171 Harrison, Johnson, and Roberts, "Space Threat Assessment 2018," 14.

172 Alexander Artamonov, "Americans Force Russia to Create Weapons Even More Powerful than S-500," *Pravda*, November 13, 2017, www.pravdareport.com/russia/economics/13-11-2017/139167-anti_missile_weapons-0/.

173 Ibid.

174 "Putin: Unilateral US Withdrawal from ABM Treaty Pushing Russia toward New Arms Race," *RT News*, June 19, 2015, www.rt.com/news/268345-putin-west-russia-relations/.

175 John A. Tirpak, "Operation Burnt Frost 'Historic,'" *Air Force Magazine*, February 22, 2008, www.airforcemag.com/Features/security/Pages/box022208shootdown.aspx.

176 Ibid.

177 Daniel R. Coats, "Worldwide Threat Assessment of the US Intelligence Community," § US Senate Select Committee on Intelligence (2018), 13.
178 Vladimir Putin, "Presidential Address to the Federal Assembly" (Office of the President of Russia, March 1, 2018), http://en.kremlin.ru/events/president/news/56957.
179 T. Grove, M.R. Gordon, and J. Marson, "Putin Unveils Nuclear Weapons He Claims Could Breach U.S. Defenses; Russian Leader Sharpens Rhetoric against West, Intensifying Arms Race with the U.S.," *Wall Street Journal*, March 1, 2018.
180 Putin, "Presidential Address to the Federal Assembly."
181 Ibid.
182 Yleem Poblete, "Remarks on Recent Russian Space Activities of Concern" (US Department of State, August 14, 2018), www.state.gov/t/avc/rls/285128.htm.
183 Ibid.
184 Ibid.
185 Ibid.
186 Vladimir Putin, "Meeting of Ambassadors and Permanent Representatives of Russia" (Office of the President of Russia, July 19, 2018), http://en.kremlin.ru/events/president/news/58037.

4 Deterrence and reassurance

The previous chapter established that a security dilemma does exist in space and that sub-optimal arms racing is occurring as a result. The three major competing space powers—China, Russia, and the US—all have misperceptions about each other's motivations. These misperceptions are leading to an accelerating arms race within an increasingly severe security dilemma; as each state seeks to increase its security further, its actions negatively impact the security of its adversary. Each nation is then forced to pursue policies to increase its security in response, and a tragic spiral of action and reaction is ensuing.

The first definitive act in this negative spiral within space was the US withdrawal from the ABM treaty in 2001. This nominally defensive act forced China and Russia to look for ways to preserve their nuclear deterrent. Each state took slightly different approaches, but both ultimately pursued their own missile defense systems. These dual-use systems are easily capable of attacking satellites as well as ballistic missiles, as the US demonstrated during Operation Burnt Frost in 2008.[1] The existence of these systems damaged established norms in space, and gradually each nation's actions in space created the intensifying sub-optimal arms race that we find ourselves in today.

Breaking out of a security dilemma-fed security spiral in space is not an easy task. Merely restoring the ABM treaty and returning to the situation that existed prior to the US withdrawal would be a positive step, but this is no longer feasible. The proliferation of these missile defense systems and the peace of mind they provide civilian populations against rogue states like North Korea means that they are here to stay. Eliminating ABMs would also do nothing to provide reassurances that other space weapons are not in development. Another approach needs to be found, one that is feasible, acceptable, and suitable. However, one variable necessary for determining strategies to reduce the severity of the security dilemma still needs to be determined, the offense–defense balance.

Space power

In this chapter, we are attempting to develop a strategy for signaling benign intent and reassuring potential adversaries that does not place the signaling nation in unnecessary jeopardy. Therefore, a theory of space power must form the basis of the selected strategy. Without that theory, it is also impossible to truly determine the offense–defense balance. The problem is that no accepted theory of space power exists, and it is outside the scope of this book to explore all aspects of existing theories or to develop an entirely new one. Despite this, the outlines of a space power theory are necessary since at sea, in the air, and on land there are theories of war that influence policy and form the paradigm for strategic thought in these domains. These theories influence national policy by generating a functional theory of war that allows for the accurate assessment of military strategies. The shape of that theory does not have to be consciously present in the mind of the policymaker or military strategist, but without that paradigm, the coherent formulation of policy and strategy lacks structure and direction. A sound space power theory allows for the controlled development and application of military power in space. Absent a controlling theory, warfare in space is nothing more than a contest to see which side can destroy or disable more enemy space assets. While a straightforward military objective, it begs the question, to what end? The form and function of military forces are a means to achieve specific goals in support of political ends. Structuring military space doctrine and acquisition around simplistic poorly understood concepts of dominance or control is only the beginning of the evolution of understanding war in space.

Military theory is not static; it is continually evolving. A new theory, or a reinterpretation of an existing theory, arises whenever the strategies derived from accepted theory fail the test of war or are challenged by the development of new technology. These new or reinterpreted theories form the basis of military strategy that attempts to apply theory to reality. Often, a supposedly new theory is nothing more than an old theory applied to new circumstances and technologies. Whether a new theory suited specifically to space is eventually formulated or an older theory is adapted to fit the newest warfighting domain is unimportant. What is important is that neither a coherent national military space strategy nor a good understanding of the offense–defense balance can exist without a broadly accepted theory of space power upon which to build upon or at least a set of guiding principles.

Before addressing existing theories of space power, it is instructive to attempt to define the term itself. What is space power? There is no accepted definition of the term. Looking to US military doctrine for a description of space power would seem to be the easiest and most straightforward way of determining a definitive space power definition. However, current US military doctrine does not include an independent definition of space power. The 2018 version of Joint Publication (JP) 3–14, *Space Operations*, includes space power in its glossary but lists "none, approved for removal from the DOD dictionary," as the definition.[2] No specific explanation is given for its removal, but looking to *US Air Force Doctrine Document 1* (AFDD-1) a possible explanation appears. In AFDD-1 the term air power is not used separately from space power but instead used as a single term, "air and space power," throughout the document.[3] Since the US Air Force controlled the bulk of US military space assets and at the time AFDD-1 was written there was the beginnings of a significant movement to separate space functions from it, this mashup of domains was probably a mis-guided attempt to conflate them for organizational reasons, further harming the development of an independent and useful space power theory.

It is possible to find a definition of space power in earlier official documents. The 2009 version of JP 3–14 defines space power as "the total strength of a nation's capabilities to conduct and influence activities to, in, through, and from space to achieve its objectives."[4] This definition seems to be broad enough to capture all elements of space power, military and civilian, and is similar to other earlier definitions of space power. Writing in 1988, David Lupton developed one of the earliest theories of space power. He advocated a policy of space control through force and described space power as "the ability of a nation to exploit the space environment in pursuit of national goals and purposes and includes the entire astronautical capabilities of the nation."[5] This definition is broadly echoed by a later RAND study which defined space power as "the pursuit of national objectives through the medium of space and the use of space capabilities."[6] These definitions also support the concept that space power is more than just the military aspects of the domain; it also includes the commercial and political aspects of space working in concert to achieve some national goal.

These definitions seem almost too broad to serve as a basis for a working theory of space power. Rather than focusing on several quantifiable aspects of space power as the basis upon which to build a useful theory, they attempt to capture all aspects of power in space. Despite this, some theorists would still call these definitions too specific. Brent Ziarnick, in *Developing National Power in Space*, criticizes definitions such as the one in JP 3–14 as "descriptions of unique cases of applied space power."[7] He argues instead for the broadest possible definition which he derives from Brigadier General William Mitchell's description of airpower as "the ability to do something in the air" into a defini-tion of space power which he simply replaces 'air' with 'space.'[8] With this, Ziarnick establishes the broadest possible definition and one that is elegant in its simplicity. What it lacks is a tie to why space is relevant to military forces that allows for the construction of a useful contemporary military theory. The ability to accomplish things in space has little relevance unless those things support national objectives. Moreover, because national objectives are so often tied to where people live, on land, or at least on Earth, that is where the impact of space power must be measured in order to gain traction with a larger audience.

Alfred Thayer Mahan is an example of a military theorist who made a previ-ously difficult to quantify military domain suddenly relevant. In his book, *The Influence of Sea Power Upon History*, Mahan argued that the contribution of mastery of the sea to victory in warfare was severely underappreciated. He clev-erly used a quote from George Washington to make his central point, "no land force can act decisively unless accompanied by a maritime superiority."[9] By demonstrating in his book how actions at sea lead to success on land Mahan allowed his readers to grasp the fundamental importance of sea power to nations. This seemingly simple realization, not entirely novel as demonstrated by Washington's quote from a century earlier, forms the core of Mahan's theory. When supported by historical examples and fleshed out into a more comprehensive analysis its impact was enormous.

While Mahan's target audience was American, he had a global impact that shaped history. German Kaiser Wilhelm stated that "I am ... not reading but devouring Captain Mahan's book ... it is on board all of my ships and constantly quoted by all of my captains and officers."[10] Germany and the other great nations latched onto Mahan's theory of sea power, and a naval arms race commenced between Britain and Germany in the years leading up to WW1. The sub-optimal naval arms race that ensued and the larger build-up leading to WW1 has become the subject of international relations research ever since.[11] The outsize influence that Mahan's book had demonstrates the impact that a military theory can have as a paradigm upon which nation's build strategy and policy.

Developing a theory of space power is made doubly difficult because, unlike the sea, space is an untested domain. Humanity lacks any empirical evidence on the nature of conflict within it. Of course this is a condition worth preserving, but it does prevent any space power theory gaining traction from historical examples as Mahan's did. This lack of domain-specific evidence leads would-be-Mahans to attempt to adapt existing theories of war to space. The most popular domains from which to adopt theories are the existing fluid domains, sea and air. Theories of sea power are particularly attractive as they revolve around actions in one domain indirectly influencing action in another. This contrasts with most air power theories as they focus on the benefits of direct application of kinetic effects from the air to contribute decisively to victory on the land or at sea. One of the more successful applications of an existing theory was John Klein's effort to adapt sea power theorist Julian Corbett's work to space.

Julian Corbett was a near contemporary with Mahan and his principal work of military theory, *Some Principles of Maritime Strategy*, was published just 20 years after Mahan's in 1911. While Mahan's thinking was an adaption of earlier work by one of Napoleon's Generals, Antoine-Henri Jomini, Corbett's treatise on sea power was an adaption of the more nuanced work of the Prussian General Carl von Clausewitz to the sea. The central point of Corbett's theory of sea power argued that the "object of naval warfare is to control maritime communications."[12] This theory contrasted with Mahan's theory of sea power in that it strongly supported the construction of vessels adapted for the pursuit of commerce such as cruisers as well as a battlefleet, whereas Mahan argued that attempts to disrupt commerce were at best secondary to the construction of a battle fleet.[13] Klein adapted Corbett's central theory of sea power to space by modifying it to argue that "command of space entails the ability to ensure access and use of celestial lines of communications when needed to support the instruments of national power—diplomatic, economic, information, and military."[14]

Klein's theory argues for more than the control of space, rather he argues for the subtler concept, "command of space."[15] Command of space is achieved through presence, coercion, and force. The concept of presence highlights the fact that nations that have few or no assets in space have little influence on the domain. The degree of a nation's space presence allows it to shape international treaties, regulations, and customary practices. Today the US has by far the largest space presence. Therefore, its actions and behavior set the baseline for other nations for better or for worse. The second piece in achieving command of space is coercion. Coercion "occurs short of open hostilities, but may be the result of the implicit or explicit threat of detrimental action."[16] Presence in space is a pre-requisite for coercion and impacts the degree to which a nation can employ it. Coercion in space may take on diplomatic, economic, or informational forms. Coercion through diplomatic means comes in the form of international agreements and other forms of norm establishing. Economic coercion can involve denying launch services, satellite construction services, or vital space technology to another nation. Informational coercion relies upon the use of space-based communications to transmit a viewpoint in opposition to a state's adversary. The US transmission of the Voice of America (VOA) broadcast into Iran using satellites is an example of informational coercion. This method of coercion is seen as disruptive enough that Iran actively jams satellites carrying VOA broadcasts.[17]

According to Klein, the final aspect of the command of space is command through force. Command through force usually only occurs when a state of open conflict exists between two nations. Returning to the core argument of Klein's work, command through force is achieved by ensuring one's own celestial lines of communication while denying those same lines to the enemy. Since the primary value of space lies in its usefulness for transmitting and gathering information, it is the ability to preserve access to information or to deny it to an opponent that provides command through force. Klein's core concepts are sound though his Corbett derived celestial line of communications approach is just one method for describing space power.

Building upon his definition of space power discussed earlier, Brent Ziarnick's work mentioned above attempts to create a structure for space power theory that echoes JFC Fuller's theoretical work in detail and scope. He borrows from Clausewitz's famous work *On War* the concept that space, like war, must have logic and grammar. He argues that space may have its own grammar but not its own logic. The specific grammar in his theory is a modification of Mahan's assertion that the basis of sea power lies in commerce, bases, and ships. He modifies this into a grammar for space power, the basis of which lies in production, shipping, and colonies centered on access. This theory is further developed with a logic that relies on economic, military, and political power. Ziarnick's theory is complex and well developed, but it suffers from being too anticipatory to truly be useful today, though it may very well stand the test of time. Much as Jomini dominated 19th-century thinking, while his contemporary Clausewitz's work suffered from anonymity, Ziarnick's work will probably age well as military and commercial space activities expand beyond Earth orbit.

Among the most recent publications dedicated to the formulation of comprehensive space strategy is the aptly named *Space Strategy* by Jean-Luc Lefebvre which was only recently translated into English. To Lefebvre, the key to space power is "acquiring the human and technical resources to increase one's freedom of action, while aiming to reduce an opponent's."[18] Toward this end, he identifies 12 principles of space warfare broken into three categories which he labels, preliminary, cardinal, and complementary. The preliminary principles center on space situational awareness, investment, public engagement, and training.[19] The cardinal principles include ensuring technical and physical access to space, avoiding the generation of orbital debris, and stealth. Finally, his complementary principles are: take advantage of the physical geography of space, promote and protect non-physical lines of communication, promote resilience, and ensure effects are designed to influence events on Earth. Lefebvre's language and descriptions are awkward and often esoteric, but the essential elements of a valid space power theory are present if poorly developed.

Former Air Force officer and NASA engineer Jim Oberg proposed a more conventional space power theory than Ziarnick or Lefebvre. He describes space power as including all aspects of civil, commercial, and military space activity.[20] The primary characteristic of space systems in Oberg's theory is their ability to view the world from orbit. This characteristic enables the most strategically relevant aspect of these space assets which is their ability to transfer and gather information.[21] Moreover, since the commercial industry controls the vast majority of systems on orbit, it will be commercial platforms that transmit and gather the majority of information. Being the primary source of most information means that "it will be the commercial manufacturers, owners, operators, and users who will contribute the larger, if less clearly perceptible, aspects of space power."[22] Oberg cites the influence of the commercial industry as the largest complicating factor in determining a clear formula for developing a comprehensive theory of space power.

Further, Oberg argues that as commercial entities become increasingly internationalized and so available for purchase by anyone "that a common level of space support will soon be available to citizens of all nations, including their armies."[23] Since the time of his writing in 1999, this objective has largely been achieved. The level of detail and ease of availability of commercial imagery from tools like Google Earth and the ubiquitous embedding of GPS in commercial devices brings a degree of space support to the average individual that even the US military was incapable of providing little more than a decade ago.

There are several additional tenets of Oberg's theory of space power that are worth considering. First, Oberg cautions that space power by itself "is insufficient to control the outcome of terrestrial conflict or ensure the attainment of terrestrial political objectives."[24] Oberg makes this point explicitly to avoid the mistakes made by early air power theorists who consistently over-promised and under-delivered. In Oberg's opinion, the control of space is only important in relation to its ability to influence events on Earth. This is something that it can only do when working in conjunction with other elements of national power and only when a nation has adequate control of space.

This need for control leads to another tenet of Oberg's theory, that "control of space is the linchpin upon which a nation's space power depends."[25] Unlike Lupton's theory of space power mentioned earlier which argued for control of space "through the destruction of the enemy's space forces," Oberg takes a broader view.[26] Oberg argues that space control and therefore space power will accrue to the nation with the largest space presence. This again reinforces the importance of commercial systems since they increasingly represent the majority of systems on orbit. The nation with the most significant commercial space industrial base will have the largest presence on orbit and as a result, the greatest degree of space control and space power. The commercial aspect of space power emphasized by Oberg does not mean that military strategy is irrelevant in space; the objective remains preserving your own information flow while disrupting an opponent's when necessary. It does mean that since the majority of information will flow over commercial satellites, any military strategy involving space must account for their presence. In the end, even with its more commercial focus, Oberg's theory has much in common with Klein's theory of space power.

Both Oberg's and Klein's theories of space power place the primacy of information at the core of their theories and its importance is highlighted in Lefebvre's. Both Oberg's and Klein's theories also develop the idea that the degree of presence in space is a large part of what gives a nation power and control over it. Oberg draws the connection between on-orbit presence and commercial systems explicitly while Klein only hints at it in his work. These theories also share the idea that actions in space are dependent on and in support of other warfighting domains. The degree of agreement between the two theories points to several ideas that taken together form an adequate foundation for a functioning theory of space power:

1   Space power is directly proportional to a nation's presence in space.
2   The strategic value of space in our current era lies in the ability to transfer information through it and to gather information from it.
3   Space is a supporting domain that is only relevant to the degree that it influences terrestrial events.[27]

With this basic theoretical framework in place, we can now determine the offense–defense balance in space.

The offense–defense balance in space

The one variable in our theoretical structure that remains undefined is the offense–defense balance. This material variable is central as it helps determine the viability and effectiveness of various reassurance strategies that can help resolve the security dilemma. If a nation misperceives the offense–defense balance, it will rule-out reassurance strategies that might otherwise be possible and instead default to sub-optimal arming policies. This sub-optimal policy of arming is occurring now largely due to a misinterpretation of the overall offense–defense balance. The common belief is that offense has a distinct advantage in space and that the offense and defense are indistinguishable because of the dual-use nature of many space systems. This misperception of offense dominance and indistinguishability is ruling out viable reassurance strategies and forcing states to pursue self-defeating policies that are further intensifying the security dilemma in space.

In the space domain, it is generally accepted that offense has the advantage. This frequently cited "fact" appears in studies, newspaper articles, and treatises on strategy, often with little support.[28] RAND studies cite it, as do prominent strategists such as Colin Gray who argues with some equivocation that "offense may appear to be the stronger form of war in space, given the absence of terrain obstacles, the relative paucity of capital assets (and targets), and the global consequences of military success or failure."[29] This is a stunning statement from Gray, a true disciple of Clausewitz, who argued that fundamentally "the defensive form of warfare is intrinsically stronger than the offense."[30] Gray's opinion is also shared by senior US policymakers. James Finch and Shawn Steen, the former Director and Deputy Director respectively for Space Policy and Strategy Development within the US Office of the Undersecretary of Defense for Policy, argue that the domain is offense dominant because "holding space targets at risk is far easier and cheaper than defending them."[31] With the notable exception of an article by Klein using a Clausewitzian-based argument, there are few serious attempts to refute the idea that space is offense dominant.

At first glance, it does seem a fairly obvious conclusion that space is offense dominant. After all, satellites are vulnerable machines.[32] They travel in predictable orbits, and every kilogram of mass devoted to their defense is one less available to perform its actual mission. The attacker is under no similar limitation and can devote all its capabilities to defeating whatever safeguards the defender has available. In addition, with many military satellites taking nearly a decade to design, its technology is already outdated upon launch.[33] During the expected ten to 15 years of lifetime that a satellite has on orbit, that technology deficit only grows with no realistic way for improvements or upgrades to occur. As a result, the attacking platform or system will almost always be newer and more capable. Even the traditional advantages of the defender do not apply. There is no terrain to leverage to a defending satellites advantage. If orbits are terrain, then the defending satellite is essentially trapped in the orbit in which it is placed. Even if it had the fuel to move, it loses its very purpose once it changes orbit and the attacker has achieved their objective merely by threatening to attack. The attacker also chooses the time and place of the attack which can occur when the defender has limited ability to observe or react. Another traditional defensive advantage that fails in space is that of interior lines. Interior lines traditionally allow a defender to mass forces and reinforce faster than an attacker. In space both the attacker and defender suffer from the same physical restrictions in achieving orbit, neutralizing any advantage to either side. Finally, the proverbial bullet is almost always cheaper than the target, assuming that the target is not another bullet. Whatever form the attacker takes, it is optimized for a single function, destroying or disabling its target. This approach will inevitably be cheaper than the function the target satellite is designed to do.

With all of these disadvantages accruing to the defending satellite, how can any argument be made that does not favor the offense in space? Returning to our theory of space power, the military utility of space lies not in the individual satellite but in the ability to transmit information through it and to collect information from it. True, the satellite is critical to this process; however, the paradigm is shifting. As recently as 15 years ago the number of satellites on orbit with service to any one region in any particular band was relatively limited. Therefore, the ability to transmit information through space and the health of the satellite were inextricably linked. In mid-2018 there were more than 1,887 active satellites on orbit, up from around 500 in 2008, and we are on the cusp of the era when active small satellite constellations and greatly reduced launch costs will significantly increase this number.[34] The space between orbital slots in the geo-belt also continues to shrink with multiple satellites now operating in the same slot. With so many satellites on orbit, a hostile entity looking to interfere with a signal will first have to contend with finding the signal. Once found, whether the attacker uses kinetic means to threaten the satellite or non-kinetic means to target the signal will not matter. The signal can move elsewhere in a matter of moments, and the attacker is again left hunting for a needle in a haystack. A competent defender will be ready for interference or attack, and just as is done with terrestrial radio interference, have a pre-planned alternate frequency. A clever defender will take things one step farther by having a plan, when threatened, to further complicate the attacker's search by switching bands or even moving from fixed satellite services to mobile satellite services.

The intermixing of military, civil, and commercial signals from a variety of sources on commercial platforms creates a further complication for an attacker. Attacking the wrong signal or satellite can involve a third party in any conflict, an undesirable situation for any attacking entity. The level of entanglement involved in commercial platforms varies, but it creates another issue that any attacker must take into account. When the variety of challenges involved in the actual mechanics of preventing the transmission of information through space is considered, the offense–defense balance is more balanced than commonly thought. While the sheer number of signals on orbit makes stopping the transmission of information extremely difficult, preserving the ability to gather information presents a more difficult problem.

Gathering information from space requires a platform to do it. This is a situation where the loss of a satellite could create a catastrophic loss of information gathering ability, although this is changing rapidly. In 2017, there were 620 satellites on orbit whose primary purpose was Earth observation in a variety of spectrums, a 66 percent increase over 2016.[35] While much of the growth is coming from small satellites, there is significant growth in larger satellites as well. Planet Labs alone can now offer 3- to 5-meter resolution of anywhere on the globe every day, with resolutions as low as .72 meters less frequently.[36] This is a capability that no one, civilian or military, ever had as recently as a year ago. It is becoming very challenging for a nation to hide anything, and even harder to prevent someone from gathering the information they need. There are simply too many commercial, scientific, or national systems imaging the Earth for any attacker to reasonably deny them the ability to image an area.

The one area where no commercial system can yet compensate is in dedicated systems that have no civilian application, such as missile warning. The satellites performing these missions are currently irreplaceable though these systems' specific association with the nuclear deterrence mission provides them with their best protection. Any attack on these systems represents an attack on a country's nuclear deterrent with attendant consequences. However, the blurring of lines with these systems toward non-missile warning conventional missions such as "battlespace awareness" represents a dangerous trend that makes these satellites legitimate targets in any conventional limited conflict.[37] Whether the intent behind an attack on these satellites is a prelude to nuclear conflict or an effort to deny an enemy information an adversary must assume the worst. Even unintentional damage by debris from another destroyed satellite could be misinterpreted as an intentional attack given the inability to directly observe the damage. Protecting this handful of expensive and irreplaceable satellites that are vital to a nation's defense is best done by avoiding a sub-optimal arms race in space.

Fundamentally, the greatest threat to a nation's space control will be an adversary's ability to disrupt or deny the information flow provided by a nation's space assets, whether commercial or military. Since a larger presence on orbit makes this task more difficult, it benefits a nation to have the largest and most resilient space architecture possible. Resiliency is the ability of a nation's assets "to continue providing required capabilities in the face of system failures, environmental challenges, or adversary actions."[38] One of the easiest ways to achieve resiliency is by dispersing the capability to gather and transmit information across as many platforms as possible, commonly called disaggregation. Since the number of commercial satellites on orbit is increasing rapidly, a nation's ability to achieve resiliency through disaggregation will depend on the size of its commercial space industry. The issue is that while there may be more commercial satellites, many commercial providers are reluctant to provide any increased protection for their satellites due to the additional costs. This reluctance means that while the individual satellites may be more numerous, they are also more vulnerable to interference or other forms of attack.

Where then does the offense–defense balance lie? The individual satellite remains vulnerable to attack and nearly impossible to defend. At this level, the tactical level of space, the advantage does lie with the offense. At the level of a constellation of similar platforms, a signal can move, or one platform can compensate for the loss of another, but the target set remains limited to a subset of satellites. A smaller constellation favors the attacker while a larger more robust constellation can shift the advantage to the defender. The balance at this level,

Level of War		Balance
Tactical	Individual satellite	Strongly favors offense
Operational	Constellation or specific architecture	Neutral depending on constellation size and architecture resiliency
Strategic	Continued national access and ability to exploit space	Slightly favors defense

Figure 4.1 Orbital offense–defense balance.

Source: author's original work.

the operational level, is then generally neutral depending on the number of satellites and the ease with which they can be replaced. At the strategic level, where the balance is measured against the aggregate ability of a nation to transmit and gather information using space, the balance begins to shift in favor of the defender (see Figure 4.1). As long as a nation can maintain access to a significant share of the commercial market, it is unlikely that another nation can deny it the use of space. Space is an environment where nations can always disrupt and degrade the capabilities of other nations. However, one nation cannot deny another nation the ability to substantially leverage space as long as a neutral commercial market exists.

Distinguishability

Determining the value of reassurance strategies also requires determining the distinguishability between the offense and the defense. This is a notoriously difficult task as most weapons are not intrinsically either offensive or defensive. Instead, it is the intent behind the weapon that determines its nature. While some space systems are more easily distinguishable than their terrestrial counterparts, others suffer from the same degree of confusion. Space also has norms of behavior established at the outset of the space race that differs from any other domain and significantly impact distinguishability. This section will evaluate the level of distinguishability in space and the degree of uncertainty that comes from dual-use systems.

Before establishing the degree of distinguishability in space, clarification on degrees of space militarization is necessary. Since the very beginning of the space age, the US has publicly supported the peaceful use of space while at the same time quietly steering the definition of peaceful toward "the non-aggressive use of space."[39] The Eisenhower Administration saw the unique benefits of space-based reconnaissance in verifying Soviet military capabilities and preserving the peace between the two superpowers. The precedent established of defining passive military use as peaceful has continued, even as the passive use of space moved beyond reconnaissance and treaty verification. Today passive systems, such as satellite communications and GPS, actively contribute to offensive military actions on the ground yet they remain classified as passive, and therefore peaceful systems. Intentionally conflating non-aggressive with peaceful from the outset established a unique domain norm that remains in effect today. In any other domain, a system that provides targeting data to weapons systems would be a legitimate military target, yet space maintains a definition of peaceful based on a precedent only tacitly agreed to during the Cold War.

The next degree of space militarization is ground-to-space weapons. These were among the first space weapons developed and have existed in one form or another since nearly the beginning of the space age, albeit in limited numbers. ASATs are the classic example of these types of systems which can also include ground-based jammers and lasers designed to degrade or damage orbiting satellites. The current arms race in space is focused on developing these capabilities, and the further testing and development of these weapons is an area of significant concern.

The next level of space militarization is space-to-space weapons. This category includes satellites designed to destroy or disable other satellites. Satellites designed to protect other satellites through offensive action would also fall in this category, one that is only just beginning to develop. Weapons at this level of militarization would be a dangerous new development in space. Even so, they would be a threat only to other space systems and remain within the accepted information-centric space power paradigm developed earlier.

The final and most dangerous level of space weaponization is the fielding of space-to-ground weapons (see Figure 4.2). No nation has crossed this proverbial Rubicon, though if it does happen, it will create a dangerously unstable situation. The advent of space-to-ground weapons would invalidate the space power theory developed above as information would no longer define the military utility of space. These space-to-ground weapons could provide nuclear effects without nuclear fallout from orbits low enough that they would give the defender minimal reaction time. Reaction time would be further limited by the likely extremely small launch signature that these weapons would have, which varies from ICBM launches which are easily detectable.[40] These space launch platforms would be individually vulnerable, and an opponent would no doubt develop ASATs designed to attack these platforms. Both nations in this situation would enter a dangerous offense-dominant environment with extremely low crisis stability. In essence, if a situation developed that increased tensions, each nation may find itself in a "use it or lose it situation." This first-strike instability would pressure leaders "to strike first in a crisis to avoid the worst consequences of incurring a first strike."[41] Thankfully, this paradigm shift is not on the immediate horizon, though many believe it is inevitable.

With the degrees of space militarization delineated it is now possible to return to the issue of determining the degree of distinguishability. Established norms determine what constitutes the peaceful use of space, and since

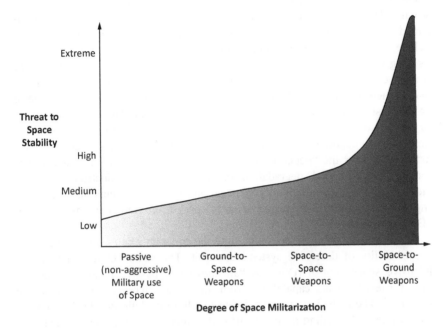

Figure 4.2 Degree of militarization and threat to space stability.

Source: author's original work.

Note
Values are for illustrative purposes only and not based on empirical analysis.

peaceful/non-aggressive and defensive are nearly synonymous terms in an environment dependent on information, anything designed to gather or transmit information is distinguishable as non-offensive. This established norm of passive military use is stretched by the dependence that conventional offensive military capabilities have on space, yet it still holds. Despite this, it is the very dependence of conventional forces on these capabilities that makes these platforms such a tempting target and drives nations to develop weapons designed to attack satellites.

The next tier of militarization, ground-to-space, is where challenges of distinguishability begin. ABMs and ASATs are the most obvious examples of systems that suffer from distinguishability problems at this level. As discussed previously, a system designed to destroy or disable ballistic missiles can easily be retargeted to strike satellites in low orbit. A dedicated ASAT is only distinguishable if it is designed to travel to altitudes beyond which ballistic missiles travel. Even if an ASAT and a missile defense system can be distinguished, the ASAT system might be a justified defensive system if there exists a legitimate reason to suspect that an adversary has placed weapons on orbit. The ASAT system then becomes indistinguishable in purpose from other weapons as it could be used defensively to deter enemy threats or offensively to attack peaceful satellites.

Other systems within this tier include ground-based lasers and jamming systems. These systems might have reversible effects, or they may cause permanent damage to the target. The intent of these systems could vary and may include causing interference with reconnaissance or navigation satellites that are assisting with enemy targeting. While this would be considered defensive usage in most contexts, any system designed to attack satellites for any reason can be classified as offensive. This simplistic categorization is only possible due to the established norms that passive satellites are peaceful. Therefore, any attack for any reason against passive satellites falls into the category of offense. This implicit understanding is present in US doctrine which defines offensive space control (OSC) as the "negation of adversary space capabilities through deception, disruption, denial, degradation, or destruction."[42] US doctrine makes no allowance for the intent behind why negation of enemy systems may be occurring which simplifies the categorization of space-to-ground weapons outside of ABMs as offensive.

The discussion of intent does enter into the equation when evaluating the distinguishability of defensive systems on orbit. The same US doctrine defines defensive space control (DSC) as "all active and passive measures taken to protect friendly space capabilities from attack, interference, or unintentional hazards."[43] The inclusion of active defense within this definition leads to confusion of intent. According to this doctrine document, active defense "consists of those actions taken to neutralize imminent space control threats to friendly space forces and space capabilities."[44] Within this definition use of the term "imminent" seems to clarify any confusion over what active defense is, however, what constitutes an imminent threat is not defined. Could a satellite belonging to a hostile nation sharing the same orbit be an imminent threat? Or does the satellite have to take aggressive action such as approaching within a given distance of a friendly satellite? If the friendly satellite possesses a defensive system capable of disabling the approaching satellite could that system be used for offensive purposes? Lack of clarity with regard to defining imminent threats blurs the line between offensive and defensive space control.

It is under these conditions that the secrecy surrounding military satellite capabilities becomes an issue. While the orbits and designations of military satellites are generally known, the purposes of these platforms are impossible to verify. Unlike terrestrial weapons systems which can be easily imaged and observed in use, it is nearly impossible to verify if a nation claims to launch a communications satellite that it is not instead launching a weapons platform or intelligence asset. The US Air Force recognized this problem and revealed a previously classified program in 2014 designed to image satellites in GEO.[45] The purpose of revealing this program according to Gen. William Shelton was to "discern when adversaries attempt to avoid detection and to discover capabilities they may have which might be harmful to our critical assets at these higher altitudes."[46] Imaging adversary satellites can provide information about a satellite's purpose, but the distance between two satellites in GEO may be tens of thousands of kilometers preventing close inspection. The adversary satellite may even possess the outward characteristics for its stated purpose while housing offensive weapons. Since military satellites are not available for general use, there is no way to verify that it can in fact perform its stated mission.

The ability to image satellites provides some reassurance, but space is large and launch platforms now have the ability to launch multiple satellites in a single launch. The Indian space agency holds the current record for a multiple satellite launch, with 104 satellites of various sizes launched at once in 2017.[47] Each time a nation launches a military satellite there exists the possibility that the official payload is not the only payload onboard. It requires an exquisite level of space situational awareness to ensure that no additional satellites are on board. If they are small enough, these additional payloads may disguise themselves as launch debris to escape detection.

A Russian launch in 2014 attempted this trick of hiding in the debris. Following the launch of three Russian Rodnik military communications satellites, a piece of supposed debris from the launch began maneuvering.[48] The object was not part of the official Russian launch declaration, and speculation on its purpose and nature ranged from experimental repair vehicle to hunter-killer satellite. Whatever the cause, it demonstrated the ease with which nations can add additional payloads to launches without declaring them, hoping to evade detection. The US has the best space tracking network of any nation and even it has substantial weaknesses that could easily allow something like this Russian action to go unnoticed.[49] Nations with much less robust tracking networks than the US can only rely on the data the US chooses to share about its satellites' purposes and capabilities, leaving significant room for suspicion and uncertainty of the type that fuels security dilemmas. With the rapidly growing number of satellites on orbit, it will be increasingly difficult to verify that every military satellite is what it is purported to be.

Military satellites are not the only ones that suffer from issues of distinguishability. While commercial communications, weather, and reconnaissance satellites create little suspicion because they are performing their intended purpose daily for a variety of users, other commercial satellites are being designed that by their very nature create suspicion. Commercial or civil ventures designed to refuel satellites, repair them, or remove debris can easily be used to damage or disable other satellites. For example, in 2018 Chinese researchers proposed a space-based laser designed to remove space debris from orbit.[50] This ostensibly civil research program would involve a satellite with a laser designed to heat targeted debris and de-orbit them. A satellite mounting a laser designed to remove debris would have obvious dual-use potential and could serve as a testbed for future space-based weapons. This proposal is just one of many attempting to deal with the problem of space debris, and any of them could easily be used to destroy active satellites. The one mitigating factor is that at this time such satellites are only in the earliest stages of development and testing.

Among existing satellites and space systems distinguishability is high relative to other fields of military endeavor. The establishment of norms early in the space-era that information gathering and transmission, even in support of military efforts, is non-aggressive and peaceful greatly aids distinguishability in space. This differentiation makes ground-to-space weapons designed to interfere with satellites inherently offensive. Some confusion of intent does exist with regard to active defenses on orbit and dual-use commercial systems. However, this is mitigated by the relative lack of known systems with these capabilities. One complicating factor is the uncertainty over the true purpose of military satellites. Distinguishability among military platforms is highly dependent on whether the satellite that a nation claimed it launched was in fact the one that it did. The Russians have shown that hiding orbital weapons among communications satellites is possible. Even with this caveat, distinguishability remains relatively high among existing space systems.

The security dilemma revisited

In Chapter 3 I determined the power balance between the three largest space powers as well as their motives and intentions relative to each other. These factors showed that a security dilemma does exist between these nations in space and sub-optimal arming is occurring as a result. Each nation is on the path to pursuing greater security through increased military capability in space. This section will show that the dangers generated by the increased level of insecurity to each nation's adversary is greater than the gains in security from arming. A negative spiral is ensuing in which each increment of increased military capability creates additional fear and uncertainty. Escaping this negative spiral requires that each state must either pursue an effective strategy of reassurance to signal its true motives in space to its potential adversaries, or it must somehow effectively deter its adversaries from attack. With the offense–defense balance and distinguishability established, all of the factors necessary for determining an effective reassurance or deterrence strategy are present. This section will investigate the challenges associated with achieving effective deterrence in space and explore the alternatives to arming using various reassurance strategies.

Deterrence in space

The current US strategy in space is focused on military means to somehow deter an adversary from attacking US space assets. The three military pillars of the 2018 US National Space Strategy are all related to deterring an adversary, and if that fails, countering threats from hostile adversaries.[51] There are various approaches to achieving deterrence in space with most focused on hardening satellites to threats, increasing resiliency, and developing methods to rapidly reconstitute disabled or destroyed military satellites. Each method has weaknesses that have proved difficult to overcome, leading to the decision that deterrence in space must somehow expand beyond space. The 2018 Space Strategy made this explicit by including language threatening to respond to attacks in space at "the time, place, manner, and domain of our choosing."[52] Expanding the threat of response beyond space makes the domain ripe for escalation and emphasizes the difficulties involved in deterrence that reinforces the need for a strategy aimed at preserving stability through reassurance.

At first glance deterrence is a simple concept; one state discourages another from aggressive or undesirable acts by adjusting the aggressor state's perceptions of the risk and reward. In practice, deterrence is a complex process that underpins all aspects of human behavior. At the strategic level deterrence ultimately involves establishing a conditional threat.[53] The conditions surrounding the threat and the credibility of it are at the core of deterrence theory. The success of the threat is dependent on the conditions in which it is applied, which can be categorized as narrow or broad, extended or central, as well as denial or punishment.[54]

Achieving deterrence in space is both narrow and broad. Narrow deterrence is the process of preventing some specific form of warfare or weapons use, while broad deterrence is the more common form of preventing armed conflict of any type.[55] Relative to other domains, deterrence in space is narrowly designed to prevent armed conflict in space while space is acting in support of armed conflict across other domains. Achieving this delicate balance requires broad deterrence within the domain. In the post-Cold-War era, no two space-faring nations have yet engaged in a significant conflict that has tested deterrence in space. One nation, usually the US, has had the full benefit of the space domain while engaging in conventional conflict with another nation or non-state adversary that enjoys only the most basic military benefits from space. When two space capable nations do engage in a limited war, space assets will be a tempting target for both parties that will test the limits of space deterrence strategy and the frayed norms of space as a sanctuary.

The problem with narrow deterrence applied broadly across the space domain is the difficulty of developing an appropriate and credible punishment that achieves the objectives of deterrence. Deterrence through punishment requires developing a suitably credible threat to an enemy that causes them to hesitate to attack for fear of the response. In the context of a limited conflict between two space powers, the threat of a response outside of the space domain rings hollow since attacks are likely already occurring outside of the domain and the effect of any punishment would be lost in the noise of that broader conflict.[56]

A further problem with deterrence through punishment occurs if the deterring nation is the US. The US cannot possibly hold its adversaries' space assets at equal risk to its own given its disproportionally large space presence and its degree of military dependence on them. Conversely, a nation with a smaller space presence upon which it is less reliant may effectively protect its space assets by threatening the more numerous US systems. The US is then at the mercy of lesser space powers that have offensive capability. Should the US attempt to rectify this imbalance by attempting to destroy an opponent's space weapons it could force the lesser space power into a dangerous use it or lose it situation with low first-strike stability.

US investment and dependence on space systems also creates opportunities for compellence. Rather than prevent or stop an opponent from taking action, compellence implies positive action to coerce an opponent. The US military is so dependent on its space assets for power projection that threats against critical space infrastructure may be used to influence political and military decision making. Recognizing this a weaker nation may threaten vital US space infrastructure as a method of compelling the US to meet broader political demands or to ensure its own security.

China, in particular, is ideally situated to use space as a domain to shape broader US decision making. The likely conditions of any conflict between the US and China place far greater demands on US space assets to support information flow and power projection than they do on China. While China will be operating close to its mainland, US forces will be operating far from theirs. Therefore, China can more easily rely on terrestrial substitutes while US Sea and Air Forces cannot. This gives China an opportunity to use threats in space to secure larger Chinese strategic objectives, something that Chinese writings suggest they are prepared to do.[57]

The one advantage the US does enjoy as the dominant space power is strategic depth. It is possible that the US may possess enough satellites to weather enemy attacks while it disables or destroys an opponent's space weapons. This scenario would be extremely costly both in monetary and environmental terms for the warring nations and for humanity. The debris generated even in a short space war that crossed the destructive threshold would likely damage other satellites and present a persistent hazard. After the shooting stopped, the US would also possess a much-reduced space presence that would likely negatively impact its conventional warfighting capabilities much more than its less dependent opponent. So, while the US might technically win in such a scenario, it would be a pyrrhic victory and a net loss to overall US warfighting capability.

The alternative to deterrence through punishment is deterrence through denial. Denial involves creating defenses strong enough that an enemy cannot hope to achieve its objectives. Due to the nature of the offense–defense balance in space, the difficulty of a denial strategy increases as the aims of the attacker become more limited. If an attacker only desires temporary tactical advantage by disabling a handful of satellites, then the balance favors the attacker's strategy. However, if an attacker has greater aims and the defender's space architecture is resilient enough, then the defender can successfully conduct deterrence through denial. The offense–defense balance and the aims of both the attacker and the defender are key in a denial strategy. If it is unacceptable that an attacker gains temporary tactical advantage, then a strategy of deterrence through denial is extremely difficult. In order for a strategy of denial to work under these circumstances the offense–defense balance must strongly favor the defense, and as we have already discussed, it is at best neutral within the space domain. If the attacker's goals are more expansive, then a sufficiently resilient architecture can achieve deterrence through denial. No nation can reasonably expect to deny an adversary the ability to temporarily degrade its space capabilities if it chooses to do so.

Another issue with effective deterrence in space is that it shares many of the challenges associated with extended deterrence. The credibility of any deterrence strategy hinges on political will.[58] Mustering the political will to follow through on a deterrent threat is often easier when the attack strikes something genuinely vital to a nation. A nations' sovereign territory is often the trigger for a response. The massive and ongoing US response to the terrorist attacks on 9/11 versus its muted response to previous attacks against its interests in Africa and the Middle East exemplifies the problems with mustering the political will for issues that do not strike close to home. Generating enough political will to credibly deter an adversary for something that is not central to that nation's interests is extended deterrence—a situation which accurately describes space.

Satellites are sovereign territory that is virtually invisible to the vast majority of a nation's populace. If an adversary attacks and destroys a satellite, there will be no images of bodies or wreckage available to incite public passions. The fact of the attack on the satellite may be made public, but it will have no visceral impact. No nation is going to muster the political will to start a conflict over a deleted entry in the satellite catalog, no matter how valuable the space asset. A nation may muster the political will to respond in kind, though here again, we run into a challenge for the dominant space power. The lack of mutual dependence causes the cost-benefit calculus to favor the nation with fewer space assets.

Deterrence is not an entirely hopeless task for the dominant space power, if it surrenders the goal of achieving broad deterrence within space, some level of narrow deterrence may be possible. Limiting the type, use, and nature of space weapons could occur because militaries find some weapons of marginal utility. Kinetic space weapons such as ASATs are incredibly destructive, and even their limited use could permanently deny humanity access to space absent some revolutionary new debris removal method. More likely, is the use of weapons that cause temporary reversible effects.[59] Satellites do not have to be permanently disabled, merely prevented from performing their mission while they are passing overhead or for some other limited duration. Employment of these weapons is also much less likely to trigger a response due to the temporary nature of the threat. The middle ground of non-kinetic weapons with destructive effects is a gray area that nations will hesitate to employ for fear of triggering a corresponding response against their own satellites. When temporary reversible methods cannot adequately achieve the objective, a nation must then weigh the marginal benefits of upping the ante or just leaving the offending satellite untouched.

The final factor to consider is that not all satellites have equal costs associated with attacking them at all levels of conflict. Certain satellites have more deterrent value by virtue of their function than others. Missile warning satellites are associated with the nuclear warning mission, and interference with these satellites could signal an intent to launch a nuclear strike. Even so, the continued association of these satellites with non-nuclear missions weakens this threshold and raises the possibility that an opponent may attempt a non-destructive reversible attack against them.[60] At the other end of the spectrum is interference with PNT signals such as GPS. These signals are weak and rather than attacking the satellite it is possible for a nation to interfere with the signal locally and receive only limited objections from the international community.[61] At the same time, the GPS satellite itself provides a global service that is vital to international commerce and so has a high threshold of deterrence.

Other dedicated military satellites have relatively weak levels of inherent deterrence. A nation's Intelligence, Surveillance, and Reconnaissance (ISR) satellites are of high military value and attacking them would have no international impact. The threshold for attacking these satellites with non-destructive attacks would be low with a proportionately low threshold for destructive threats.[62] Military satellite communications (MILSATCOM) would have the same threshold for non-destructive attacks as ISR platforms, with possibly a slightly higher threshold for destructive attacks. An attacker might hesitate to destroy MILSATCOM platforms if they can effectively deny their use to an adversary through non-destructive means. This hesitation would stem primarily from a desire to preserve the option of allowing an opponent to communicate with forward-deployed military forces and de-escalate a conflict.

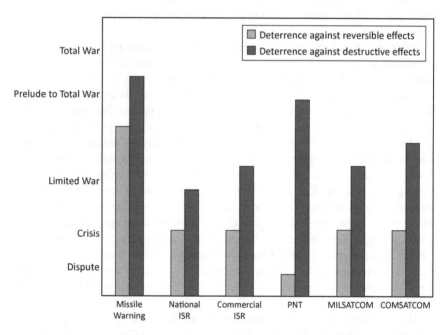

Figure 4.3 Relative deterrence capabilities of different space systems at various conflict levels.

Source: author's original work. For a different opinion on deterrence levels of various satellite classes see Forrest E. Morgan, "Deterrence and First-Strike Stability in Space: A Preliminary Assessment" (Santa Monica, CA: RAND Project Air Force, January 2010), 19.

Note
Values are illustrative only.

In contrast to MILSATCOM, commercial satellite communications would have a higher degree of protection from destructive attacks. While commercial satellites carry substantial military traffic, they also carry traffic from other nations and may be owned by third parties. Attacks on these satellites would require more precision and caution to avoid involving third parties in the conflict. It may even be possible that a commercial satellite is carrying traffic from both parties in a conflict which could effectively prevent an attacker from carrying out destructive attacks.

A final category to consider would be commercial ISR capabilities. For most of the space age reconnaissance satellites belonged exclusively to governments. In the last decade this paradigm has shifted. Driven by small satellites, more commercial imaging satellites now exist on orbit than national platforms.[63] The national affiliation of these satellites may be the primary determiner of the threshold for attack against them. Commercial imaging platforms owned by a US company are likely to have only marginally more deterrent value than national assets, while those owned by a neutral party will enjoy a similar deterrent value to commercial SATCOM. The estimated deterrent threshold for various categories of satellites is shown in Figure 4.3.

The preferred outcome for the dominant space power is continued stability in space. Since effective deterrence is nearly impossible for the dominant space power, it must avoid igniting an arms race that gives lesser space powers greater comparative advantage due to the lack of mutual dependence. Avoiding an arms race requires a solution to the security dilemma that further attempts at deterrence are only exacerbating. Given the offense–defense balance and the misperception of motives and intent between the great space powers, the pre-requisite conditions are present for a reassurance strategy to possibly succeed.

Strategies of reassurance

Reassurance relies on building enough trust to achieve cooperation. The key to building trust lies in accurately determining the motives and intent of an adversary state while clearly signaling your own benign intent. Returning to the analysis of misperceptions of motivation determined in Chapter 3, the US is at the center of the arms competition among the three major space powers which is being fueled by misperceptions of intent, uncertainty, and fear. Therefore, a reassurance strategy pursued by the US will have an impact on both Russia and China while any reassurance strategy these nations pursue independently would only impact the US. To preserve stability in space and avoid an arms race, the US must adopt a reassurance strategy that is costly enough to clearly signal benign intent and spark cooperation between the three competing space powers without overly jeopardizing its security in space.

There are three general strategies for signaling benign intentions: pursuing arms control agreements, unilaterally adopting a defensive posture, and unilateral disarmament.[64] Pursuing arms control agreements is the least costly strategy for the signaling state as it places the pursuing nation at very little strategic risk, which means that it is also the weakest method of signaling of intent. Unilaterally adopting a defensive posture is costlier to the signaling state and so sends a clearer signal of intent than pursuing arms control agreements. Finally, the most dangerous and costly method of signaling intent is through unilateral disarmament. If a state adopts this approach and has incorrectly interpreted its adversary's intentions, then it will leave itself dangerously exposed to attack.

Pursuing arms control agreements is the least costly method of signaling intent and one that has generally proved ineffective in space. The negative US reaction to the repeated attempts by Russia and China to gain approval for the "Treaty on Prevention of the Placement of Weapons in Outer Space and of the Threat or Use of Force Against Outer Space Objects" (PPWT) is a case in point. Russia and China have jointly pursued this treaty for over a decade, and it has not resulted in meaningful negotiations with the US or a reinterpretation by the US of their motivations or intentions in space. Both nations have expressed frustration with the lack of constructive US response, arguing that the US is misinterpreting their actions through "the appalling attempts of the United States of America to impose on the international community its politicized assessment of the space programs of certain states."[65] The lack of a US counter-proposal can only strengthen China and Russia's view that the US has malign intentions of dominating space at their expense, even as they acquire space weapons that further reinforce US views of their malign intent.

A US counter-proposal for a treaty for the prevention of weapons in outer space would be a positive first step toward signaling benign intent. Opening good faith negotiations with both states would signal a departure from the currently perceived US policy of space dominance. Sadly, it is doubtful that this would be enough to defuse the growing security dilemma in space. Perceptions of state behavior are too entrenched among all parties for a low-cost signal to succeed in building enough trust to lessen the severity of the security dilemma. China and Russia would very likely enter negotiations with great caution suspecting that the US is attempting to somehow slow down their weapons development programs just as the US perceives their attempts at negotiations as malicious attempts to prevent it from defending itself.

In order to break the entrenched perceptions among the space powers of malign intent, the US needs to send a costly signal of reassurance. The signal must be costly enough to break the entrenched perceptions of malign intent among the great space powers. The other two broad categories discussed above are useful guides that provide a pathway to specific strategies that can possibly accomplish this lofty objective.

Unilateral disarmament is the costliest category of signaling, but it is difficult to envision what form successful unilateral disarmament would take in space. Since neither the US, nor any other state, currently possess space-based weapons, removing them or pledging not to emplace them would be no more costly than pursuing an arms control agreement. Perhaps it could take the form of unilaterally disbanding US missile defenses. The US has already demonstrated that these systems can double as ASAT weapons.[66] The challenge with this approach is that it would be extremely costly in terms of security. General disarmament of this type would need to involve all US missile defenses seen as capable of reaching orbital altitudes. This would include a broad spectrum of US missile defenses, including theater level defense systems such as THAAD. These systems are vital to US conventional defenses, especially given the proliferation of theater ballistic missiles and their use by lesser US rivals North Korea and Iran. This approach to unilateral disarmament then falls into the trap of dual-use technologies and multi-party verification. It would dangerously expose the US to other nations outside those that we can be reasonably sure are security seekers and not merely contained greedy states.

Going a step further, would unilateral disarmament require the US to completely stop or at least sharply curtail the passive military use of space? Beyond the nearly inconceivable negative impact this would have on US military capability, existing norms in space have already established that the passive use of space is peaceful. This seemingly disposable course of action does have some merit. As discussed earlier, each dedicated military launch and platform creates uncertainty, that breeds fear and feeds the security dilemma. An approach that sharply curtailed the number of military space launches would not only strongly signal benign intent, it would also reduce the level of uncertainty in space.

The middle approach to signaling benign intent is unilaterally adopting a defensive posture. This method is best applied under specific conditions. If the offense–defense balance strongly favors the defense, then the risk of adopting this approach is low and therefore less costly, and less costly signals are less effective at reassuring an adversary.[67] Alternatively, if the offense–defense balance strongly favors the offense and is clearly distinguishable from the defense then adopting a defensive posture would send the clearest signal of intent. The difficulty with this approach is that it would place the signaling nation at extreme risk. This strategy is best applied when there is a high level of distinguishability between the offense and defense and the balance between the two is neutral. Since this closely resembles the current situation in space, this approach has significant potential. The challenge is determining what form this strategy would take and how it would be different from the current US posture.

The current US posture is defensive in the form of non-aggressive passive military use. As some significant change is required to signal benign intentions, maintaining the status quo would be an ineffective approach. The US needs to change its posture enough to effectively signal intent without jeopardizing the advantages it gains from space or harming its overall space power. Resolving these seemingly conflicting objectives has only one clear answer, commercialization.

Commercialization offers a blend of the reassurance strategies of unilateral disarmament and defensive posture adoption that would reassure the US's adversaries without sacrificing national space power. Commercialization would preserve US national space presence and therefore national space power while at the same time reassuring potential adversaries that the US is not building or deploying disguised space weapons. The capabilities of commercial satellites are also unclassified and available for verification through use by any paying customer, reducing uncertainty and increasing the level of entanglement between US military signals and those of other nations.

Choosing commercialization as a signaling strategy also shares many of the desirable signaling aspects of unilateral disarmament. Depending on the degree of successful commercialization it would curtail the number of US military launches, thereby further reducing uncertainty. Reducing active US military presence in space would reduce the need for military space forces, inevitably leading to a reduction in their numbers. These combined effects from commercialization would likely result in reduced costs for the US military with relatively little loss in capability.

This strategy raises the immediate question of which aspects of the US national security architecture could be commercialized and to what degree? The national space architecture can be subdivided into five categories: remote sensing, launch, positioning, timing and navigation (PNT), communications, and missile warning. Each category of the national security space architecture has unique issues and challenges that determine the difficulty of pursuing a commercialization strategy.

Launch systems are already effectively commercialized, especially now that the US government has broken its exclusive relationship with the United Launch Alliance (ULA) for national security payloads. The entry of SpaceX, Blue Origin, and other companies into the launch market has expanded the US launch market from a ULA driven monopoly to a competitive marketplace. As a result, the US now has the lowest cost and most reliable access to space of any nation. The emergence of this competitive market place for launch services has driven innovation and real decreases in the cost of access to space in what was previously a staid and unexciting industry. The positive impact this has had on US space power cannot be overstated, and it was made possible by a NASA commercialization effort that expanded beyond the traditional launch providers.[68] A carryover from the success of the launch industry to other categories of space could yield similar benefits.

Another category that is not in need of commercialization is PNT. Epitomized by GPS, the PNT mission may be better served by transitioning from a military to civilian-led organization. GPS is currently controlled by the US military mainly by virtue of its origins. Having expanded far beyond its original military purpose, GPS would be relatively easy to shift to a civilian managed organization as Europe has done with the Galileo constellation. Transitioning GPS from military to civilian management would have only a moderate impact on the security dilemma because it would not prevent military use of GPS or the development of active jamming systems by adversary states. Also, GPS satellites are launched infrequently as existing satellites reach their end of life and the constellation is optimized for better coverage.[69] Moreover, because of their unique orbit and the ability of any user to verify the function of GPS satellites, their continued presence and ongoing replenishment contributes relatively little to the uncertainty driving the security dilemma.

The final three categories have varying degrees of commercialization potential. SATCOM is an area where the US military already has a substantial commercial presence, though it continues to rely primarily on dedicated MILSATCOM systems for a variety of reasons. Remote sensing as an industry is only just beginning to reach a capability level where commercialization could represent a realistic option. Finally, missile warning and other platforms directly associated with the nuclear deterrent would seem to be the last category where any degree of commercialization could reach, though some level of commercialization may be possible as demonstrated by the Commercially Hosted Infrared Payload (CHIRP) demonstration program launched in 2011.[70]

Commercialization is a potentially viable solution as a signaling approach to building trust and resolving the growing security dilemma. It reduces uncertainty and equates to adopting a defensive posture in space without overly jeopardizing the security of the signaling state. The US is the state best positioned to adopt this strategy with the greatest incentive for ensuring stability in space and avoiding a costly and self-defeating arms race. As the dominant space power with the most substantial commercial industry, the US is also in the best position to implement a commercialization strategy. Finally, the US is the one nation engaged in a space arms race with both Russia and China. These two nations see the US as their primary adversary rather than each other despite sharing an extensive land border and a historically adversarial relationship. The above discussion leads to the first hypothesis of this book:

Hypothesis 1: Full commercialization by the status quo dominant space power (the US alone) would significantly reduce the security dilemma.

As discussed earlier, of the five categories of national security space, launch, and PNT are either commercialized or contribute relatively little to the security dilemma. Of the three remaining categories missile warning and communications satellites directly associated with the nuclear deterrent present the greatest challenge to a commercialization strategy. Pursuing a commercialization strategy for these systems may be infeasible. It is also possible that if these systems are limited to supporting the nuclear deterrent mission their continued militarization would only have a moderate impact on the security dilemma. Retaining these systems under military control then forms the second hypothesis of this book:

Hypothesis 2: Commercialization of all functions except those directly associated with the nuclear deterrent would somewhat reduce the security dilemma.

A last point of investigation is if the commercialization effort is broadened beyond just the US. Should a US commercialization effort prove successful in building trust and reducing the security dilemma, other states may pursue a similar strategy. If it is possible for the US to fully or mostly commercialize while preserving or even increasing its national space power other states may follow suit. A demilitarized space architecture that relies on commercial capabilities for the passive use of space forms a final hypothesis:

> **Hypothesis 3:** Commercialization by all great powers in space would effectively eliminate the security dilemma.

Summary

The willingness to cooperate is based on trust, whereas building security through competition is based on fear. The challenge is to determine how states signal a willingness to cooperate and build trust in a world driven by the uncertainties of the security dilemma. Of the three strategies for signaling reassurance the one strategy that best fits the conditions that currently exist is adopting a defensive posture. Differentiating a defensive posture from the current US non-aggressive posture is necessary to signal benign intent. Commercialization of as much of the US national security space structure as possible is the closest the US can come to adopting a defensive posture and signaling benign intent.

Adopting a commercialization strategy to signal intent must avoid overly jeopardizing the security of the signaling state. The functioning theory of space power used for this research has three primary tenets that apply as long as space militarization falls short of the development of space-to-ground weapons. Commercialization directly supports all aspects of this space power framework. First, commercialization will have little negative impact on a nation's presence in space and may have a positive impact. Second, commercialization will not significantly decrease the ability of the signaling nation to transfer information through space or gather information from it. Also, it will not impact the supporting domain aspect of space or the ability of the military to benefit from its passive use. Commercialization could, in fact, have a positive impact on all aspects of national space power if it results in increased innovation and industrial development.

A strategy of commercialization is also dependent on the offense–defense balance and degree of distinguishability meeting specific conditions. As a strategy that most closely resembles adopting a defensive posture with some aspects of unilateral disarmament, it is best executed when the offense–defense balance is neutral and distinguishability its high. Since both of these conditions currently exist in the space domain, commercialization could prove to be an effective signaling strategy that does not jeopardize national space power.

An arms race is currently in progress in space that does not benefit the US. Adopting a commercialization strategy that focuses on the three categories of space systems that would most impact the security dilemma is possibly the best approach to avoiding a sub-optimal arms race. The question remains of what degree commercialization is possible for the US? The next three chapters of this book are a unit level analysis of each category beginning with SATCOM, followed by remote sensing, and finishing with the most challenging commercialization case, missile warning.

Notes

1 John A. Tirpak, "Operation Burnt Frost 'Historic,'" *Air Force Magazine*, February 22, 2008, www.airforcemag.com/Features/security/Pages/box022208shootdown.aspx.
2 *Joint Publication 3–14: Space Operations* (US Department of Defense, 2018), GL-6.
3 "Air Force Doctrine Document 1 (AFDD-1)" (US Air Force Curtis E. Lemay Center for Doctrine Development and Education), Chapter 3, accessed October 12, 2018, www.au.af.mil/au/awc/awcgate/afdc/afdd1-chap3.pdf.
4 *Joint Publication 3–14: Space Operations* (US Department of Defense, 2009), GL-9.
5 David E. Lupton, *On Space Warfare* (Maxwell Air Force Base, AL: Air University Press, 1988), 7.
6 Dana J. Johnson, Scott Pace, and C. Bryan Gabbard, *Space: Emerging Options for National Power* (Santa Monica, CA: RAND Corporation, 1998), xi.
7 Brent Ziarnick, *Developing National Power in Space: A Theoretical Model* (Jefferson, NC: McFarland, 2015), p. 14.
8 Ibid., p. 13.
9 Alfred Thayer Mahan *The Influence of Sea Power Upon History 1660–1783* (New York, NY: Dover Publications, 1987), 400.
10 Quoted in Peter J. Hugill, "German Great-Power Relations in the Pages of 'Simplicissimus,' 1896–1914," *Geographical Review* 98, no. 1 (2008): 2.
11 Charles L. Glaser, *Rational Theory of International Politics* (Princeton, NJ: Princeton University Press, 2010), 238–243.
12 Julian S. Corbett, *Some Principles of Maritime Strategy* (The Perfect Library, 2016), 81.
13 Alfred Thayer, *The Influence of Sea Power Upon History 1660–1783*, 589.
14 John J. Klein, *Space Warfare: Strategy, Principles and Policy*, 1 edition (London; New York, NY: Routledge, 2006), 60.
15 Ibid.
16 Ibid., 63.
17 "Iran Jams VOA's Satellite Broadcasts," *Voice of America News*, October 5, 2012, www.voanews.com/a/iran-jams-voa-satellite-broadcasts/1521003.html.
18 Jean-Luc Lefebvre, *Space Strategy* (Hoboken, NJ: John Wiley & Sons, 2017), 214.
19 The terms used here to summarize Lefebvre's are not literal from his work but rather the author's summary in order to remain concise. For instance, rather than simply using the term "engagement" Lefebvre literally says, "elicit a sense of wonder turned toward the stars."
20 James E. Oberg, *Space Power Theory* (Colorado Springs, CO: USAFA Government Printing Office, 1999), 125.
21 Ibid.
22 Ibid.
23 Ibid., 126.
24 Ibid., 127.
25 Ibid., 130.
26 Lupton, *On Space Warfare*, 118.

27 This applies only to conditions short of space-to-ground weapons and economic conditions in space short of substantial resource extraction and production.

28 See Forrest E. Morgan, "Deterrence and First-Strike Stability in Space: A Preliminary Assessment" (Santa Monica, CA: RAND Project Air Force, January 2010), 2; and James P. Finch and Shawn Steene, "Finding Space in Deterrence," *Strategic Studies Quarterly* 5, no. 4 (2011): 11 and; Paul Scharre, "The US Military Should Not Be Doubling down on Space," *Defense One* (blog), August 1, 2018, www.defenseone.com/ideas/2018/08/us-military-should-not-be-doubling-down-space/150194/.

29 Colin S. Gray, *Weapons Don't Make War: Policy, Strategy, and Military Technology* (Lawrence, KS: University Press of Kansas, 1993), 14–15.

30 Carl Clausewitz, *On War*, ed. Michael Howard and Peter Paret (Princeton, NJ: Princeton University Press, 1984), 358.

31 Finch and Steene, "Finding Space in Deterrence," 11.

32 Edward Ferguson and John Klein, "The Future of War in Space Is Defensive," *Space Review*, December 19, 2016, www.thespacereview.com/article/3131/1.

33 Cristina T. Chaplain, Director Acquisition and Sourcing Management, "Space Acquisitions: DOD Continues to Face Challenges of Delayed Delivery of Critical Space Capabilities and Fragmented Leadership," § Testimony before the Subcommittee on Strategic Forces, Committee on Armed Services, US Senate. (2017), www.gao.gov/assets/690/684664.pdf.

34 "Union of Concerned Scientists: Satellite Database" (Union of Concerned Scientists, May 1, 2018), www.ucsusa.org/nuclear-weapons/space-weapons/satellite-database#.W7eDtPZFxEZ.

35 Pixalytics ltd, "Earth Observation Satellites in Space in 2017?" (Pixalytics Ltd., November 29, 2017), www.pixalytics.com/eo-sats-in-space-2017/.

36 "Defense and Intelligence," *Planet*, July 18, 2018, https://planet.com/markets/defense-and-intelligence/.

37 Lockheed Martin Corp., "SBIRS Fact Sheet" (Lockheed Martin Corp., 2017), www.lockheedmartin.com/content/dam/lockheed-martin/space/photo/sbirs/SBIRS_Fact_Sheet_(Final).pdf.

38 "Resiliency and Disaggregated Space Architectures" (Air Force Space Command, April 14, 2016), 4.

39 "Draft Position Paper for UN Ad hoc Committee on Peaceful Uses of Outer Space: Legal Problems Which May Arise in the Exploration of Space" (White House Office of the Staff Secretary: Records, 1952–1961, April 22, 1959), 8, Box 24, Space Council (7), Eisenhower Library.

40 "Defense Support Program Satellites" (US Air Force Space Command Public Affairs Office, November 23, 2015), www.af.mil/About-Us/Fact-Sheets/Display/Article/104611/defense-support-program-satellites/.

41 Glenn A. Kent and David E. Thaler, "First-Strike Stability and Strategic Defenses: Part II of a Methodology for Evaluating Strategic Forces" (RAND Corporation, October 1990), xviii.

42 *Joint Publication 3–14: Space Operations*, 2018, II–2.

43 Ibid.

44 Ibid., II–3.

45 Amy Butler, "USAF Space Chief Outs Classified Spy Sat Program," *Aviation Week*, February 21, 2014, http://aviationweek.com/defense/usaf-space-chief-outs-classified-spy-sat-program.

46 Gen. William Shelton quoted in Irene Klotz, "US Air Force Reveals 'neighborhood Watch' Spy Satellite Program," *Reuters*, February 22, 2014, www.reuters.com/article/us-space-spysatellite-idUSBREA1L0YI20140222.

47 Samantha Mathewson, "India Launches Record-Breaking 104 Satellites on Single Rocket," *Space.Com*, February 15, 2017, www.space.com/35709-india-rocket-launches-record-104-satellites.html.

48 Sam Jones, "Object 2014–28E—Space Junk or Russian Satellite Killer?," *Financial Times*, November 17, 2014, www.ft.com/content/cdd0bdb6-6c27-11e4-990f-00144fe abdc0#axzz3JPZDZk6I.

49 Brian Weeden, "Space Situational Awareness Fact Sheet" (Secure World Foundation, May 2017), https://swfound.org/media/205874/swf_ssa_fact_sheet.pdf.

50 Kyle Mizokami, "China Proposes Orbiting Laser to Combat Space Junk," *Popular Mechanics*, February 20, 2018, www.popularmechanics.com/military/weapons/a18240128/china-orbital-laser-space-junk/.

51 "President Donald J. Trump Is Unveiling an America First National Space Strategy" (The White House, March 23, 2018), www.whitehouse.gov/briefings-statements/president-donald-j-trump-unveiling-america-first-national-space-strategy/.

52 Ibid.

53 Lawrence Freedman, *Deterrence* (Cambridge: Polity Press, 2004), 32.

54 Ibid.

55 Ibid., 32–33.

56 Morgan, "Deterrence and First-Strike Stability in Space," xiii.

57 Dean Cheng, "Evolving Chinese Thinking About Deterrence: What the United States Must Understand about China and Space" (Heritage Foundation, March 29, 2018), 5–6.

58 Freedman, *Deterrence*, 35.

59 Finch and Steene, "Finding Space in Deterrence," 12.

60 Lockheed Martin Corp., "SBIRS Fact Sheet."

61 US National PNT Advisory Board, "Russia Undermining Confidence in GPS," May 17, 2018, www.gps.gov/governance/advisory/meetings/2018-05/goward.pdf.

62 Morgan, "Deterrence and First-Strike Stability in Space," 19.

63 "Union of Concerned Scientists: Satellite Database."

64 Charles L. Glaser, "The Security Dilemma Revisited," *World Politics* 50, no. 1 (1997): 181.

65 Permanent Representative of China to the Conference on Disarmament, "Letter from the Permanent Representative of China to the Conference on Disarmament and the Charge d'affaires a.i. of the Russian Federation Addressed to the Secretary-General of the Conference Transmitting the Comments by China and the Russian Federation Regarding the United States of America Analysis of the 2014 Updated Russian and Chinese Texts of the Draft Treaty on Prevention of the Placement of Weapons in Outer Space and of the Threat or Use of Force against Outer Space Objects (PPWT)," September 11, 2015, 6, https://documents-dds-ny.un.org/doc/UNDOC/GEN/G15/208/38/PDF/G1520838.pdf?OpenElement.

66 Tirpak, "Operation Burnt Frost 'Historic.'"

67 Andrew Kydd, "Trust, Reassurance, and Cooperation," *International Organization* 54, no. 2 (2000): 326.

68 Ashlee Vance, *Elon Musk: Tesla, SpaceX, and the Quest for a Fantastic Future* (New York, NY: Harper-Collins, 2015), 210.

69 "GPS Space Segment" (US government), accessed October 23, 2018, www.gps.gov/systems/gps/space/.

70 "CHIRP" (Hosted Payload Alliance), accessed October 25, 2018, www.hostedpayload alliance.org/Hosted-Payloads/Case-Studies/Commercially-Hosted-Infrared-Payload-(CHIRP)-Fligh.aspx#.W9dhIfZFxEZ.

5 Satellite communications

Current US government use of satellite communications is categorized into two major areas, military satellite communications (MILSATCOM) and commercial satellite communications (COMSATCOM). MILSATCOM is made up of satellites built by contracted commercial companies for dedicated military use. Often these platforms include advanced security features, operate in unique frequency ranges, and are designed to interface with specific ground systems. Examples of MILSATCOM include the Military Strategic and Tactical Relay (Milstar) and the Wideband Global SATCOM Satellites (WGS) which both operate in geosynchronous orbit, providing dedicated communications for US and selected allied military users. COMSATCOM platforms, like MILSATCOM, are usually located in geosynchronous orbit and operate by leasing bandwidth on transponders for a contracted length of time. Examples of COMSATCOM providers include INTELSAT which operates 53 geosynchronous satellites spanning the globe.[1] In contrast, WGS, which is designed to be the backbone of the US MILSATCOM constellation, currently consists of just ten satellites with a planned total of 12.[2] The 65 current dedicated US government communications satellites represent the second largest category of satellites belonging to the US government. It is a category that could grow dramatically if the Space Development Agency succeeds in developing its concept of a space transport layer, a global constellation of LEO communications satellites.[3] Eliminating these satellites in favor of commercial systems would send a powerful signal of benign intent and remove a current and growing source of uncertainty.[4]

The security dilemma has a fundamentally defensive character stemming from the fear and uncertainty of being attacked. In space this fear translates into two primary outcomes. First, the more valuable a satellite system is to a nation's defense the more it fears its loss. Second, the more numerous satellites are the more uncertainty they create in an adversary that some of them could be a threat to their space systems. Currently, the first outcome applies more to MILSATCOM though the future adoption of constellations of satellites could shift this in favor of the second. Despite being neither the most numerous or arguably the most valuable of US satellite capabilities, SATCOM still contributes to the security dilemma due to the impact its loss would have if it faced interference or destruction over a theater of conflict.

Of the three categories of satellites analyzed in this research, MILSATCOM satellites are arguably the least valuable collectively and individually. Satellite communications are still critical to national security and nearly invaluable, but unlike remote sensing or missile warning, it is a category whose loss could be somewhat mitigated. Some degree of global military communications capability could remain by relying on undersea cables and by bouncing radio signals off the ionosphere. Of course, these mitigation factors are also vulnerable to adversary interference and not nearly as versatile as space-based capabilities. Naval assets are perhaps the most reliant on SATCOM for coordination and information flow and would suffer the most if SATCOM is denied due to enemy attack. Since any conflict between China and the US would lean heavily on naval and air capabilities, attacks against SATCOM could prove extremely disruptive. Scenarios such as this increase the fear of loss and drive a need to find ways to preserve SATCOM capability.

The number of US government satellites performing the SATCOM mission is only slightly less than the number of active remote sensing satellites, though this could change in the near future. OneWeb and SpaceX are launching LEO constellations of thousands of small and cheap satellites designed to provide global high-speed internet access. DOD is watching their progress closely to determine if they could serve as models for future military constellations.[5] Communications constellations consisting of hundreds or thousands of inexpensive satellites are more resilient than individual satellites to attack. They would shift the current dynamic away from slightly favoring the offense toward somewhat favoring the defense at the operational level. Instead of being able to target a handful of satellites to obtain a tactical advantage an attacker must disable or destroy dozens of platforms to disrupt a small satellite constellation. While constellations reduce the fear of loss, they significantly increase the uncertainty in an adversary that some of these numerous satellites may be weapons designed to destroy their satellites or block access to orbit.

The threshold for attacking SATCOM also influences the fear of loss that drives the security dilemma. A rare capability with a low threshold for attack would generate the greatest fear of loss. For SATCOM the likely threshold for attacking it is relatively low, though it varies slightly between MILSATCOM and COMSATCOM. Both forms of SATCOM are likely to face non-destructive attacks short of conflict to disrupt communications and military preparations. There is variation between them when it comes to destructive attack as MILSATCOM will likely be faced with the threat of destructive attack before COMSATCOM once conflict begins. The additional deterrent capability granted to COMSATCOM is due to the possibility that the attacking nation could disrupt third party or even friendly communications in addition to its adversary's.

A capable adversary will almost certainly target US SATCOM with non-destructive attacks once it becomes likely that conflict is about to occur. These attacks will disrupt military command and control systems to varying degrees hampering military operations. The ability to quickly leverage the large number of alternative commercial options available will depend on the flexibility of the

US government's acquisition system and its relationship with commercial industry. Adopting a commercial architecture in an effort to signal benign intent is unlikely to prevent an opponent from attempting to interfere with US Military communications. What it can do is lessen the security dilemma and avoid the further development of destructive space capabilities. It will also allow the US military to leverage the latest commercial technology and disaggregate its signals architecture, increasing the difficulty of effectively disrupting military communications.

COMSATCOM and the US military

In 1964 the Tokyo Olympics was broadcast via Syncom III, a NASA experimental communications satellite launched in August of that year into geosynchronous orbit. The Syncom III signal reached the United States and then was rebroadcast from the US across the Atlantic Ocean to Europe via another NASA experimental satellite, Relay I.[6] The age of global satellite-enabled communications had begun.

Just two years before, Congress passed the Communications Satellite Act which sought to "establish ... a commercial communications satellite system, as part of an improved global communications network."[7] The goal of this act was to facilitate private enterprise in global satellite communications and promote competition in the sector. The act created the Communications Satellite Corporation (COMSAT) which was authorized to "plan, initiate, construct, own, manage, and operate itself or in conjunction with foreign governments or business entities a commercial communications satellite system."[8]

The Communications Satellite Act created the legal structure for the initial commercialization of space. At that time there was no commercial space capability; neither was there a method of transferring the technology developed by the US government to a private corporation. Space access was still very new and the workforce and technology to access space was controlled entirely by major national governments. The act overcame these difficulties by directing NASA to assist COMSAT with research and development as well as with satellite launching and services on a reimbursable basis.[9] By doing this, the US government created a legal venue for the commercialization of much of the space technology that NASA was developing to support manned space flight.

The act also placed several restrictions on COMSAT. Though foreign participation and relations by COMSAT were encouraged in the act, supervision by the president was required to ensure any foreign contacts were consistent with existing US foreign policy and interests.[10] Under the provisions of the act, COMSAT helped create the International Telecommunications Satellite Consortium (INTELSAT) in August 1964 which successfully launched the first commercial communications satellite, Early Bird, in April 1965.[11]

Even after the creation of COMSAT, military communications in space remained separate from civilian communications satellites due to cost and development concerns. In 1962, then Secretary of Defense Robert McNamara

Figure 5.1 President Kennedy signs the Communications Satellite Act, August 31, 1962.

Source: John F. Kennedy Library, accessed 14 August 2019, www.jfklibrary.org/asset-viewer/archives/JFKWHP/1962/Month%2008/Day%2031/JFKWHP-1962-08-31-A?image_identifier=JFKWHP-AR7444-G.

canceled a joint Army and Air Force program, Advent, that aimed to develop a military communications satellite constellation. McNamara canceled the program due to ballooning costs and technical concerns and opened discussions with newly formed COMSAT to lease bandwidth from them at a lower cost.[12] McNamara's idea was sound, but COMSAT was only a few months old and had not yet founded INTELSAT or launched its first satellite. Negotiations failed because the DOD and COMSAT could not agree on costs or the need for dedicated military transponders aboard COMSAT's satellites. As a result of these disagreements with COMSAT, in July 1964 McNamara ended negotiations with COMSAT and opted for the development of a dedicated military satellite constellation under the directions of the Air Force called the Initial Defense Communications Satellite Program.[13]

From that point until Operation Desert Storm the military was able to meet its satellite communications needs with its own resources. That the MILSATCOM constellation could do this was primarily the result of timing. The Vietnam war ended before the US military could develop a significant dependence on satellite communications capabilities and peacetime usage did not stress the available bandwidth to the point at which it would drive the purchase of commercial bandwidth. This changed with the first Gulf War in 1991 when demand spiked, and satellites carried approximately 80 percent of communications.[14] This percentage was achieved despite demand exceeding supply in both bandwidth and satellite capable ground equipment. Air Force Space Command (AFSPACECOM) in a review of lessons learned from the conflict identified that communications plans had underestimated the level of demand and recommended that it acquire more military communications satellites to support future operations.[15] Nowhere in the lessons learned document was using COMSATCOM as a backup or developing a permanent relationship with a commercial provider mentioned as an alternative despite the extensive reliance on it during Desert Storm.

The role of COMSATCOM in enabling the communications architecture in the Gulf War cannot be understated. Just before the start of the conflict the total bandwidth usage in the Central Command (CENTCOM) area of operations was 4.54 Mbps which was entirely provided by MILSATCOM.[16] Within the first month there was no longer any available MILSATCOM bandwidth, and the DOD was forced to transfer satellites from other global locations and adopt other extreme measures to support the growth of demand. At the height of the conflict demand had increased to 67.65 Mbps carried over MILSATCOM and 31.39 Mbps carried over COMSATCOM for a total of 99.04 Mbps.[17] COMSATCOM, provided entirely by INTELSAT, was carrying 31.6 percent of all military satellite traffic and nearly 20 percent of all traffic in the entire theater. Interestingly, the military had returned to COMSAT founded INTELSAT to carry the majority of data that traveled back to the continental United States because INTELSAT possessed both the constellation of satellites and the ground transfer stations to support it whereas the military did not.

Desert Storm serves as a benchmark for SATCOM usage.[18] During Desert Storm, SATCOM usage was 140 bits per second (bps) per deployed soldier. Future conflicts would see exponential growth. In 1999, for example, usage in Kosovo had increased to 3,000 bps per soldier. This growth accelerated and reached 8,300 bps per soldier in the opening days of Operation Enduring Freedom in Afghanistan in 2001 and a further 13,800 bps per soldier by 2004 in Operation Iraqi Freedom.[19] Total bandwidth used during the invasion of Iraq in 2003 was 3.2 Gbps compared to the 99 Mbps used for a force more than twice as large in Desert Storm.[20] This exponential growth in SATCOM usage came at significant cost to the US government and drove substantial change in how COMSATCOM was acquired.

A troubled history of acquiring COMSATCOM

The sudden increase in the tempo of operations during the Gulf War and the associated increase in demand for satellite bandwidth created a free-for-all in acquiring COMSATCOM to meet the surge in demand. After the conflict ended, the DOD addressed this by mandating that the Defense Information Systems

Agency (DISA) would manage the process for acquiring commercial band-width.[21] DISA managed the process in accordance with federal regulations and standards, but the process was slow and demand was always immediate. Users of DOD commercial satellite services were dissatisfied with DISA's process, claiming that it was too slow for military operations and too expensive.[22] For this reason, many users circumvented the DISA process. The General Accounting Office (GAO) in 2003 estimated that at least 20 percent of DOD's purchased bandwidth was acquired without going through DISA.[23] Subsequent analysis by a GAO report in 2015 raised this estimate to 55 percent.[24]

Lack of compliance with government mandates to rely on DISA processes for securing bandwidth stemmed from the DISA's high costs and slow process. In 2003, the average time for DISA to award a task order was 79 days after receiving the request, and further time was required after that for a service provider to be selected and for service to begin to the requesting customer.[25] In contracts where users acquired bandwidth directly, the process could be completed in a few weeks and was significantly cheaper than what it cost to go through DISA. In an example cited by the GAO in its 2003 report, the US Army was able to acquire COMSATCOM directly from the commercial provider for $34,700 per month as compared to a price estimate from DISA of $139,000 per month.[26] A later upgrade of a ground terminal for this same contract was priced

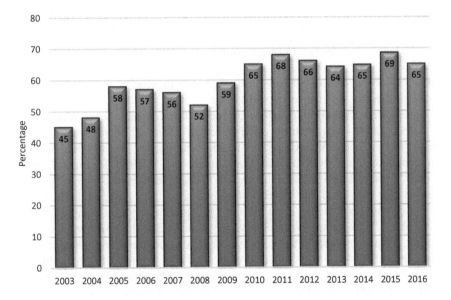

Figure 5.2 Percentage of DOD fixed satellite services acquired by DISA, 2003–2016.

Sources: United States General Accounting Office, *GAO-15-459 Defense Satellite Communications: DOD Needs Additional Information to Improve Procurements*, (Washington, DC: GAO, July 2015); "Defense: DOD Commercial Satellite Communications Procurements (2016–01)," US Government Accountability Office, 2016; Fiscal Year 2012, 2013, and 2015 Commercial Satellite Communications Usage Reports: In Response to Chairman of the Joint Chiefs of Staff Instruction 6250.01E." (US Strategic Command, April 6, 2015, March 30, 2016, and September 7, 2017).

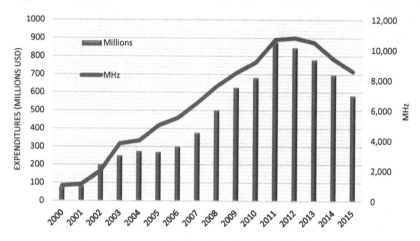

Figure 5.3 DOD fixed satellite service bandwidth cost and usage (excludes MSS).

Sources: Department of Defense. *Satellite Communications Strategy Report: In Response to Senate Report 113–44 to Accompany S.1197 NDAA for FY14*. Washington, DC: Office of the Chief Information Officer, 14 August 2014; Fiscal Year 2012, 2013 and 2015 Commercial Satellite Communications Usage Reports: In Response to Chairman of the Joint Chiefs of Staff Instruction 6250.01E." (US Strategic Command, April 6, 2015, March 30, 2016, and September 7, 2017).

Note
Fixed satellite services (FSS) are communication services between fixed earth stations at specific locations; mobile satellite services (MSS) are communications between mobile ground-based stations at varying locations by means of one or more satellites, Iridium is an example of MSS.

at $579,000 by DISA, but the Army was able to acquire it at $240,000 by working directly with the service provider.[27]

There were a number of reasons for the cost discrepancy between acquiring COMSATCOM through DISA vice purchasing it directly. First, DISA charged requestors an 8 percent surcharge on the contract price to cover its costs incurred in conducting the work as authorized under the Defense Working Capital Fund.[28] Second, DISA acquired bandwidth through vendors who charged a small fee, usually between 1 and 4 percent of the total contract. Also, while the DISA contract with its vendors included the "termination for convenience" clause typical in US government contracts, which should allow the government to stop paying for the bandwidth after it was no longer needed, the service providers did not. The vendors' contracts reflected typical industry practice of being liable for the remaining cost of the lease upon cancelation. The result is that the vendor, and not the service provider, is the entity that the government would cancel the contract with and under government contract law the vendor is entitled to recoup reasonable costs and a fair profit.[29] Since the vendor is liable for the full cost of the lease to the service provider, these costs become a factor in the negotiated termination settlement with DISA, from which DISA ends up bearing some portion of these costs that are in turn passed on to the service requestor. The final factor in inflated DISA prices can be blamed on poor estimates based on old or inaccurate data.[30] Whatever the cause, a requesting user must still have the money budgeted to meet the initial cost estimate, and this leads users to pursue alternative acquisition methods.

DOD users are still bypassing DISA and acquiring COMSATCOM directly, and neither the DOD nor DISA have firm control over the process. The GAO criticized the DOD in 2003 for poorly managing the entire process from acquisition and performance metrics to the enforcement of existing policies.[31] In 2015, the GAO criticized the DOD for almost exactly the same things as in 2003.[32] The legacy of the poor process that DISA was using in 2003 is that today users continue to circumvent it and go directly to the provider to acquire bandwidth even though the GAO estimated that using DISA was 16 percent cheaper than going directly to the vendor in 2015 as the result of improved contracting processes.[33]

Despite improvements in achieving cost competitiveness the DOD remains inefficient at acquiring COMSATCOM and has faced continued criticism. In 2013, the chief executive officers (CEOs) and presidents of five of the largest satellite communications companies in the world published an open letter to the DOD titled, *Seven Ways to Make the DOD a Better Buyer of Commercial SATCOM*.[34] The CEOs criticized the DOD purchasing model as inefficient and its cost comparison methodology between MILSATCOM and COMSATCOM as deeply flawed. They pointed out that the DOD showed no ability ever to be able to meet its own satellite communications needs again. As a result, continuing to purchase COMSATCOM using spot market and indefinite-delivery/indefinite-quantity (IDIQ) short-term leases resulted in much higher prices for the DOD and created the possibility that the needed bandwidth might not be available in times of crisis due to the crowded commercial market. Instead, they recommended that DOD adopt a long-term baseline approach with a dedicated Program Objective Memorandum (POM) line for acquiring COMSATCOM. They also recommended that the DOD adopt commercial technology and standards and stop building architectures that are incompatible with commercial infrastructure. The industry leaders also cited the Civil Reserve Air Fleet (CRAF) model as a way for the DOD to cover the marginal cost of building additional desired protective features into commercial satellites in order to increase the pool of commercial satellites available for use. Finally, they proposed replacing DISA with a single office that could integrate both MILSATCOM and COMSATCOM requirements.

In the same month of the open letter's publication, the Defense Business Board (DBB) released a report on DOD COMSATCOM that echoed many of the things proposed by the CEOs. The DBB found that the DOD purchased COMSATCOM through a mixture of one-year leases, 75 percent of the total, and "spot market" purchases, 25 percent of the total, for which the DOD paid a premium. Commercial satellite development also was progressing significantly faster than military satellites. Commercial satellite development timelines were 3–4 years and at the same time substantially cheaper while the military development timeline was 5–15 years.[35] Partnering with the COMSAT industry would allow the DOD to define requirements, leverage commercial technology, and establish a compatible architecture. The DBB also found that military users were loath to plan for COMSATCOM purchases because those costs came from the requesting user's budget whereas MILSATCOM was perceived as free by most users due to the funding structure. Thus, COMSATCOM was purchased for the shortest interval possible and only upon critical need, driving up costs. The DBB conducted an analysis of alternative contracting models and found obstacles to adopting each of them, see Figure 5.4 below.

	Alternative approaches to COMSAT acquisition	**Obstacles to each approach**
Buy to Lease	Make offer to a commercial operator for system use and obtain quid pro quo global service access for discount or zero charge	Funds derived from DOD asset must go to the national treasury vs. global service access deal
Capital Lease	Long term lease for satellite life (>10 years)	Programmers resist O&M dollars for investment; Procurement dollars ineligible for these deals; existing regulation limited to 5 year maximum lease
Anchor Tenancy	NASA/NOAA ability to enter into multi-year contracts to serve as the anchor tenant for commercial space ventures	Termination liability concerns; statute limited to NASA/NOAA; cannot be used for COMSAT unless authorized by Congress
Indefeasible Right of Use	Pays for up-front costs; signs agreements with others to get services and pays a large up-front fee, followed by annual charges for maintenance and upkeep	Failed providers pulling out early; poor pricing methods
Multiyear/ Long term lease	Opportunity to reduce costs with longer leases	Congress uncomfortable committing dollars beyond first year; multi-year contracts limited to five years; termination liability concern
Hosted Payloads	DOD furnished payload; special needs; short timeframe	Requires prior funding; delays to timeline in acquiring new bandwidth; US launch vehicle requirement per Space Transpiration Policy
Pathfinder	Finding optimal approach to leverage COMSAT technologies, a long-term solution	Near term budget issues with the added challenge of accepting large scale non-traditional approaches; acquisition, policy, and legal concerns generate risk

Figure 5.4 Alternative approaches to COMSAT acquisition.

Source: adapted from Defense Business Board. Report to the Secretary of Defense: Taking Advantage of Opportunities for Commercial Satellite Communications Services, Report FY13-02. January 2013, 8.

Following the DBB report and the industry letter, congressional attention on the cost of COMSATCOM resulted in the draft Senate version of the National Defense Authorization Act (NDAA) for FY14 included a requirement for the DOD to explore methods of long-term leasing and to determine the appropriate mix of military and COMSATCOM.[36] The DOD response described their current, as of 2014, acquisition process as a "multiple year contract with a (single) base year and price-negotiated option years."[37] The report stated that this mitigated the risk of excessive cancelation charges in case of "termination for convenience" while still ensuring below market prices. Long-term contracts, like those requested by the Senate, required stability in requirements, predictable funding, and substantial savings to the government in order to be viable. Related to achieving this goal, the DOD identified four related issues that needed to be addressed:

1   accurate and timely prediction of demand,
2   appropriate and stable sources of funding,
3   managing capacity use to realize any apparent cost-benefit of long-term bulk leasing,
4   statutory authority for multi-year contracting.[38]

The DOD cited multiple reasons why it could not resolve the issues that it had identified. Developing an accurate demand prediction requires projecting past trends onto future events. While this is a difficult and inherently inaccurate process, it is also something that the DOD is required to do when developing MILSATCOM. DOD also argued that appropriate and stable funding required centralized procurement to eliminate overlap between multiple theater commanders with separate funding sources, often Overseas Contingency Operation (OCO) funds outside the base DOD budget. The DOD would then need to increase its funding levels in the baseline budget to cover the increased cost if OCO funds were no longer authorized for SATCOM purchases. Managing capacity required understanding user needs, and again the DOD argued that it needed a centralized authority which it had years before dictated through internal policy but not effectively enforced. DOD concluded by pointing out that the multi-year contracting authority options recommended by the Senate were not authorized under US law for purchasing COMSATCOM.[39]

In its 2014 report, the DOD had identified three conditions under which commercial services were acquired: when military bandwidth was unavailable, when user demand exceeded military capability, or when user ground terminals were incompatible with MILSATCOM.[40] The technical aspects of this problem are interrelated. Incompatible ground terminals are the result of the DODs resistance to adopting technology compatible with commercial standards. Therefore, when a user needs SATCOM bandwidth for immediate operational purposes, the user is forced to also acquire a compatible ground system at substantial cost which encourages the user to remain on commercial bandwidth, both to recoup the investment in ground hardware and ensure against future

MILSATCOM bandwidth availability limitations. This creates a self-reinforcing cycle where large bandwidth users, such as UAV systems, must plan to rely on COMSATCOM in order to ensure availability of the SATCOM necessary to function.

In recognition of this limitation, Air Force SMC (Space and Missile Systems Center) is pursuing an enterprise approach to the ground terminal problem. SMC Commander, Lt. Gen. John Thompson, has argued that "not every satellite or satellite system, whether it's communications or not, needs to have its own independent ground system."[41] With more than 17,000 ground terminals in the DOD, replacing or upgrading these systems is a monumental effort.[42] One proposed solution is a software upgrade for those military systems which already have the capability to receive and transmit in the necessary frequency bands. For those systems that do not have the required hardware, it will be necessary to replace them with more modern equipment. No accountability of the scale or the cost of this effort currently exists, though SMC has started laying the groundwork for a future upgrade. A further complication is that commercial providers are developing managed network services that use proprietary waveforms and modems that limit the ability of a user to move between providers. SMC has identified this as a significant obstacle to meeting cost savings objectives by globally trading bandwidth between providers and various DOD customers and is attempting to develop a ground terminal capable of adapting to multiple providers' needs.[43] All of these efforts take time, as a result of the likely extended timeline required to replace or reduce US military dependence on MILSATCOM, Congress authorized the purchase of two more WGS satellites at a cost of $600 million in 2018.[44]

Historically the DOD strategy remained focused on maximizing the use of MILSATCOM and efforts to move to commercial platforms have come only under threat from Congress.[45] This enforced reliance on MILSATCOM was due to DOD analysis that showed that MILSATCOM bandwidth cost $14,200 per MHz while the cost to the DOD for comparable COMSATCOM was $56,220 per MHz in 2013.[46] Numbers showing such a stark cost differential like these were contested by the CEOs of the COMSATCOM companies in their 2013 letter, where they argued that the DOD did not account for all of its costs in acquiring MILSATCOM and had an inefficient model for acquisition that distorted the facts.[47] Relying on its flawed cost analysis, the DOD strategy has remained focused on MILSATCOM utilization while minimizing the use of COMSATCOM.

DOD currently acquires most of its COMSATCOM through the custom SATCOM solution 3 (CS3) contract vehicle formerly managed by DISA and transitioned to Air Force Space Command.[48] This contract vehicle is valued at $2.5 billion over the course of its potential ten-year life.[49] It is substantially longer than its predecessor contract, CS2, which was valued at $3.4 billion and had a three-year base period with two one-year options. Unlike CS2, CS3 has a five-year base period that ends in 2022 and two optional periods which extend the contract out to 2027 with the last task order issued under CS3 expiring not later than 2032.[50] The primary purpose of this extended contract period under CS3 is to realize the cost savings associated with long-term contracts, yet it could have the unintended effect of limiting the DOD's ability to adopt new emerging commercial technologies.

The CS3 contract represents the evolution of DOD commercial SATCOM acquisition over more than 30 years, yet it is clearly not satisfactory for Congress. Continued congressional pressure on the alternative methods of COMSATCOM acquisition demonstrates that a desire exists to find cost savings and leverage rapidly emerging commercial technologies. Despite these efforts, the existing legacy ground infrastructure and caution continue to limit the use of COMSATCOM.

Possible future pathways for commercialization of SATCOM

Reducing the security dilemma in space requires reducing the number of military satellites. The current US military and government SATCOM architecture consists of 65 satellites of various types and capabilities.[51] If the DOD chooses to develop a more disaggregated space architecture relying on a constellation of dedicated military small satellites this number could dramatically increase and further fuel the security dilemma. The research above demonstrates that the US military already relies heavily on COMSATCOM even though it has a troubled history of developing efficient acquisition strategies for acquiring it. Even with its poor record of effectively leveraging COMSATCOM, there are three primary pathways to pursuing a commercialization strategy. First, the DOD could simply purchase all of its bandwidth needs from the commercial market and rely on commercial ground terminals and encryption. Alternatively, the DOD could subsidize increased resiliency of US and allied flagged commercial systems through the development of more robust commercial satellites. Another approach would be a mixed method of subsidized commercial platforms combined with the use of hosted payloads on some commercial platforms for specific strategic communications requirements.

Commercial reliance approach

Under this approach the DOD and the US government transition from a mixed architecture to a purely commercial structure using contract vehicles similar to CS3. The DOD would no longer possess any dedicated MILSATCOM and instead rely on US and allied commercial providers to meet its communications needs. Ground terminals and other infrastructure could be commercially procured or developed under contract to meet military requirements for cybersecurity and form factor. The full commercial reliance approach creates the largest reduction in on-orbit military infrastructure and so would have the most substantial impact on reducing the security dilemma.

Relying on the commercial market for SATCOM requirements has benefits beyond its impact on the security dilemma. Among the largest non-security benefits would come in the form of cost savings. Historically, DOD analysis has found the cost of MILSATCOM to be cheaper than COMSATCOM, though industry leaders claim this is most likely because DOD underestimates the cost of MILSATCOM and over-pays for COMSATCOM resulting in a skewed cost-benefit analysis.[52] Congress recognizes that DOD cost analysis is flawed and in the 2017 NDAA it directed the DOD to capture costs associated with MILSATCOM more accurately.[53] Shifting to a fully commercial architecture would remove the friction between MILSATCOM and COMSATCOM that drives inefficient and fractured purchasing processes and force the DOD to budget appropriately for future SATCOM usage. This approach would allow the DOD to achieve costs much closer to the market baseline. Moving to a commercial structure would also allow for a reduction in military space personnel and infrastructure further signaling benign intent and increasing cost savings.

The benefits of a full commercialization strategy extend beyond cost savings and include the ability to leverage the latest technical advancements in the commercial sector. While military satellites produced under contract average between 5–15 years in development, commercial platforms take approximately 3–4 years.[54] This disparity in developmental timelines allows commercial platforms to integrate the latest technology more rapidly. Technology advancement in the commercial industry also outpaces what the government can produce under contract given that technology for military satellites is often baselined at the beginning of an extended acquisition process.

There are also considerable challenges to relying entirely on the commercial sector. One of the largest is that military communications are essential and the DOD's market leverage is small. Unlike in the ISR industry where DOD dominates the market, in the COMSATCOM industry the DOD represents a relatively small fraction of the total industry revenue. In 2015, the DOD spent $581.6 million on FSS and $11.8 million on MSS.[55] The total market for FSS and MSS type services in 2015 was $17.9 billion and $3.4 billion, respectively.[56] DOD then represented only 2.7 percent of the global market for COMSATCOM. Since this expenditure is not concentrated regionally but dispersed globally, in line with DOD responsibilities, the DOD does not provide enough economic incentive for the commercial industry to flex to meet unique military needs.

Entirely relying on the commercial industry also means that, for better or for worse, the US military must also mostly rely on civilian anti-jam technology. While SMC is currently developing a Protected Tactical Waveform (PTW) that uses a specialized modem that a user can insert into a commercial ground terminal for use on military or commercial satellites, it has no impact on the ability of the satellite itself to resist jamming.[57] Current military communications satellites typically use GEO orbits to achieve maximum coverage using the minimum number of satellites. This reduces cost but requires additional investment in expensive anti-jam technology for what is undoubtedly a high-value target.[58]

Commercial high-throughput satellites (HTS) offer a potential solution to this problem. Initially designed to meet other market needs, these HTS satellites use

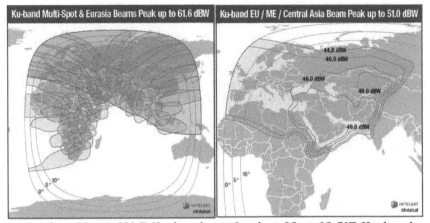

Intelsat 33e at 60° E Ku-band   Intelsat 20 at 68.5°E Ku-band

Figure 5.5 HTS footprint compared to a traditional transponder.

Source: © Intelsat License Holdings LLC, All Rights Reserved, used with permission.

dozens of small steerable spot beams to achieve maximum throughput for customers and in some cases use MEO, rather than GEO orbits. This closer orbit
allows the systems to provide lower latency to users and higher data rates to
satisfy growing customer demand.[59] HTS satellites differ from the traditional
approach of regional spot beams with near hemispheric coverage, making HTS
satellites much harder to jam because the jammer has to be within the same spot
beam as the jamming target.[60] As a result, a jammer based safely away from the
battlefield cannot interfere with signals carried on HTS satellites. The geographic impact of this change is shown in Figure 5.5 where the multiple small
spot beams of a new HTS satellite are juxtaposed with the standard large spot
beam used by non-HTS satellites. Modern commercial HTS platforms also have
data rates several orders of magnitude higher than current military satellite communications platforms with effectively similar anti-jam capability. HTS is a case
of a commercial solution naturally emerging that effectively supports both
military and civilian needs.

Another issue with a pure contract approach to acquiring SATCOM is the
uncertain reaction of the commercial market to a hostile space environment. If
satellites carrying US military signals come under attack, commercial providers
may be hesitant to lease the DOD bandwidth for fear of damage to their satellite.
Even the use of temporary non-kinetic attacks such as jamming may discourage
commercial providers from accepting business from the US military. These
attacks may disrupt other commercial signals on the satellite and lead to customers fleeing to satellites that do not carry military signals. As demonstrated
earlier, commercial providers have so far shown no hesitation to carry US
military traffic, but the US has yet to face a near-peer competitor in space.

Should this occur, commercial providers may hesitate to accept the DOD as a customer, potentially creating a catastrophic scenario for military operations.

Relying entirely on a contract approach to pursuing a commercialization strategy has several evident weaknesses. Since the DOD does not come close to dominating the market for SATCOM, it cannot dictate terms or ensure access to bandwidth in times of conflict. As a high-risk customer, it may even find itself shunned by some commercial companies during times of conflict. Efforts to legislate access to US COMSATCOM in times of crisis may drive commercial companies to rebase themselves outside of the US to avoid burdensome regulations. Driving COMSATCOM overseas would harm US space power and leave the nation weaker than before it attempted to defuse the space arms race.

Subsidized commercialization

An alternative to pure contract-driven commercialization is a subsidized approach. Rather than pursuing an expanded version of the existing CS3 contract vehicle, the US military could invest in a commercial satellite using variations of the Pathfinder approaches that SMC and DISA are experimenting with at congressional direction. The US military could purchase transponders in advance of a satellite being built, or it could promise to subsidize the cost of construction in exchange for future access. Using an approach similar to the Civil Reserve Air Fleet (CRAF) is also a possibility. No matter the exact approach, subsidizing commercial providers could resolve the customer risk issue while at the same time sending nearly the same signal of benign intent as a contract-driven method.

The CRAF model is a promising model for the basis of a space equivalent effort. CRAF is a voluntary cooperative program between the Department of Transportation, DOD, and US Airlines. The CRAF program was established following the Berlin airlift in 1951 as a way to provide "supplemental airlift to support a major national defense emergency."[61] The idea behind it is a public-private partnership where the government guarantees a minimum level of annual business to civilian airlines in exchange for the ability to mobilize them to support the military in times of crisis. Mobilization proceeds in stages based on aircraft need and has only been activated twice in its history, to support the 1991 Gulf War and secondly to support Operation Iraqi Freedom from February 2003 to June 2003.[62] Essentially "the CRAF program meets the military's mobilization requirements while saving taxpayers billions of dollars by foregoing the cost of procuring a government fleet to meet those requirements."[63] A 1994 RAND study estimated that the CRAF program had saved the US government $128 billion from 1951 to 1994.[64] Prior to 9/11, the US military did $600 million a year in business with CRAF members, increasing after 9/11 to over $3 billion a year.[65] This program has been a valuable partnership for US-flagged carriers, supporting the industry while meeting US military wartime surge requirements.

Adapting a CRAF model to space would solve a number of issues with a commercialization strategy. First, it would give US commercial providers an incentive to consider military security and compatibility requirements when developing satellites. Second, it would ensure the availability of US-flagged satellites that could be quickly leveraged during periods of national crisis. Third, a CRAF model would also help overcome any potential reluctance to carry US military signals due to threat and give the US government leverage over the industry. Fourth, it would provide a stable and predictable subsidy for US commercial providers that would encourage the further development of the US space industrial base. Finally, it would make little difference to the security dilemma as a reassurance strategy compared to the pure contract approach as improving the security of commercial platforms could only be seen as defensive and prudent.

Applying the CRAF model to space is not a revolutionary idea, and the senior leadership in OSD recognizes its potential benefits.[66] DOD has explored adopting the CRAF model but historically has not had the level of broad, high-level government support necessary to make it happen.[67] With the recently renewed focus on space, it may be possible to build the broad coalition in Congress required for the concept of a Civil Reserve Space Fleet (CRSF) to have a chance at success. It is certainly less controversial than the concept of creating a separate Space service and could have real economic benefits to the country. Creating a bipartisan coalition to support the concept of a CRSF is a real possibility even in a polarized political environment.

A subsidized CRSF style approach to commercialization would still not resolve all of the problems associated with the pure commercial approach. It would likely cost more than adopting a contract-driven approach even if it did have a positive impact on the US space industrial base. A CRSF style approach would also likely not be enough to encourage industry to meet all unique military communications needs. The US military still would have to accept the limitations of commercial coverage and capabilities under certain circumstances. For example, commercial providers have little incentive to provide coverage in extreme northerly or southerly latitudes that have little commercial value. These areas are best served by satellites operating in highly elliptical orbits (HEO) that are extremely uncommon in civilian applications. The primary commercial alternative to dedicated FSS platforms is MSS provided by small satellite constellations operating in LEO which do not yet exist outside of Iridium but may provide a solution to this unique problem in the near future.

Market subsidies are a potential solution to some of the challenges associated with a commercial strategy for signaling benign intent that resolves many, but not all, of the issues associated with such a strategy. The military would remain limited to applications and technologies that are primarily designed for civilian use and could be adapted to military use. If the military requires a specific orbit or application that has no commercial value, then it would either need to find alternatives to space-based capabilities or pursue some other approach to meeting its communications requirements. One alternative approach to building dedicated military satellites that could help resolve some of the application issues is using hosted payloads.

Hosted payloads

Hosted payloads are government sensors or communications packages that are integrated into a commercial satellite. These payloads share the basic support functions of the parent satellite while providing a unique capability that the commercial provider would otherwise not include on the satellite. This approach varies from the previous two because instead of accepting the difficulty of finding providers to meet unique military needs, it relies on developing dedic-ated payloads. A hosted payload approach could rely exclusively on government developed payloads, or it could be used in a mixed manner with either of the two previous approaches by using hosted payloads to accommodate unique military capabilities. Compared to the previous two approaches, relying on hosted payloads represents a "soft" form of commercialization which would send the weakest signal of benign intent, though it still preserves the overall message while providing increased military control.

Before 2018, the US military successfully demonstrated the viability of hosted payloads on three separate occasions.[68] The driving impetus behind the existing hosted payload efforts is that their purported benefits are similar to the advantages that the US would gain from a commercialization strategy: cost savings, faster on-orbit capability, increased deterrence, and resilience, and benefits to the industrial base of increased government contracts.[69] In comparison to either of the two commercialization strategies discussed earlier, hosted payloads provide smaller gains in each category. The benefit they do have for the strategy of commercialization is that they do provide a solution to the inability of commercial providers to meet unique military needs.

If the US military chose to pursue a hosted payload approach only for unique capabilities, such as strategic communications, in combination with either of the two previously discussed approaches it could preserve some of the advantages of signaling through commercialization. The damage this approach would have to signaling benign intent would depend on the extent of hosted payload distribution. Should the US pursue a policy of only placing hosted payloads on a small number of satellites expressly for providing unique strategic communications capabilities then it could have minimal impact on signaling. If instead the US vigorously pursued hosted payloads as a path to disaggregation while preserving its existing MILSATCOM architecture, it would only further fuel the security dilemma. Over-diluting the signaling of benign intent associated with a commercialization strategy by including hosted payloads is a real possibility.

There are significant challenges to using hosted payloads that makes them difficult to develop as well. Currently, the DOD considers the challenges of matching up a government payload with a commercial satellite "too difficult for programs to overcome."[70] The reasons for this include the difficulty of matching up size, weight, and power concerns, matching government development timelines to commercial ones, and ongoing concerns over payload control.[71] Difficulties overcoming these challenges are the primary reason why the US military has developed so few hosted payloads despite the apparent advantages over the current architecture. These challenges would likely limit any broad deployment of hosted payloads as part of a disaggregation approach designed to increase deterrence and resiliency, reinforcing the idea that this approach is best used for unique capabilities only to augment either of the two previously proposed commercially-centric approaches.

Exploring the possibilities of commercialization: the case of Iridium communications

Iridium is an example of a functioning relationship between a US-based commercial partner and the US government that preserved a capability for commercial users while at the same time benefiting the military. With 66 cross-linked satellites in low-Earth orbit providing global coverage using lightweight mobile handsets, Iridium is an example of a successful public–private partnership model.[72] The company has a uniquely close association with the DOD, and its ongoing dependence on government business makes it a suitable lens through which to predict the future relationship between the DOD and other commercial providers when pursuing a commercialization strategy.

Motorola Corporation initially conceived Iridium in the mid-1980s when various mobile phone technologies were competing to dominate the emerging mobile phone market. The Iridium constellation was an attempt by Motorola to solve the early coverage issues associated with mobile phones by deploying a global satellite constellation. The final cost of the launch and development of the system was $5 billion, and it was not on orbit and available as a service until 1998.[73] The company held an opening ceremony in November of that year where Vice President Al Gore made the first phone call to inaugurate the launch of the service.[74] Despite being a marvel of technology, Iridium had only 20,000 customers by August the following year and was in severe financial trouble.[75] Without an adequate customer base, Iridium defaulted on $1.5 billion in debt in August 1999 and was forced to file for Chapter 11 bankruptcy.[76] Despite having global coverage, Iridium had failed to compete with terrestrial alternatives that focused on providing coverage in cities and other population dense areas using smaller devices that cost substantially less to purchase and use.

There were many reasons for the failure of Iridium. The system was under development for 11 years, and during that time the spread of cellular technology removed the original need for it. The significant leap in technology required to develop the Iridium constellation resulted in as many as 1,000 patent filings.[77] The technological innovation required to launch Iridium and the long lead time that satellite and system hardware development required put Iridium at a significant disadvantage compared to terrestrial competitors. By the time Iridium launched, rapidly advancing cell phone technology had rendered the Iridium handset design obsolete. The system relied on large bulky phones that cost $3,000 each and charged between $6 and $30 a minute.[78] Cellular coverage area also increased substantially during the time that it took Iridium to develop its satellites. By the time Motorola launched in 1998, terrestrial carriers already served most of its original target audience of business users in the major cities of the world. Iridium's failure was primarily due to the cost and time required to develop and put its system in place. When Motorola initially conceived the business model it was valid, but the extended development timeline allowed competing terrestrial technology to develop.

Dan Colussy, a veteran aviation industry executive, orchestrated the buyout of Iridium in 2000.[79] Putting together a consortium of four buyers, including the DOD, Colussy was able to buy out Iridium for just $25 million. This buyout offer was contingent on the US government indemnifying Motorola against any damage from future debris that might result when the constellation de-orbited. In order to achieve this indemnification, Colussy leveraged his relationship with the DOD. Iridium entered into an indemnification contract with the US government in December 2000 whereby Motorola was not liable for future debris damage, and the new buyers agreed to maintain adequate insurance while subject to being ordered at any time to de-orbit the entire constellation at the US government's discretion.[80] The de-orbiting agreement satisfied all parties, and Iridium communications was purchased for a small fraction of its development cost in December 2000.[81]

The DOD had maintained an interest in Iridium since its inception. Of the 18 original ground stations developed for the system, one was entirely owned by the DOD. This gateway, located in Hawaii and dedicated to DOD use, was one of only two that remained open after the buyout of Iridium.[82] Colussy also negotiated a $36 million a year contract with the DOD, locking in a lucrative customer with considerable leverage before the buyout was even complete.[83] Today, this gateway remains open and dedicated to US government use. The DOD has invested substantial amounts in upgrading and maintaining this gateway since it was originally built in the late 1990s. Iridium highlights in its annual investor reports that this gateway and the significant investments that DOD has made in voice and data systems are only compatible with Iridium satellites.[84] This investment in dedicated hardware is common with MSS which means that any future effort to incorporate a distributed commercialized architecture based on small satellites will require an investment by the DOD in cross-system compatible ground hardware.

The US government was Iridium's largest customer after its buyout and remains so today. In 2017 the US government accounted for 24 percent of all Iridium revenue at $106.1 million.[85] The DOD also has a fixed price contract with Iridium for its services valued at $400 million with a five-year term that it has extended through 2019.[86] This most recent contract continues the US government relationship with Iridium that has existed since its launch.

The long association with Iridium by the DOD has resulted in the development of DOD hardware dependent on the Iridium constellation. The Distributed Tactical Communications System (DTCS) is an example of this. DTCS is a system managed within DISA that provides over the horizon tactical and voice communications.[87] DISA advertises the system as a solution to line-of-sight communications issues that only requires customers using the system to have appropriate hardware to take advantage of it.[88] This system represents a substantial investment on the part of DOD and creates ecosystem lock-in that encourages resistance to the adoption of other commercial platforms that may be launched in the future.

Despite the continued survival of Iridium, it has remained dependent on the US government. Since the buyout of Iridium from bankruptcy, the company has remained financially viable only with DOD support. This support, in the form of long-term fixed price contracts, has allowed Iridium substantial time to develop a customer base outside of the DOD. Even with its broadened customer base, in the seven-year period from 2010 to 2017 the DOD actually increased from 23.6 percent of Iridium's revenue base to 24 percent.[89] This minuscule change obscures the fact that total revenue increased during the same period from $76 million to $448 million.[90,91] The growth in total revenue demonstrates that there is strong demand for a global satellite communications network that functions independently of fixed ground stations as long as the DOD exists as an anchor customer.

The DOD's early investment in Iridium ensured the survival of the company and its technology; it could also serve as an example of how a commercialization strategy might assist in the development of a robust national space industrial base. Iridium continues to thrive as a company today and with its updated Iridium Next constellation currently being deployed, it is prepared to continue to provide a valuable service to the DOD at a relative bargain for the foreseeable future. Iridium provides the DOD with a unique capability that would cost billions for the US military to launch and maintain at a fraction of the price. The relationship between Iridium and the DOD shows how promising space technology can benefit from having the government as an anchor customer and is a strong argument for the benefits that commercialization can provide to both the military and commercial industry.

Summary

The US government built the modern commercial satellite communications market when it established COMSAT Corp during the Kennedy Administration. From there the commercial market advanced rapidly, at first adapting government developed technology for use in the commercial world and then developing its own innovations. Today the pace of commercial communications technology innovation far exceeds the government acquisition cycle, and the US military is becoming ever more dependent on it for essential operations. Acceptance of the long-term dependence on COMSATCOM has led the US government to attempt multiple methods of acquiring it that have met with mixed success. With emerging small satellite constellations poised to disrupt the SATCOM market and provide a generational leap forward in capability, the commercial market will soon possess the capability to meet almost all US military SATCOM needs.

There are three primary strategies that the US, or any of the major space powers, could adopt in commercializing their on-orbit SATCOM architecture.

From directly purchasing its SATCOM on the market, to subsidizing national systems using a CRSF approach or relying on hosted payloads, there is a range of options. All of them are viable strategies that likely would positively impact the security dilemma by reducing military space presence while preserving capability.

Iridium serves as a good example of how government support can help make an innovative but risky capability viable in the commercial market. Iridium provides a global MSS communications architecture that the US military has come to rely on that it would otherwise not be capable of developing on its own. The blending of commercial and military interests represented by Iridium provides a template for how future partnerships between the military and commercial companies could work.

The current reliance on COMSATCOM demonstrates that it is possible to transition all or most of the US military's use of SATCOM to commercial systems. Doing so would significantly reduce the total number of US government satellites on orbit and could even have a positive impact on US capability as well as its latent space power. Transitioning would take time, as the ground system architecture distributed across the various military services represents a significant drag on technology adoption. Once transitioned, though, the US would find itself able to leverage the latest technology using a distributed architecture that a potential adversary would have difficulty effectively disrupting.

Notes

1 "Intelsat Global Network," Official Company Page, Intelsat, accessed November 5, 2018, www.intelsat.com/global-network/.
2 Sandra Erwin, "Boeing to Accelerate Production of WGS Satellites," *Space News*, April 18, 2018, https://spacenews.com/boeing-to-accelerate-production-of-wgs-satellites/.
3 Sandra Erwin, "Space Development Agency Releases Its First Solicitation," *Space-News.Com*, July 4, 2019, https://spacenews.com/space-development-agency-releases-its-first-solicitation/.
4 "Union of Concerned Scientists: Satellite Database" (Union of Concerned Scientists, May 1, 2018), www.ucsusa.org/nuclear-weapons/space-weapons/satellite-database#.W7eDtPZFxEZ.
5 Douglas Loverro, Deputy Assistant Secretary of Defense for Space Policy-interview by author, Phone, January 13, 2017.
6 Delbert D. Smith, "The Legal Ordering of Satellite Telecommunication: Problems and Alternatives," *Indiana Law Journal* 44, no. 3 (1969): 339–340, www.repository.law.indiana.edu/ilj/vol.44/iss3/1.
7 "Communications Satellite Act of 1962," Pub. L. No. 87–624, § 102a, HR11040 (1962).
8 Ibid., sec. 305a.
9 Ibid., sec. 201b.
10 Ibid., secs. 201a, 4.
11 David J. Whalen, "Communications Satellites: Making the Global Village Possible," NASA History Division, accessed November 13, 2018, https://history.nasa.gov/satcomhistory.html.
12 David N. Spires and Rick W. Sturdevant, "From Advent to Milstar: The U.S. Air Force and the Challenges of Military Satellite Communications," NASA History, accessed November 13, 2018, https://history.nasa.gov/SP-4217/ch7.htm.

13 Ibid.

14 "AFSPACECOM Desert Shield/Desert Storm Lessons Learned" (Air Force Space Command, July 12, 1991), 3.

15 Ibid., 1,8.

16 E. Bedrosian et al., "Tactical Satellite Orbital Simulation and Requirements Study," RAND Study (Santa Monica, CA: RAND Corp., 1993), 9.

17 Ibid., 9–10.

18 Benjamin D. Forest, "An Analysis of Military Use of Commercial Satellite Communications" (Naval Post-Graduate School, 2008), 10.

19 Greg Berlocher, "Military Continues to Influence Commercial Operators," *Satellite Today*, September 1, 2008, www.satellitetoday.com/publications/via-satellite-magazine/supplement/2008/09/01/military-continues-to-influence-commercial-operators/.

20 "Space Systems Bandwidth," Global Security, accessed December 30, 2016, www.globalsecurity.org/space/systems/bandwidth.htm.

21 "Satellite Communications: Strategic Approach Needed for DOD's Procurement of Commercial Satellite Bandwidth" (Washington, DC: United States General Accounting Office, December 10, 2003), 3.

22 Ibid., 2.

23 Ibid.

24 "Defense Satellite Communications: DOD Needs Additional Information to Improve Procurements" (Washington, DC: United States General Accounting Office, July 2015), 9.

25 "Strategic Approach Needed," 13.

26 Ibid., 15.

27 Ibid.

28 Ibid.

29 "Federal Acquisition Regulation Part 49-Termination of Contracts," § 49.202 a (n.d.).

30 "Strategic Approach Needed," 16.

31 Ibid., 19.

32 "Defense Satellite Communications: DOD Needs Additional Information to Improve Procurements," 19.

33 Ibid., 8.

34 Ron Samuel et al., "Seven Ways to Make the DOD a Better Buyer of Commercial SATCOM," January 14, 2013, https://ses-gs.com/press-release/seven-ways-to-make-the-dod-a-better-buyer-of-commercial-satcom/.

35 "Report to the Secretary of Defense: Taking Advantage of Opportunities for Commercial Satellite Communications Services" (Defense Business Board, January 2013), 6.

36 "National Defense Authorization Act for Fiscal Year 2014" (Washington, DC: US Senate, 2013).

37 "Satellite Communications Strategy Report: In Response to Senate Report 113–44 to Accompany S.1197 NDAA for FY14" (Washington, DC: Department of Defense Office of the Chief Information Officer, August 4, 2014), 3.

38 Ibid., 4.

39 Ibid., 7.

40 "Satellite Communications Strategy Report: In Response to Senate Report 113–44 to Accompany S.1197 NDAA for FY14" (Washington, DC: Department of Defense Office of the Chief Information Officer, August 4, 2014), 8–9.

41 Kimberly Underwood, "Air Force Pursues SMC 2.0 Effort," *Signal Magazine*, October 30, 2018, www.afcea.org/content/air-force-pursues-smc-20-effort.

42 Sandra Erwin, "Defense Budget Bill Creates Path for Future Network of Military, Commercial Communications Satellites," *Space News*, June 16, 2018, https://spacenews.com/defense-budget-bill-creates-path-for-future-network-of-military-commercial-communications-satellites/.

43 "Pathfinder 3 Request for Information: Solicitation Number 16–076" (Air Force Space Command, May 20, 2016).

44 Erwin, "Defense Budget Bill Creates Path for Future Network of Military, Commercial Communications Satellites."

45 "Satellite Communications Strategy Report: In Response to Senate Report 113–44 to Accompany S.1197 NDAA for FY14," 13.

46 Ibid., 8.

47 Ron Samuel et al., "Seven Ways to Make the DOD a Better Buyer of Commercial SATCOM," January 14, 2013, https://ses-gs.com/press-release/seven-ways-to-make-the-dod-a-better-buyer-of-commercial-satcom/.

48 Erwin, "Defense Budget Bill Creates Path for Future Network of Military, Commercial Communications Satellites."

49 Laura Moreno-Davis, "UltiSat Named Awardee on $2.5 Billion Complex Commercial SATCOM Solutions (CS3) IDIQ Contract," *Globe Newswire*, August 28, 2017, https://globenewswire.com/news-release/2017/08/28/1101038/0/en/UltiSat-Named-Awardee-on-2-5-Billion-Complex-Commercial-SATCOM-Solutions-CS3-IDIQ-Contract.html.

50 "Complex Commercial SATCOM Solutions (CS3): CS3 Customer Ordering Guide" (US General Services Administration, October 2017), 5.

51 "Union of Concerned Scientists: Satellite Database" (Union of Concerned Scientists, May 1, 2018), www.ucsusa.org/nuclear-weapons/space-weapons/satellite-database#.W7eDtPZFxEZ.

52 Samuel et al., "Seven Ways to Make the DOD a Better Buyer of Commercial SATCOM."

53 "National Defense Authorization Act for Fiscal Year 2017," Pub. L. No. 114–328 (2016), sec. 1605.

54 "Taking Advantage of Opportunities for Commercial Satellite Communications Services: Report FY 13-02" (Defense Business Board, January 2013), 6.

55 Kerry Kelley, "Fiscal Year 2015 Commercial Satellite Communications Usage Report: In Response to Chairman of the Joint Chiefs of Staff Instruction 6250.01E." (US Strategic Command, September 7, 2017), ES-1.56.

56 "State of the Satellite Industry Report, Satellite Industry Association" (Tauri Group, September 2016), 11, https://a3space.org/wp-content/uploads/2017/09/tauri-satellite.pdf.

57 Mike Gruss, "US Air Force Award Contract for New Waveform Demonstrations," *SpaceNews*, August 25, 2016, https://spacenews.com/u-s-air-force-awards-contracts-for-new-waveform-demonstrations/.

58 Former Deputy Assistant Secretary of Defense Douglas Loverro described the trade-off in the development of military satellites as one where "launching military satellites into GEO, the coverage the military wanted came at a lower price tag, but with an increased risk of jamming." Ryan Schradin, "Overcoming the Largest Threats to Military Satellites and Increasing Resiliency," *The Government Satellite Report*, April 4, 2018, https://ses-gs.com/govsat/defense-intelligence/overcoming-largest-threats-military-satellites-increasing-resiliency/.

59 Sanjeev Bhatia, "Understanding High Throughput Satellite (HTS) Technology," June 2013, www.intelsat.com/wp-content/uploads/2013/06/HTStechnology_bhartia.pdf.

60 Schradin, "Overcoming the Largest Threats to Military Satellites and Increasing Resiliency."

61 "Civil Reserve Airfleet Allocations," US Department of Transportation, US Department of Transportation, accessed November 13, 2018, www.transportation.gov/mission/administrations/intelligence-security-emergency-response/civil-reserve-airfleet-allocations.

62 Christopher Bolkcom, "Civil Reserve Air Fleet" (Congressional Research Service, October 18, 2006), 3.

63 Jerry F. Costello, "Hearing on the Economic Viability of the Civil Reserve Air Fleet Program," Pub. L. No. 111–130, § House Subcommittee on Aviation (2009).
64 Costello. Values given in 2009 dollars.
65 Duncan J. McNabb, "Hearing on the Economic Viability of the Civil Reserve Air Fleet Program," Pub. L. No. 111–130, § House Subcommittee on Aviation (2009).
66 Joseph Koller to author, "Office of the Under-Secretary of Defense for Space Policy—Civil Reserve Air Fleet," January 23, 2017.
67 Ibid.
68 "Military Space Systems: DOD's Use of Commercial Satellites to Host Defense Payloads Would Benefit from Centralizing Data" (United States Government Accountability Office, July 2018), 10.
69 Ibid., 4–5.
70 Ibid., 15.
71 Ibid.
72 "United States Securities and Exchange Commission Form 10-K: Iridium Communications Inc.," SEC 10-k (Iridium Communications Inc., February 22, 2018), 1.
73 Craig Mellow, "The Rise and Fall of Iridium," *Air and Space Magazine*, September 2004, www.airspacemag.com/space/the-rise-and-fall-and-rise-of-iridium-5615034/.
74 S. Finkelstein and S.H. Sanford, "Learning From Corporate Mistakes: The Rise and Fall of Iridium," *Organizational Dynamics* 29, no. 2 (2000): 138.
75 Ibid.
76 Reuters, "Iridium Declares Bankruptcy," *New York Times*, August 14, 1999.
77 Finkelstein and Sanford, "Learning from Corporate Mistakes: The Rise and Fall of Iridium," 139.
78 Ibid., 138.
79 Craig Mellow, "The Rise and Fall of Iridium."
80 "United States Securities and Exchange Commission Form 10-K: Iridium Communications Inc.," SEC 10-k (Iridium Communications Inc., March 16, 2010), 10.
81 Craig Mellow, "The Rise and Fall of Iridium."
82 Craig Mellow, "The Rise and Fall of Iridium."
83 Craig Mellow, "The Rise and Fall of Iridium."
84 "United States Securities and Exchange Commission Form 10-K: Iridium Communications Inc.," SEC 10-k (Iridium Communications Inc., December 2015), 2.
85 "United States Securities and Exchange Commission Form 10-K: Iridium Communications Inc.," February 22, 2018, 2.
86 Annamarie Nyirady, "Iridium Awarded $44 Million DISA Contract Extension," *Via Satellite*, October 15, 2018, www.satellitetoday.com/government-military/2018/10/15/iridium-awarded-44-million-disa-contract-extension/.
87 "The Distributed Tactical Communications System: Fact Sheet" (Defense Information Systems Agency, n.d.), www.disa.mil/~/media/Files/DISA/Services/DTCS/DTCS-Overview.pdf.
88 Ibid.
89 "United States Securities and Exchange Commission Form 10-K: Iridium Communications Inc.," February 22, 2018, 2.
90 "United States Securities and Exchange Commission Form 10-K: Iridium Communications Inc.," March 16, 2010, 1.
91 "United States Securities and Exchange Commission Form 10-K: Iridium Communications Inc.," February 22, 2018, 1.

6   Remote sensing

The Chernobyl disaster on April 26, 1986, was the catalyst that first brought home the potential impact that commercial satellite imagery could have. Images taken of the Chernobyl reactor by the satellite Landsat-5 just three days after the incident showed the extent of the damage, allowing the news media to pierce the Iron Curtain and obtain information in a way never before possible.[1] The single image taken by Landsat-5 revealed more about the extent of the accident than anyone had been able to piece together up to that point and was the single greatest source of evidence available to the public.[2] Information previously limited to superpowers was now available for publication and analysis by anyone with enough money. It was possible for almost anyone to purchase imagery from the US Landsat program or the newly launched French SPOT satellite. The image resolution was low by modern commercial standards, 10-meters for SPOT and 30-meters for Landsat, but it was good enough to make out details of major disasters and to see places and things to which the public previously had no access.[3]

Today commercial satellite imagery of significantly higher quality is easily available through a variety of tools and venues. It has revolutionized everything from how we see the world to how we navigate. Satellite-based observation, which began as a reconnaissance tool for the US and Soviet governments, has progressed to a commercial tool that is increasingly necessary to the economy, allowing the monitoring of commercial activity ranging from port activity to agriculture, and offering high-quality imagery and video of the entire world.

The world is in the middle of revolutionary changes to the remote sensing market that make commercializing all or most of the national remote sensing architecture a real possibility. As recently as five years ago, the idea that everywhere on Earth could be imaged in high resolution every day was science fiction. Now it is reality. Soon it may even be possible to obtain continuous real-time video of almost anywhere on the planet as well. Until recently, commercial providers largely copied the government approach of building a small handful of expensive and exquisite imagery satellites placed in relatively high orbits and designed to last a decade or more. Companies like Planet completely overturned this model by using large numbers of cheap small satellites operating close enough to the Earth that their small cameras could still obtain high-quality imagery at the cost of short on-orbit lifetimes. Advances in small satellites and the fall in launch costs have made their business model viable and present an opportunity to shift away from government-provided imagery services.

Responsibility for developing and operating US imaging satellites falls to the National Reconnaissance Office (NRO) supported by the National Geospatial-Intelligence Agency (NGA). As of 2018, the NRO controls at least 36 satellites, representing the bulk of the more than 60 remote sensing satellites that the US government has on orbit performing a variety of missions from environmental monitoring to intelligence collection.[4] Plans by the US military and the NRO to develop their own small satellite constellations in the next decade would dramatically increase this number, potentially adding hundreds of new satellites in a variety of orbits. These new satellites would only further fuel uncertainty and fear in space, especially given the highly secretive nature of the NRO. Placing large numbers of new government platforms on orbit would also provide additional justification for developing dual-use defensive capabilities to protect them. Avoiding this outcome may be possible if the US chooses instead to pursue a policy of commercializing as much of its existing imagery architecture as possible and relying on commercial small satellite constellations rather than developing its own.

Remote sensing and the security dilemma

The fundamentally defensive character of the security dilemma has two primary drivers in space, fear of loss and uncertainty of purpose. Fear of loss drives the dilemma from the perspective of the owning nation and uncertainty of purpose drives it from the perspective of potential adversaries. Uncertainty of purpose begins with a lack of trust between nations and increases the more numerous that nations satellites are and the more secrecy that surrounds them. Remote sensing platforms are currently the most numerous category of US government satellites. They also operate under the highest level of security with the majority belonging to the highly secretive NRO. This combination of relatively large numbers and secrecy means that, unlike other categories of satellites, those labeled as remote sensing contribute to the security dilemma as much through uncertainty of purpose as from fear of loss.

Even though secrecy drives uncertainty of purpose for remote sensing systems operated by US intelligence agencies and the military, the fear of loss is still significant. As a group, the US government has invested more in remote sensing than in any other category of satellites.[5] Unlike missile warning or even satellite communications, the benefits of taking images or collecting other data from orbit without interference are immediately apparent. If this capability is removed, then it cannot easily be replaced. Aerial reconnaissance is a suitable replacement for satellite-based remote sensing in permissive environments, but as shown by the downing of a US U-2 spy plane piloted by Gary Powers over Russia in 1960, nations do not generally allow foreign powers to collect intelligence in this manner even during peacetime.[6] Without satellite reconnaissance, activities within adversary nations become inscrutable, contributing to speculation, insecurity, and instability.

The peacetime benefits to strategic stability of allowing an adversary to verify military activity disappear during periods of crisis. Allowing an opponent to observe military activities provides them with critical intelligence once tensions between nations develop into a crisis verging on conflict. When a nation determines that military action is likely, the strategic stability incentives for allowing an opponent to verify its military activities disappears along with its peaceful intent. For this reason, remote sensing and intelligence satellites have the same low threshold for non-destructive attack as SATCOM. However, unlike SATCOM, where an opponent might desire to avoid destroying an adversary's systems to allow them to communicate surrender or withdraw orders, remote sensing systems have little lingering value to an adversary. Also, it may be more difficult to interfere with the ability of these satellites to collect data, forcing an opponent to consider destructive attacks earlier than they might for SATCOM.[7] This proportionally low threshold for destructive attack drives the fear of loss that helps this category of satellites fuel the security dilemma.

Uncertainty driven by secrecy is already in the process of being compounded by increasing numbers of US government remote sensing and intelligence satellites on orbit. The Defense Advanced Research Projects Agency (DARPA) is exploring possibilities for greatly expanding that number through the Blackjack program. DARPA's Blackjack program envisions a 20-satellite test constellation that can fly multiple types of national security payloads and demonstrate the same performance as GEO systems at the cost of less than $6 million per satellite.[8] While it is cost-effective to develop a flexible small satellite bus capable of operating in constellations and supporting multiple national security payloads, efforts such as Blackjack create uncertainty and fuel the security dilemma.

Remote sensing and intelligence satellites are a large and growing contributor to the security dilemma in space. This is the result of secrecy driven uncertainty as well as increasing numbers of these types of platforms. The low threshold of attack for this critical capability also drives fear of loss, especially for the US which is highly dependent on technical intelligence from space. Finding a way to preserve some measure of the capability provided by these satellites through commercial means is strategically valuable; it is also largely dependent on the health of the US commercial remote sensing market. A market that has proven highly vulnerable to US remote sensing policy and acquisition needs.

Remote sensing policy

In 1984, Congress passed the Land Remote Sensing Commercialization Act, designed primarily intended to privatize the Landsat program.[9] The act specifically targeted developing a private industry to exploit the data available from Landsat. The act expressed doubt that the private sector had the ability to develop a remote sensing system "because of the high risk and large capital expenditure involved."[10] Despite this, the legislation also included a provision to allow the Secretary of Commerce to issue licenses for commercial remote sensing satellites. The Department of Commerce then delegated licensing authority to the National Oceanic and Atmospheric Administration (NOAA).[11]

The conditions for operation placed upon a company obtaining a license under the 1984 act are similar to those that exist today, with a few notable exceptions. Among those requirements that remain in effect today, a licensee is required to:

a   operate the system in such manner as to preserve and promote the national security of the United States and to observe and implement the international obligations of the United States;

b   upon termination of operations under the license, make disposition of any satellites in space in a manner satisfactory to the president;

c   furnish the Secretary with complete orbit and data collection characteristics of the system, obtain advance approval of any intended deviation from such characteristics, and inform the Secretary immediately of any unintended deviation;

d   notify the Secretary of any agreement the licensee intends to enter with a foreign nation, entity, or consortium involving foreign nations or entities;

e   furnish the Secretary with complete orbit and data collection characteristics of the system, obtain advance approval of any intended deviation from such characteristics, and inform the Secretary immediately of any unintended deviation.[12]

None of the requirements listed above are particularly onerous, and all exist in current law unchanged from 1984. However, several additional requirements were included in the 1984 act which made developing a commercial remote sensing platform an extremely risky investment. These included a condition allowing the Secretary of Commerce to inspect the licensee's "equipment, facilities and financial records" with no limitations.[13] The act also allowed the Secretary to "terminate, modify, condition, transfer, or suspend licenses" without any judicial recourse provided for the licensee.[14] A final aspect of the law that made private investment risky was that a system operator had the right to sell exclusive data only "for a period to be determined by the Secretary."[15] These requirements placed absolute power over commercial remote sensing in the hands of the Commerce Secretary with virtually no limitations, except for those that were self-imposed.

The news media led the first effort to take advantage of this new licensing arrangement. The Radio-Television News Directors Association (RTNDA) saw the obvious advantages in having access to high-quality independent satellite imagery that far exceeded what was available for purchase from the privatization of Landsat, which at the time provided only a 30-meter resolution.[16] The RTNDA concept was dubbed "Mediasat" and proposed a three-meter resolution that would have been far beyond anything then available for purchase. Combining resources, the major news outlets possessed the ability to fund the development of Mediasat, but had to engage in a protracted legal battle with the US government to clarify regulations. The news association believed that the vagueness of regulations "chill(ed.) commercial interest in remote sensing."[17]

The RTNDA accusation that vague and arbitrary regulations were preventing investment was valid. No US commercial remote sensing platform had been launched or was in design, but the French had their SPOT-1 satellite on orbit beginning in 1986, and it offered better resolution than the US Landsat program. Regulations governing the licensing process were finally released by NOAA in 1987 in the form of 15 Consolidated Federal Regulations (CFR) Part 960.[18]

The long delay in formulating the appropriate regulations by NOAA did not result in the necessary clarification called for by the commercial market. Criticism of the regulations included their failure to define which national security obligations warranted the denial of a license, and there were serious concerns that the entire set of regulations violated the First Amendment.[19] Legal literature at the time also noted the irony of denying a license for a commercial platform that could provide additional sources of data to national intelligence in order to protect national security.[20]

National Security Presidential Directive 3 (NSPD-3), signed on February 11, 1991, attempted to fix the problems with the original 1984 law.[21] NSPD-3 included several significant changes. It listed remote sensing as one of five major commercial space sectors; US government agencies were directed to use commercial space products and services to the "fullest extent feasible"; and it directed that the US government avoid regulation that deterred investment in commercial space.[22] The law also encouraged the establishment of cooperative agreements with commercial space firms. An example included in the law was the use of "anchor tenancy." The idea behind anchor tenancy is that the government could purchase enough of the service or product to provide the industry venture with a viable initial customer base. NSPD-3 led directly to the passage of an updated law on land remote sensing, the Land Remote Sensing Policy Act of 1992.

This legislation repealed the 1984 law and deleted the most egregious portions of it. The Act placed limitations on the Secretary of Commerce's power to revoke a license. It required the Secretary of Commerce to obtain an injunction from Federal Court to exercise the previously unlimited powers of license termination and modification.[23] The warrantless-inspection authority of the Commerce Secretary also was now subject to the normal judicial process of obtaining subpoenas and warrants in order to conduct inspections. The law also removed the ability of the government to decide the period of exclusivity for images taken with remote sensing platforms. With the exception of a renumbering of the section as part of its inclusion in the 2010 National and Commercial Space Programs legislation, the 1992 Act remains the foundation of commercial remote sensing law, though various legislative efforts are underway to update it.[24]

The 1992 Act finally created enough certainty in government regulation to encourage the development of a true commercial remote sensing industry. One of the largest US commercial-imagery providers today, DigitalGlobe/Maxar and its founder, Walter Scott, identify the 1992 law as the beginning of the industry.[25] Scott founded his first company, named WorldView, in January of that year.[26] In 1993, WorldView received the first license under the 1992 Act to operate a high-resolution commercial satellite.[27]

The motivation for the new act was not due to its predecessor's failure to promote the development of commercial remote sensing platforms, but rather the failure of the attempted privatization of the Landsat program. Under the 1984 Act, the Landsat program was managed by the Earth Observation Satellite Corporation (EOSAT) with a ten-year contract.[28] Under EOSAT management the price of data from Landsat increased sharply, which led to a corresponding decrease in demand.[29] The launch in 1986 of the government subsidized French SPOT-1 satellite, which had superior resolution, exacerbated this problem.[30] The combination of foreign competition and unsuccessful commercialization resulted in Congress returning the Landsat program to NASA and the Department of Defense (DOD) management.[31] The Department of Commerce and NOAA did retain the licensing authority granted to them in the 1984 law with the updates made to the licensing agreement mentioned above.

The next significant modification to remote sensing policy and law was Presidential Decision Directive 23 (PDD-23) signed by President Bill Clinton in 1994. This PDD had the stated intent of allowing "US firms to compete aggressively in a growing international market."[32] The PDD recognized the increasingly competitive commercial market, which now included Russia, China, and Japan as well as France, with several other nations developing their own capabilities.[33] To accomplish its stated objective, the PDD relaxed restrictions on the export of remote sensing technology; and, following its release, the Department of Commerce approved the sale of 1-meter-resolution imagery. This was significant, as imagery resolution had been progressing rapidly, and the industry needed this authorization to be competitive in the global marketplace. The loosening of restrictions also represented a significant shift in risk calculus for the US national security community. In the late 1980s the RTNDA had lobbied for 3-meter resolution as part of Mediasat, but security concerns had prevented efforts to gain approval for such a "high resolution."[34]

In combination with relaxing restrictions on imagery sales, PDD-23 introduced the concept of "shutter control." Under PDD-23, commercial-imagery providers might be required "during periods when national security ... may be compromised, as defined by the Secretary of Defense or the Secretary of State, respectively, to limit data collection and/or distribution by the system to the extent necessitated by the given situation."[35] Shutter control represented a clear and substantial way that the license granted by NOAA could be used to protect national security. The PDD did leave it up to each of the concerned departments to develop its own internal statutory mechanisms for implementing this guidance and placed responsibility for resolving disagreements directly on the president.

On July 22, 1998, the Secretary of Commerce announced that US commercial companies would no longer be allowed to sell high-resolution imagery of

Israel.[36] Movement in this direction began the previous year with the Kyl-Bingaman amendment to the 1997 National Defense Authorization Act.[37] This amendment prevented the sale of imagery of Israel that was more detailed than imagery available from non-US sources. The enforcement of this limitation contradicted earlier assurances given by the Clinton Administration that the policy would not be enforced beyond the then 1-meter-imagery restriction, as it harmed US commercial-imagery industries' business interests in the Middle East.[38] The policy established a new precedent that the president had the power to designate any geographic area as limited by the same law. Today this specific exemption remains a part of 15 CFR 960 but has been applied only to Israel.[39]

Policy toward remote sensing satellites remained static for the remainder of the Clinton Administration. This was primarily due to the lag in the growth of the commercial market. NOAA issued several licenses after the 1992 law was passed, but they had yet to result in a successful on-orbit satellite. World-View, the company that received the first license after the 1992 law passed, had been renamed EarthWatch; and its first satellite failed on orbit in 1997 shortly after launch.[40] In 1999, the Ikonos satellite built by Lockheed Martin became the first US commercially built and funded imaging satellite to achieve orbit.[41] It was quickly followed by a second and third failed attempt by EarthWatch and another attempt by a company called Orbimage, that also failed.[42] Despite these failures, by 2002 there were two US commercial remote sensing platforms on orbit that had one-meter or better resolution capability, though US policy still prevented them from selling anything better than one-meter imagery.

As a result of the availability of commercial imagery, in 2002 the Director of Central Intelligence issued a memorandum that created a valuable new market for the emerging industry. The memorandum made it the policy of the US "intelligence community to use US commercial space imagery to the maximum extent possible."[43] This memorandum specifically sought to stimulate the US commercial-imagery market. The imagery was not being used for intelligence purposes but primarily for mapping purposes. This memorandum led directly to the ClearView contracts issued in January 2003 to DigitalGlobe and Space Imaging. These contracts represented the first between the US government and commercial-imagery providers. The ClearView contract, worth $500 million, between DigitalGlobe, Space Imaging, and the National Imagery and Mapping Agency (NIMA) provided valuable financial support and legitimacy to commercial providers, though many other agencies had yet to embrace commercial imagery.[44]

The ClearView contract was quickly followed by the first formal action related to remote sensing from the new Presidential Administration. The 2003 Commercial Remote Sensing Policy superseded PDD-23.[45] The policy reinforced the 2002 memorandum from the CIA Director and broadened it to include all government agencies. It also encouraged the development of long-term sustainable relationships between government and industry. As a result of the policy, imagery up to a .5-meter resolution was now authorized for sale. The first satellite capable of utilizing the new resolution limit was Worldview-1, launched in September 2007.[46] Its launch and the new regulations made US commercial providers the highest-resolution vendors available on the civilian market.

Despite the stated goals of promoting industry, the policy still sought to strike a balance between national security and commercial viability. Exports of remote sensing data and components were limited to what was already available in the global commercial marketplace.[47] Private companies did not receive the ability to determine what was available independently. PDD-23 required case-by-case review of all exports of remote sensing data and technology. Satellite components were added to the International Traffic in Arms Regulations (ITAR) in the fiscal year (FY) 1999 NDAA, and the 2003 policy was not accompanied by legislative action that could change that.[48] Not until 2013 would legislative measures in the FY 2013 NDAA authorize the president to remove most satellite technology from ITAR to the much less restrictive commerce control list.[49]

Obtaining a resolution that followed the currently permitted limit exactly was extremely difficult to do. Commercial companies quickly developed satellites capable of taking images at higher resolutions than they were authorized to sell. US government agencies naturally wanted access to the best possible imagery, but national security concerns still dictated a lower resolution than was available. From this dilemma, a two-tier arrangement evolved. Under it, commercial operators were allowed to sell .5-meter imagery on the commercial market, and up to .25-meter imagery to recipients individually authorized by the US government.[50] This arrangement allowed commercial providers to develop the most capable platforms available in anticipation of a future authorization to sell still better imagery on the commercial market.

At the same time that these changes were occurring in the Administration, NOAA established the Advisory Committee for Commercial Remote Sensing (ACCRES) in 2002. This committee included representatives from government, academia, and industry with the purpose of providing advice to the Secretary of Commerce on issues relating to commercial remote sensing.[51] The establishment of ACCRES was an important development for a mature industry; previous efforts to influence remote sensing policy could be accomplished only through expensive lobbying of Congress. While the establishment of ACCRES as a forum for public involvement in the licensing process was a positive step forward that has remained relevant NOAA has struggled to keep pace with the commercial industry. The current version of 15 CFR 960 governing commercial remote sensing was published in 2006, and it is becoming increasingly dated. The 2006 version included all of the previous developments in policy and law, but it was written prior to the small satellite revolution. The legal regime within 15 CFR 960 is built around the 1992 remote sensing act with its dated process of site inspections and license approval. Within its regulatory framework, the agencies involved in the licensing process have sought to clarify their processes and keep up with the increase in licensing applications while Congress develops a new legal basis for the remote sensing industry.

In 2017, the various agencies involved in the complicated process of approving a remote sensing license developed a new memorandum of understanding (MOU) that sought to improve the process of license approval. The MOU gives NOAA 120 days to review a license application and provide feedback to the applicant.[52] Within this timeline, the Department of State (DOS), Department of Defense (DOD), and the Department of the Interior (DOI) have 30 days to review the application with additional review provided by the Office of the Director of National Intelligence (ODNI) and the Joint Chiefs of Staff (JCS). Despite this bureaucratic soup of agencies involved in the approval process, NOAA has managed to reduce their average time for license approval from 210 days in 2015 to just 91 days in 2017.[53] The key to this success is a provision in the MOU which provides for a series of escalating dispute resolution boards between the agencies that are each only given five days to resolve their issues before escalating to the next level, ultimately ending with a decision required by the president.[54] NOAA was forced to use these dispute resolution boards in 14 of the 16 licenses it issued in 2017, demonstrating that the various agencies involved in the approval process are still far from consensus on what constitutes national security risk for commercial remote sensing.[55]

Congress attempted to improve the process through in 2018 with the *American Space Commerce Free Enterprise Act*. This act placed the burden of proof on NOAA and the Secretary of Commerce to demonstrate why an application for a remote sensing platform should not be approved. Under the proposal, if the Secretary of Commerce does not personally provide written justification to the remote sensing applicant within 90 days, the application is automatically approved.[56] The bill also proposed preventing the placing of operating restrictions on remote sensing systems "for which the same or substantially similar capabilities, derived data, products, or services are already commercially available or reasonably expected to be made available in the next three years in the international or domestic marketplace."[57] This bill represented a significant change that would have had a dramatic impact on the future competitiveness of the US remote sensing market but it failed to pass the Senate before the congressional session ended. Despite this failure, aspects of this bill may be incorporated into future space legislation as space assumes a larger role in the global economy.

While the Congressionally driven efforts to reform remote sensing regulation failed to pass, the National Space Council effectively reinvigorated efforts at reform with Space Policy Directive-2 (SPD-2). SPD-2 called for streamlining regulations governing commercial space activities and specifically directed the Commerce Secretary to develop new regulations which would "encourage expansion of ... commercial remote sensing activities."[58] The Commerce department appears to be taking an approach similar to that proposed by the failed *Space Commerce Free Enterprise Act* of differentiating systems based on availability. This new approval process would accelerate the timeline for approval to 60 days from the current 120 and reduce the "regulatory burden imposed by individual interagency review" for technologies that are or will soon be available on the commercial market.[59] The details of how this new regulatory structure will be implemented are still unclear though it is apparent that NOAA has taken to heart the guidance in SPD-2 to reduce regulatory burdens on commercial remote sensing. This relaxing of regulatory controls over the commercial remote sensing market is not without opponents within the DOD.[60] The final outcome of this new regulatory structure is still unclear but its impact on the remote sensing industry will be profound as effective government regulation remains instrumental to the health of the industry.

Exploring the evolution of commercial remote sensing

In 2016, the commercial-imagery industry in the US briefly consolidated down to just one company, DigitalGlobe (now Maxar). At that time the director of the NGA referred to the company as his "mission partner" that was essential to his "mapping, charting, geodesy, and intelligence missions."[61] With five satellites offering resolutions between .31-meters and .5-meters that cost as much as $750 million and weigh more than 2,400 kg each, the company is vastly different from new entrants to the industry like Planet, but until late 2016 it represented the sole remaining commercial entity providing imagery to the US government. How DigitalGlobe became the only significant operator of commercial-imagery satellites under the regulations and policies established by the government reflects the health of the industry and the dependent nature of the industry on government support. It is an illustrative example of how the US government is both the primary customer for commercial remote sensing companies and the biggest obstacle to US dominance of the commercial market.

DigitalGlobe today represents the final merger of all remaining commercial-imagery companies formed after the 1992 changes to remote sensing law. Incorporated initially as WorldView Imaging Corporation in January 1992, it anticipated the 1992 law passing later that year and received the first license to operate high-resolution imagery.[62] The company then merged with Ball Aerospace's imagery division to become EarthWatch in September 1993.[63] The company's first two launches, Earlybird 1 and Quickbird 2, failed in 1997 and 2000, respectively, nearly destroying the company.[64] Earthwatch finally succeeded in launching its first satellite in October 2001 when QuickBird 2 successfully achieved orbit from Vandenberg AFB.[65] The company changed its name to DigitalGlobe the following year in August 2002.[66]

The successful launch of Quickbird 2 increased the number of US commercial-imagery satellites in orbit to two and prompted the US government to take the nascent industry seriously. The other imagery satellite was owned by Space Imaging which had successfully launched a .8-meter resolution satellite in 1999, making it the first commercial company to provide better than 1-meter imagery.[67] Following the CIA director's 2002 guidance, in 2003 NIMA awarded its first commercial-imagery contract, ClearView.[68] The ClearView contract awarded up to $500 million each to Digital Globe and Space Imaging over a five-year period. A third US commercial provider, Orbimage, was added to the contract a year later after it successfully launched its first satellite in June 2003.

The ClearView contract created a vehicle for the purchasing and sharing of imagery within the government, removing the need for government agencies to send orders for specific images directly to the commercial providers. Under the contract, NIMA, renamed the National Geospatial-Intelligence Agency (NGA) in 2003, guaranteed a minimum revenue in the form of imagery-purchase commitments.[69] The contract also replaced a cumbersome licensing strategy for sharing the imagery purchased by the NGA among other government agencies.[70] Guaranteeing a certain amount of income was a boon for the fledgling commercial companies, and in 2004 public filings, Orbimage reported that the government was its primary customer.[71]

The successor contract to ClearView, NextView, drove the next phase of industry consolidation. The NGA selected DigitalGlobe and Orbimage under the NextView contract, worth $1 billion, in late 2004. These contracts went beyond the simple purchasing of imagery under the ClearView contract and subsidized the construction of high-resolution satellites by both companies. In the case of Orbimage, this contract provided $237 million for the construction of OrbView-5 which the company projected to have a total cost of $502 million.[72] The third US commercial provider, Space Imaging, a joint venture between Lockheed Martin and Raytheon, was not awarded a contract under NextView. This company was founded with $750 million in capital by the two major aerospace giants and was the first to launch a commercial imaging satellite successfully, but the lack of a contract from the NGA doomed the company.[73] Orbimage purchased the company for the "fire sale" price of just $58.5 million in early 2006 and renamed itself GeoEye.[74]

By 2006 there were just two US commercial-imagery providers, DigitalGlobe and GeoEye. Despite a forecast by Merrill Lynch in 2000 that the market for commercial imagery could be as large as $2.5 billion per year by 2005, the companies in the market at that time were unable to develop enough commercial demand to support themselves outside of government contracts.[75] That this financial prediction covering such a relatively short time horizon would prove so wrong was remarkable. As a result of failures and commercial costs, GeoEye emerged as the largest provider of satellite imagery with sales of $160 million in 2005. The government share grew from 49 percent of revenues for GeoEye in 2004 to 61 percent in 2005.[76] This further dependence on government funding would again drive the next round of industry consolidation when the demand for commercial imagery tapered off as the wars in the Middle East scaled back.

In 2008, the NGA transitioned to a Service-Level-Agreement (SLA) structure under NextView with GeoEye and DigitalGlobe that continued in future contracts. Under this SLA structure, the NGA agreed to purchase $12.5 million in imagery from GeoEye and DigitalGlobe each month.[77] The SLA provided a greater amount of revenue predictability for both companies. The Enhanced-View contract superseded NextView in 2010. Under this contract, both companies received a combined $7.3 billion over ten years. This included the investment in another satellite from GeoEye, GeoEye-2.[78] The NGA agreed to subsidize $337 million of the satellite's a projected total cost of between $750 and $800 million.[79]

The two companies did not face any domestic competition, but the French company Astrium Services, owner of the SPOT satellites, did compete for the contract. The SPOT satellites had inaugurated the commercial imaging era with the first launch in 1986.[80] Since then the company has placed a total of five satellites in orbit, all subsidized by the French government. However, the French declined to support the latest generation of SPOT satellites, forcing the company to look to the US government, which remained the largest consumer of commercial imagery. Astrium's CEO, Eric Beranger, wondered publicly after the contract award if the industry as a whole was capable of surviving without government support.[81]

Under the EnhancedView contract, the government offered to co-invest in any new satellites as long as the company could file a letter of credit that demonstrated its ability to fund its portion of the cost. The letter of credit ensured the government would not be left without a final product if the company was financially unable to support its end of the bargain. GeoEye was forced to take on a new investor to obtain the required letter of credit.[82] Despite GeoEye's efforts, the government was compelled to drop this requirement as it drove up the total cost of GeoEye-2 to unacceptable levels. Instead, the NGA decided to tie funding for the satellite to milestones in GeoEye-2's construction.[83] By adopting this method, the contract looked very much like a traditional acquisition program for government satellite construction. The government was not, however, funding the entire project. GeoEye was responsible for the majority of the cost and now subject to the risk of the government budgeting process. In effect, the NGA had shifted the satellite's development risks to GeoEye.

In June 2012, the NGA informed GeoEye that it would not provide additional funds toward the development of GeoEye-2.[84] The reason cited by the NGA was that it simply did not have any additional funds to provide.[85] Despite this, in its memo to GeoEye, the NGA wished to explore the ability of the company to complete the satellite using already-obligated funds. The memorandum noted that the government continued to reserve its rights "under the Termination for Convenience Article."[86] This article under FAR 49.202a allowed the NGA to terminate the contract without providing funding for "any consequential damages" to GeoEye since the funding was limited to what was already obligated.[87] The final act of ruin, however, came with another NGA memo sent the same day informing GeoEye that the government would not be exercising the next SLA for imagery purchases beginning later that year "due to funding shortfalls."[88]

The Senate Armed Services Committee (SASC) attempted to rescue the company by providing enough funding in the FY 13 National Defense Authorization Act (NDAA) to continue the SLA with GeoEye as well as DigitalGlobe.[89]

Before the SASC could add the provision to the NDAA, DigitalGlobe and GeoEye announced plans to merge, leaving just one US commercial-imagery provider. Congress noted that if the merger failed to obtain approval from the Justice Department, it would reconsider providing the requested funds.[90]

In a July 2012 press release, DigitalGlobe and GeoEye publicly announced their merger.[91] This announcement came just one month after GeoEye received notice that it would no longer be receiving funding under the EnhancedView SLA or for the construction of GeoEye-2. Neither in press releases nor other public-investor documents did the two companies cite GeoEye's loss of government funding as the impetus for the merger. Instead, they touted the efficiencies and talent-gathering aspects of the merger, with a particular focus on advantages in cost savings for the government.[92]

The revenues for the combined companies for 2012 totaled more than $600 million, even after accounting for the decrease in public funds.[93] DigitalGlobe acknowledges in their annual filings with the Securities and Exchange Commission (SEC) that their government funding is "subject to Congressional appropriations and the right of the NGA to terminate or suspend the contract at any time."[94] This creates obvious uncertainties for a company whose revenue is so dependent on government funds. In 2015, 63.7 percent of total revenue was from the government, with the EnhancedView contract from the NGA making up 48 percent of that total, see Table 6.1 below.[95] This level has fluctuated in the past but remains by far the single largest source of funding for the company. Since 2012, when GeoEye and DigitalGlobe merged, the total amount of revenue generated has increased at only a moderate rate. Before 2009, the industry demonstrated substantial growth, but this leveled off when government contracts stabilized following the merger of the two companies. Non-government customers remain a small portion of the business, and, of these, the international market accounts for the largest share. In 2015, 28.9 percent of total revenue resulted from international sales, compared to 7.4 percent from domestic non-governmental sources.[96] This sourcing imbalance makes the industry extremely vulnerable to regulatory and geopolitical fluctuations.

With a product long associated with surveillance and intelligence, political implications can have significant impacts on the industry. DigitalGlobe's deep association with the US government has negative consequences for sales when sensitive foreign-policy issues arise. In 2014, the company saw a $14.5 million decline in Russian business from a high of $23 million in 2013.[97,98] The company cited several potential causes for the downturn in Russian business. These included the downturn in the Russian economy due to sanctions, although this did not affect DigitalGlobe's legal ability to sell imagery to Russian customers. Potentially the most significant reason cited by the Digital-Globe CEO was the "very public use of DigitalGlobe imagery, by the U.S. government and The NATO alliance, showing Russian troop locations and, more recently, purporting to prove that missile strikes in Ukraine came from batteries located in Russian territory," (see Figure 6.1 for the image referenced).[99] This assertion by the company cannot be proven, but it neatly

Year	Revenue (million USD)		All Gov. Funding (% of total)		NGA (% of total)	
	GeoEye	DigitalGlobe	GeoEye	DigitalGlobe	GeoEye	DigitalGlobe
2007	$183.0	$151.7	55%	68.20%	37.10%	No Data
2008	$146.6	$275.2	39%	80.80%	26%	No Data
2009	$271.1	$281.9	67%	81.90%	46%	75%
2010	$330.3	$322.5	66%	78.20%	45%	62.20%
2011	$356.4	$339.5	64%	77%	41.20%	60.10%
2012		$421.4		76.20%		60.80%
2013		$612.7		58.40%		37.10%
2014		$656.6		60.40%		38.90%
2015		$702.4		63.70%		48%
2016		$725.4		63.71%		46.50%

Figure 6.1 Percentage of revenue for DigitalGlobe and GeoEye provided by government and NGA 2007–2016.

Source: compiled by author from Securities and Exchange Commission 10K and 10Q filings from both companies from 2007 to 2016. In 2017 Digital Globe Merged with MDA.

explains the nearly complete disappearance of revenue from Russian sources following the publication of the photos by NATO. Political exposure represents a unique risk that US-based companies take when providing imagery to the military and government, potentially jeopardizing its non-governmental business.

The emergence of competition from a variety of companies relying on constellations of small satellites presented a further challenge to DigitalGlobe. In 2016 the first of a new generation of imagery providers, Planet, entered the commercial market with the successful launch of its first "flock" of Dove Cube-Sats.[100] Using a business model that relied on "good enough" imagery refreshed daily from hundreds of satellites, Planet won its first contract with the NGA in 2016, further threatening DigitalGlobe's business model. The pressure from this emerging competition possibly drove one final round of consolidation as the Canadian space company MacDonald, Dettwiler and Associates Ltd. (MDA) purchased DigitalGlobe for $2.4 billion in 2017.[101] Following the purchase MDA, a leader in satellite radar imaging rebranded itself as a nominally US company called Maxar Technologies in order to retain access to the lucrative US government market.[102]

In sum, the brief consolidation in the commercial-imagery market to just one major supplier resulted from two major trends. First, the commercial market

Figure 6.2 DigitalGlobe image showing Russian military units within Ukraine on August 21, 2014.

Source: NATO, Supreme Headquarters Allied Powers Europe, News Release 28 August 2014, "New Satellite Imagery Exposes Russian Combat troops inside Ukraine" http://shape.nato.int/new-satellite-imagery-exposes-russian-combat-troops-inside-ukraine.

proved too weak to provide the funding to support the high capital costs of building and launching a satellite using traditional launch and design methodologies. This weakness, therefore, limited the number of market players from the beginning while creating a dependence on government funding for the remaining companies. Second, the dependence on government contracts for support exposed commercial providers to the uncertainties of government budgeting and demand. The result is that each change in government funding drove industry consolidation. In 2004, when Space Imaging was not awarded a contract under NextView, it was purchased at a massive discount by GeoEye.[103] GeoEye, in turn, was forced to merge with DigitalGlobe in 2012 when the NGA abruptly cut off its funding. Changes in a single government contract were the catalysts for each of these phases of industry consolidation. Emerging competition from companies with radically different business models finally brought about the end of DigitalGlobe as an independent company. Even so, its successor company felt that US government revenues were still vital enough to rebrand itself as a US-based company.

Exploring the development of Synthetic-Aperture Radar

Synthetic-Aperture Radar (SAR) remote sensing satellites represent a greater challenge to national security than typical electro-optical imaging satellites. They have the ability to image at all hours and through any weather condition, whereas traditional electro-optical imaging satellites typically capture images between 10:00 and 14:00 local time to ensure the best daylight and minimize shadows and cannot image through cloud cover. The ability of SAR to image areas independent of lighting conditions or weather makes it nearly impossible for military forces to conduct operations unobserved when a SAR satellite is in view range. As a result of its versatility, commercially available SAR represents a greater national security risk than normal imaging systems preventing approval for domestic SAR systems and allowing foreign suppliers to effectively control the industry.

Twelve separate US companies had licenses to operate remote sensing satellites by 1997, but none of those were granted for radar satellites.[104] By 1997, the lack of a US commercial SAR remote sensing platform, even in the planning stages, became a concern to Congress. Then Senator Dennis DeConcini argued that "if Commerce does not license a radar satellite system, then a foreign-owned radar system, with a one-meter or less capability, will enter the market leaving the U.S. government with no effective control in this area."[105] Canada had already launched Radarsat-1 in 1995 with a maximum resolution of eight-meters, with plans in place to launch a second satellite with a resolution of three-meters.[106] This case study explores how, unlike with electro-optical imaging where US commercial regulations, while burdensome, were still relatively reasonable and updated fast enough to allow for commercial growth potential, national security concerns ceded the commercial market for SAR remote sensing satellites to foreign entities. These national security concerns hamstrung the US industrial base and prevented the US from having any control over the SAR market.

Following Senator DeConcini's arguments, the DOD announced that it would oppose granting any commercial license that allowed for better than five-meter resolution imagery.[107] This restriction, when combined with other additional restrictions, such as limits on the release of "phase history data" which allows for more accurate interpretation of spectral data, created obstacles to market entry that discouraged a successful US effort.[108] Though NOAA granted a handful of licenses over the ten years following Senator DeConcini's statements, no US company successfully produced a SAR remote sensing platform. Only under pressure from US companies to match existing foreign competition did the Department of Commerce gain approval from the interagency process to relax the resolution restrictions over time. In 2000, the requirement was dropped to three-meters, and in 2009 it was decreased further to one-meter.[109] Germany had already launched a commercial platform, TerraSar-X, with one-meter resolution in June 2007 so the US authorization was too late for US companies to establish market leadership.[110]

The consistent application of overly stringent restrictions on US companies' sale of SAR imagery effectively prevented the entry of US providers into the market until recently. In October 2015, NOAA granted a Virginia-based company, XpressSAR, a license to operate four satellites and sell X-band SAR imagery at unspecified sub-meter resolutions to private, public sector, and government customers.[111] This placed a US company firmly at the leading edge of the market for the first time. However, in 2018 XpressSAR was forced to purchase its satellites from an Israeli company due to cost and capability concerns with other manufacturers.[112] It also has significant foreign competition. Urthe-Cast, a Canadian company, plans to build a SAR constellation as part of a grouping of non-US commercial-imagery providers called the PanGeo Alliance which includes companies based in China, Belarus, and the UAE among others operating a variety of optical and SAR platforms with resolutions as good as .7 m.[113] UrtheCast's plans involve launching 16 satellites in pairs over two orbital planes. One satellite in the pair will be a SAR platform and the other an optical platform with 1-meter or better capability in various bands.[114] Urthe-Cast's constellation will allow the company to guarantee customers some form of imagery, no matter the weather conditions, every day.

Unlike in the optical-imagery realm, US regulatory restrictions ceded the SAR market to foreign companies. Ceding the market prevented the United States from achieving the goals set out in various national policies to "maintain the nation's leadership in remote sensing space activities."[115] It also forced US government agencies to turn to foreign companies to obtain unclassified SAR imagery for mapping and other purposes. The NGA alone spent $85 million from 2010 to 2015 to obtain SAR imagery from Canadian, German, and Italian sources.[116] Interestingly, the NGA has no issue with this. Karyn Hayes-Ryan, the then NGA director of the commercial imagery, data, and programs group, has been quoted as saying that "we do purchase SAR [radar] imagery from several foreign sources as there is not a US source for this at present," and "we have no problem with this."[117]

The lack of US government agency support combined with national security concerns has historically overcome US commercial desire to develop SAR. The imagery market is small, and so easily saturated, and an early effort in SAR could have prevented the growth of foreign competition. Instead, the US government must pay foreign companies for access to unclassified radar imagery and has no control over what foreign customers those companies sell to. In addition, lack of presence in the market harms the US commercial satellite industry by ceding technological development and commercial growth in this field to Canada and various European nations. The temporary national security gain from restrictions on the development of advanced commercial SAR systems was outweighed by the loss of control over the market. If the US controlled the SAR market, it could benefit from private investment, and through its licensing process, influence the sale and release of imagery directly. That NOAA was eventually able to grant a license to XpressSar for sub-meter-resolution imagery demonstrates that the DOD and other government agencies finally recognize that there is little to be gained from further restrictions on the US commercial SAR market.

Figure 6.3 Synthetic-aperture radar image of several fields and buildings at Kirtland AFB. Taken by Sandia's FARAD radar.

Source: courtesy of Sandia National Laboratories, Radar ISR, www.sandia.gov/radar/imagery/index. html.

Impact of the licensing process on the commercial industry

Since the 1984 Land Remote Sensing Commercialization Act was passed, the legal structure to license and regulate space-based remote sensing platforms has existed in various forms. That licensing structure has been a source of continued tension with industry, as the need to balance national security concerns competed with the desire to develop a robust commercial industry. After the passage of the Land Remote Sensing Policy Act of 1992, which resolved some of the largest concerns with the 1984 law, the industry was finally able to grow, and commercial companies finally began to apply for licensing under the Act and build satellites. Despite this early growth and the later stabilization of the industry, albeit with just one major provider, the licensing structure still possesses several flaws that make the regulation burdensome, convoluted, and unnecessary.

For much of the history of the license-security-review process, only a single factor was considered. According to former Deputy Assistant Secretary of Defense for Space Policy Douglas Loverro, the only national security concern considered when reviewing a licensing application prior to 2014 was "the harm they could do when used by an adversary."[118] Three factors are now considered when granting a license according to Mr. Loverro. These two additional factors make a significant difference in the evaluation process by expanding the strategic scope of the security review. The first factor added was a consideration of what additional potential benefits the launch of a system could provide to the resiliency of the US national space architecture. Resiliency is defined as "the ability of a system architecture to continue providing required capabilities in the face of system failures, environmental challenges, or adversary actions."[119] Any addition to the number of US remote sensing satellites effectively increases resiliency through redundancy, even if they are only US-licensed commercial systems. The emergence of Planet and others building constellations of remote sensing small satellites take this idea a step further and achieve resiliency through disaggregation, another term meaning "the dispersion of space-based missions, functions or sensors across multiple systems spanning one or more orbital plane, platform, host or domain."[120]

The second factor Mr. Loverro succeeded in adding was whether the technology under consideration for a license was controllable by the US government. If the technology did not fall under existing export controls, or a foreign entity could easily develop it, then it was not "logical to presume you could control the development of the system going forward."[121] While not as clear as the provision in H.R. 2809, adding these two additional factors to the security review rebalanced the process in favor of commercial growth and made the process more realistic in light of developing foreign capabilities. Mr. Loverro also attempted to add a fourth factor to the consideration process. This was the idea that there is a "presumption of no harm unless harm can be proven."[122] If this factor were included, the bar for granting a license to a commercial system would have been lowered, and commercial providers denied a license would have had significant legal grounds to challenge any denial.

Along with the licensing application process, licensees are also required to notify NOAA of "any significant or substantial agreement that they intend to enter into with any foreign nation, entity, or consortium."[123] The agreement is then reviewed by the DOD, the Department of State, the Department of the Interior, and "any other Federal agencies determined to have a substantial interest in the foreign agreement."[124] Review through any interagency process with multiple stakeholders is not timely or easy though the 2017 MOU has improved the process even if it has required elevation through the conflict resolution process almost every time. Once the licensing process is complete, the burden of regulation does not cease. NOAA also requires each licensee to produce a quarterly and annual report on compliance with the terms of the license agreement. As recently as 2016, all of these reports, inspections, and license approvals were managed by just a single overworked civil servant within NOAA.[125] The process for getting and maintaining a license is a burden for both the applicant and the government, with the security-review requirements achieving little in the way of added national security.

Further, the inability to approve licenses for new capabilities such as SAR has ceded portions of the market to foreign competition. Despite plans by XpressSar to launch a SAR constellation, the US lacks a domestic option for commercialization. The responsibility for this shortfall in the industry is entirely the fault of agencies within the US government that prioritized short-term gains in security over long-term space power. As a result of this prioritization, the US has no direct control over commercial providers, and any new US company faces stiff market competition. Should the US shift to a reliance on commercial capabilities this might encourage the DOD and other agencies that have stood in the path of license approval for security reasons to instead push for approval and innovation, potentially paving the way for new technology development and market dominance.

Resolution limitations

Maintaining a resolution limit provides the US government with no significant advantages. The Commerce Department sets resolution limits on US-based commercial-imagery providers in response to national security concerns. These limits have decreased from one-meter panchromatic resolution to .5-meters, and then to .25-meters. Foreign competition has been the driving motivator behind each decrease in the authorized resolution, and the current limit is set beyond the capability of any commercial platform on orbit or any that are planned. Ultimately, resolution is determined by a combination of optics and orbital altitude; small satellites that cost significantly less than the satellites launched by more traditional providers like DigitalGlobe trade orbital altitude and lifespan on orbit for a lower altitude to obtain better imagery with inferior optics. This tradeoff still puts the resolution of these satellites around 1-meter, making the panchromatic resolution restrictions irrelevant to them at this time.

It is hard to see what is protected by limiting the resolution of commercial satellites. If national security was endangered by images with better than one-meter resolution, then what changed—besides foreign competition—to lower that resolution to .5-meters and again to .25-meters? At that resolution, relatively little is obscured by the current limitation. Mr. Loverro stated that "there were those of us who believed that no restrictions were necessary, others disagreed, so we set a limit at the (reasonable) physical limits and satisfied those who felt they had defended some turf."[126] This statement demonstrates that the limitation is not the result of objective security concerns but the result of competition among stakeholders within the US government. Those whose primary concern is the health and resiliency of the US commercial space industry supported dropping restrictions altogether, while those whose primary concern is security pushed back. The .25-meter compromise represents a point beyond which there is no current commercial need for higher-resolution imagery, and the cost of developing a commercial platform becomes prohibitive. In the words of Deputy Secretary Loverro, the US government has "created boundaries that are really no boundary at all."[127]

While the limit on panchromatic imagery is .25-meters, possibly beyond the limit of commercial interest for the foreseeable future, the limit for multispectral imagery remains 1-meter.[128] Industry uses multispectral imagery for everything from determining soil quality to mineral exploration; even normal color images are considered multispectral. A one-meter limitation for multispectral that is out of synch with the panchromatic imagery limit is an issue that the US government will need to address in the near future. For now, this limit is not an issue as the most advanced commercial satellite on orbit, WorldView-4, is currently capable of only 1.24-meter multispectral imagery. However, unlike for panchromatic imagery, a demand for sharper multispectral commercial imagery is likely.

A potential solution to future multispectral concerns may be possible by using an authorized buyer arrangement. In order to allow US companies to build and operate satellites with better-than-authorized capabilities for future proofing, NOAA has in the past allowed operators to sell higher-than-authorized resolution imagery to the US government under a two-tier arrangement formalized in 15 CFR 960.[129] Under this arrangement, satellite operators can sell their best-available imagery to the US government but cannot sell beyond the current limit to the public. An agreement like this was in place when the imagery limitation was originally one-meter. Imagery companies could sell one-meter imagery to the public and .5-meter imagery to the government. When the public limit was reduced to .5-meters, the government limit was adjusted to .25-meters. There is no evidence that a two-tier arrangement currently exists, but this may be due to the fact that no commercial system in 2018 can image better than the authorized limit of .25-meters.

Existing spectral resolution restrictions are reasonable for the commercial market at the moment but if the US government shifted toward a reliance on these companies more fidelity might be required. Here again, the US government could drive commercial innovation by subsidizing a portion of the cost of development of certain platforms or at least providing a stable market for higher-resolution imagery. Once developed for the government, the industry would have time to find and develop a commercial market. Though this has proven difficult for DigitalGlobe, especially since it has been undercut by "good enough" imagery provided cheaper and faster by small satellite providers who focus less on quality and more on quantity of imagery.

Resolution restrictions focused purely on the ability of an individual system fail to account for the threat posed by high-revisit rate systems that currently sacrifice quality for global coverage. This new approach is demonstrated by Planet's ability to image the entire globe daily and represents a new threat vector for national security. A single satellite is individually not an operational threat to military forces as its orbit would limit it to passing over any single location only once per day. As constellations are developed with the ability to pass over a single point multiple times per day a new factor, temporal resolution, becomes an issue. This varies from spectral resolution measured in physical distance; instead it is measured in units of time and represents the period between revisit to a single point on the Earth's surface. As temporal resolution decreases alongside spectral resolution an imagery singularity will eventually be reached where satellite systems will be able to image the entire Earth in real time at high resolution. An imagery singularity will have profound implications for future military operations that present opportunities for a commercialization approach.

Other methods of control

Since 1984, when the first licensing requirements were laid out, the US government has required licensees to operate their system "in such a manner as to preserve and promote the national security of the United States."[130] The specific controls put in place to meet this obligation are defined in 15 CFR 960. This section briefly explores the various methods of controlling remote sensing platforms and data that are available by law or have been used in the past by the US government in addition to the resolution restrictions discussed above.

Shutter control

The term shutter control describes the ability of the US government to restrict the operation of a satellite over a designated geographic region. PDD-23 granted this ability in combination with increasing the authorized resolution of commercial satellites to one-meter. The specific language contained in a license granted by NOAA contains the provision that:

> During periods when national security or international obligations and/or foreign policies may be compromised, as defined by the Secretary of Defense or the Secretary of State respectively, the Secretary of Commerce may, after consultation with the appropriate agency(ies), require the Licensee to limit data collection and/or distribution by the system to the extent necessitated by the given situation.[131]

At the time it was created, this restriction was extremely controversial but has since never been used for two primary reasons. First, invoking the stipulation in the license would likely trigger a legal battle unless both the government and the commercial provider agreed that the reason for invoking it was truly a demonstratable threat to national security. Even then, the legal challenge would probably not come from the commercial provider but rather from news agencies or other organizations seeking access to imagery. A second concern with invoking shutter control is that it could have long-term repercussions on the health of the commercial remote sensing industry. Invoking shutter control would damage the goal of fostering a healthy commercial satellite industry by demonstrating the vulnerability of US providers to government interference.[132] Combined with the likelihood of a legal challenge, the political cost of implementing this provision has proven to be too high, even in the days following the 9/11 attacks.

Buy-to-deny

The Department of Defense (DOD) has executed a buy-to-deny form of shutter control referred to as "checkbook shutter control" just one time since the advent of commercial-imagery satellites.[133] This buy-to-deny strategy was invoked in Afghanistan during Operation Enduring Freedom in the early days of the conflict during the bombing campaign in October 2001.[134] The reported reason for this strategy was that news outlets were seeking images from the single commercial satellite available at the time with high enough resolution, Space Imaging's Ikonos, to verify reports of massive civilian casualties.[135] Attempting to invoke the legal powers granted under PDD-23 would likely have been challenged in court, so the DOD quietly entered into an exclusive contract with Space Imaging for all of the imagery over Afghanistan. Exclusivity worked when there was only a single high-resolution provider available; but, as the number of commercial providers increases, the cost of executing a strategy like this will become increasingly prohibitive.

Diplomacy

Another option that could have more success on an international scale would be to deny access to imagery through international diplomatic channels. This option was exercised during the Gulf War when the United Nations mandated an embargo on satellite imagery sales to Iraq.[136] The only available non-US imagery was from France's SPOT satellite, and the agreement required SPOT to forgo sales to media companies in order to avoid the inadvertent release of imagery to Iraq through third parties. SPOT had a relatively low 10-meter resolution at the time but could still have provided valuable overhead intelligence to the Iraqi government, which also had lost access to aerial reconnaissance.[137] Denying access to up-to-date commercial imagery of an area during times of conflict using diplomatic means could prove effective for short periods with support from the United Nations or as the result of carefully targeted diplomacy toward the handful of nations that host commercial satellite providers.

Denied-parties screening

Denied-parties screening is a control developed and maintained by the State Department where commercial-imagery companies receive a list of individuals and entities to whom they cannot sell their products. Since US-based remote sensing companies are subject to licensing through NOAA from the Department of Commerce, they are subject to this list. A publicly available and searchable version of this consolidated screening list is maintained online. It is designed to make it easier for companies doing business with foreign entities to ensure they are not under an export restriction. For any company dealing with foreign entities, this list is an important tool for ensuring compliance with US law and sanctions. However, since commercial-imagery products are data, they are easily shareable, and this is a poor tool for preventing the release of sensitive imagery into the public domain or through third parties.

Summary

Unlike in the COMSATCOM market, the US government is a large market player in remote sensing, determining winners and losers through regulation, policy, and contract purchases. Changes in any of these factors have had significant effects on the commercial remote sensing industry since Congress first authorized it under the 1984 Land Remote Sensing Act. Policy and regulation have proven to be consistent limiting factors in the development of a healthy domestic remote sensing industry. These regulatory limitations are driven by a balancing act between promoting commercial competitiveness and preserving national security. Foreign competition has consistently been the factor that has driven the relaxation of restrictions. When the relaxing of restrictions is not accomplished fast enough to allow US commercial providers to maintain a technical lead over foreign competition in the field, the market is nearly irretrievably ceded to foreign competition. US government regulation has been relaxed fast enough for the US to maintain dominance of the optical market but not the SAR market. The industry also has demonstrated that it is not yet able to remain commercially viable without government support. DigitalGlobe's history demonstrated that changes to a single contract managed by the NGA have been the cause of industry consolidation from three companies to one since 2003.

The leverage the US government does have over the commercial industry and its technical sophistication makes it very probable that almost all remote sensing for security purposes could be outsourced. The size of the Global remote sensing market in 2016 was estimated at $9.7 billion. Shifting half of the combined budget from the NRO and the NGA toward purchasing commercial capabilities would nearly double the global market for imagery.[138] Linking US funding to US corporate presence and regulation would no doubt incentivize the bulk of the global industry to shift to the US. MDA demonstrated how attractive the incentive of US government contracts are when it moved from Canada to the

US so that it could retain DigitalGlobe's government business following its acquisition.

US financial incentives also would be enough to drive technological development to suit government needs while allowing the US to consolidate the commercial industry under US regulatory rules. The leverage that large US government contracts would provide in this much smaller industry would encourage commercial providers to develop remote sensing capabilities and systems, such as high-resolution hyperspectral imaging systems, that might not yet have enough commercial demand to justify their development otherwise. Also, if the methods of control applied under licensing agreements were used sparingly and only in times of conflict, the US might find itself in the position of actually being able to exercise some control over who has access to critical imagery. Exercising this control is something that the US has almost no ability to do today given the successful proliferation of commercial providers. By incentivizing the shifting of the bulk of remote sensing development and ownership back toward the US, national security would greatly benefit from the increase in material space power and the increased oversight of the commercial market.

Notes

1 NASA GSFC Landsat/LDCM EPO Team, "Landsat Image Gallery," accessed December 6, 2016, http://landsat.visibleearth.nasa.gov/view.php?id=40535.
2 Robert P. Merges and Glenn H. Reynolds, "News Media Satellites and the First Amendment: A Case Study in the Treatment of New Technologies," *Berkeley Technology Law Journal* 3, no. 1 (January 1988): 1.
3 "SPOT Satellite Pour Observation Terre," *GIS Geography*, July 30, 2016, http://gisgeography.com/spot-satellite-pour-observation-terre/.
4 "Union of Concerned Scientists: Satellite Database" (Union of Concerned Scientists, May 1, 2018), www.ucsusa.org/nuclear-weapons/space-weapons/satellite-database#.W7eDtPZFxEZ.
5 The size of the US budget devoted to imagery and analysis is revealed in Wilson Andrews and Todd Lindeman, "The Black Budget," *Washington Post*, August 29, 2013, www.washingtonpost.com/wp-srv/special/national/black-budget/?noredirect=on.
6 The Open Skies Treaty of which the US is a member state allows for a limited number of observation overflights by member countries. While Russia is a signatory, China is not. This limited number of overflights is not a substitute for the daily global coverage provided by existing remote sensing platforms.
7 SATCOM interference is a common commercial issue that is usually the unintentional result of careless or improperly calibrated transmissions. The same cannot be said for remote sensing platforms.
8 "Broad Agency Announcement: Blackjack HR001118S0032" (Defense Advanced Research Agency Tactical Technology Office, May 25, 2018), 6–7, www.darpa.mil/attachments/HR001118S0032-Amendment-01.pdf.
9 "Land Remote-Sensing Commercialization Act of 1984," Pub. L. No. 98–365, 15 USC 4201 (1984), sec. 101.
10 Ibid.

11 Dorinda Dalmeyer and Kosta Tsipis, "USAS: Civilian Uses of Near-Earth Space," *Heaven and Earth* 16 (1997): 47.
12 Land Remote-Sensing Commercialization Act of 1984, sec. 402a.
13 Ibid., sec. 402a(8).
14 Ibid., sec. 403a(1).
15 Ibid., sec. 602e.
16 Robert J. Aamoth, J. Lauerent Scharf, and Enrico C. Soriano, "The Use of Remote Sensing Imagery by the News Media," *Heaven and Earth* 16 (1997): 141.
17 Robert A. Weber and Kevin M. O'Connell, "Alternative Futures: United States Commercial Satellite Imagery in 2020: Research Report for Department of Commerce and NOAA" (Washington, DC: Innovative Analytics and Training, November 2011), 14.
18 Ibid., 13.
19 Merges and Reynolds, "News Media Satellites and the First Amendment: A Case Study in the Treatment of New Technologies," 21.
20 Ibid., 27.
21 "National Security Presidential Directive (NSPD) 3, US Commercial Space Guidelines" (White House, February 1991).
22 Ibid.
23 "Land Remote-Sensing Act of 1992," Pub. L. No. 102–588, 15 USC 5623 (1992), sec. a2.
24 "National and Commercial Space Programs," Pub. L. No. 111–314, 51–3328 (2010).
25 Walter Scott, "U.S. Satellite Imaging Regulations Must Be Modernized, Op-Ed by Digital Globe Founder," *Space News*, August 29, 2016, http://spacenews.com/op-ed-u-s-satellite-imaging-regulations-must-be-modernized/.
26 "Commercial Remote Sensing: A Historical Chronology" (Digital Globe, April 9, 2010), 1.
27 Ibid.
28 Atsuyo Ito, *Legal Aspects of Satellite Remote Sensing* (Boston, MA: Marinus Nijhoff Publishers, 2011), 75.
29 Aamoth, Scharf, and Soriano, "The Use of Remote Sensing Imagery by the News Media," 148.
30 Ito, *Legal Aspects of Satellite Remote Sensing*, 76.
31 "Landsat Science: Landsat 5," NASA.gov, accessed February 8, 2017, http://landsat.gsfc.nasa.gov/landsat-5/.
32 Samuel L. Berger to the President of the United States, "Subject: US Policy on Foreign Access to Remote Sensing Space Capabilities," *Memorandum*, March 3, 1994.
33 Ibid., 2.
34 Aamoth, Scharf, and Soriano, "The Use of Remote Sensing Imagery by the News Media," 141.
35 "Presidential Decision Directive 23: US Policy on Foreign Access to Remote Sensing Space Capabilities," March 9, 1994.
36 Shawn L. Twing, "U.S. Bans High-Resolution Imagery of Israel," *Washington Report on Middle East Affairs*, September 1998, www.wrmea.org/1998-september/u.s.-bans-high-resolution-imagery-of-israel.html.
37 "National Defense Authorization Act for Fiscal Year 1997," Pub. L. No. 104–201, 15 USC 5621 (1996), sec. 1604.
38 Twing, "U.S. Bans High-Resolution Imagery of Israel."
39 "Kyl–Bingaman Amendment—Licensing of Private Land Remote-Sensing Space Systems," 15 CFR 960, Vol. 71, No. 79 § (2006).

40 "Commercial Remote Sensing: A Historical Chronology," 1.
41 "Press Release: IKONOS Imaging Satellite Achieves 15 Years of On-Orbit Operation" (Lockheed Martin Corp., September 24, 2014), www.lockheedmartin.com/us/news/press-releases/2014/september/0924-space-IKONOS.html.
42 "Commercial Remote Sensing: A Historical Chronology," 2.
43 George J. Tenent to Director, National Imagery and Mapping Agency, "Director Central Intelligence Agency: Expanded Use of US Commercial Space Imagery," *Memorandum*, June 7, 2002.
44 Ron Stearns, "ClearView Contract and Greater Imagery Availability Move U.S. Satellite Commercial Imaging Market Forward," *Frost & Sullivan*, February 21, 2003, www.frost.com/sublib/display-market-insight.do?id=GLEN-5JYVUJ.
45 "Press Release: US Commercial Remote Sensing Policy" (White House, April 25, 2003), www.whitehouse.gov/files/documents/ostp/press_release_files/fact_sheet_commercial_remote_sensing_policy_april_25_2003.pdf.
46 "Commercial Remote Sensing: A Historical Chronology," 2.
47 "Press Release: US Commercial Remote Sensing Policy."
48 "Strom Thurmond National Defense Authorization Act for Fiscal Year 1999," Pub. L. No. 105–261, 22 USC 2278 (1998), sec. 1261.
49 "National Defense Authorization Act for Fiscal Year 2013," Pub. L. No. 112–239, H.R. 4310–4314 (2013), sec. 1261.
50 Christian J. Kessler, "Leadership in the Remote Sensing Satellite Industry: Report Prepared for US Department of Commerce and NOAA" (North Raven Consulting, 2009), 8.
51 "Minutes of Advisory Committee on Commercial Remote Sensing (ACCRES)" (NOAA, September 30, 2002), www.nesdis.noaa.gov/CRSRA/accresMinutes.html.
52 "Memorandum of Understanding Among the Departments of Commerce, State, Defense, and Interior, and the Office of the Director of National Intelligence, Concerning the Licensing and Operations of Private Remote Sensing Satellite Systems" (Department of Commerce, April 25, 2017), 1.
53 Samira Patel, "NOAA's Commercial Remote Sensing Regulatory Affairs" (ACCRES Meeting, April 3, 2017), 4.
54 Ibid., 2–6.
55 Ibid., 4.
56 "H.R. 2809, American Space Commerce Free Enterprise Act," Pub. L. No. 2809, Title 51 (2018), sec. 80202 (2).
57 Ibid., sec. 80202 (2.e.1).
58 "Space Policy Directive-2, Streamlining Regulations on Commercial Use of Space" (2018), www.whitehouse.gov/presidential-actions/space-policy-directive-2-streamlining-regulations-commercial-use-space/.
59 "Remote Sensing Regulatory Reform ACCRES Recommendations" (NOAA, July 15, 2019), 2, www.nesdis.noaa.gov/CRSRA/pdf/NOAA_Proposed_Remote_Sensing_Regs-FINAL_ACCRES_Recommendations_7-15-19.pdf.
60 Theresa Hitchens, "DoD, Commerce Wrangle New Commercial Remote Sensing Regs," *Breaking Defense*, December 4, 2019, https://breakingdefense.com/2019/12/exclusive-dod-commerce-negotiate-commercial-remote-sensing-regs/.
61 Jeffrey R. Tarr, "DigitalGlobe CEO: Letter to Investors," Shareowner Letter, 2015.
62 "Commercial Remote Sensing: A Historical Chronology," 1.
63 "Securities and Exchange Commission Form 10-k" (DigitalGlobe, December 31, 2012), 3.

64 "Commercial Remote Sensing: A Historical Chronology," 1.

65 Ibid., 2.

66 "DigitalGlobe 2012 SEC Form 10-k," 3.

67 "Ikonos Satellite," GeoImage, accessed December 12, 2016, www.geoimage.com. au/satellite/ikonos.

68 Stearns, "ClearView Contract and Greater Imagery Availability Move U.S. Satellite Commercial Imaging Market Forward."

69 "Orbimage Securities and Exchange Commission Form 10-Q" (Orbimage, September 30, 2004), 13.

70 Dan Caterinicchia, "NIMA Seeks 'ClearView' of World," *FCW: The Business of Federal Technology*, January 16, 2003, https://fcw.com/articles/2003/01/16/nima-seeks-clearview-of-world.aspx.

71 "Orbimage Securities and Exchange Commission Form 10-Q," 6.

72 Ibid., 7.

73 Joseph C. Anselmo, "A New Image," *Aviation Week & Space Technology*, January 30, 2006, 55.

74 Ibid.

75 Joseph C. Anselmo, "Commercial Space's Sharp New Image," *Aviation Week & Space Technology*, January 31, 2000, 52.

76 "Orbimage: Securities and Exchange Commission Form 10K" (Orbimage, 2004), 6; "Orbimage: Securities and Exchange Commission Form 10K" (Orbimage, 2005), 8.

77 "GeoEye: Securities and Exchange Commission Form 10K" (Dulles, VA: GeoEye, 2009), 6.

78 Peter Selding, "EnhancedView Contract Awards Carefully Structured, NGA Says," *Space News*, September 10, 2010, http://spacenews.com/enhancedview-contract-awards-carefully-structured-nga-says/#sthash.7K7GA9Q4.dpuf.

79 "GeoEye: Securities and Exchange Commission Form 10K," 8.

80 "SPOT Satellite Pour Observation Terre."

81 Selding, "EnhancedView Contract Awards Carefully Structured, NGA Says."

82 Ibid.

83 Ibid.

84 Marc M. Lessor to GeoEye Imagery Collection Systems Inc., "Subject: EnhancedView Other Transaction For Prototype Project (OTFPP) Agreement HM0210-10-9-0001," *Memorandum*, June 22, 2012.

85 Ibid.

86 Ibid.

87 "Federal Acquisition Regulation Part 49-Termination of Contracts," § 49.202 a (n.d.).

88 Marc M. Lessor to GeoEye Imagery Collection Systems Inc., "Subject: Enhanced-View Contract HM0210-10-C-0003," *Memorandum*, June 22, 2012.

89 Marcia S. Smith, "SASC Adds Funds for ORS, STP and Commercial Imagery Purchase," *Space Policy Online*, May 25, 2012, www.spacepolicyonline.com/news/sasc-adds-funds-for-ors-stp-and-commercial-imagery-purchase.

90 Marcia S. Smith, "House Passes Final Version of NDAA, Goes Home for Now— UPDATE," *Space Policy Online*, December 21, 2012, www.spacepolicyonline.com/news/house-passes-final-version-of-ndaa-goes-home-for-now-update.

91 "Press Release: DigitalGlobe and GeoEye Agree to Combine to Create a Global Leader in Earth Imagery and Geospatial Analysis" (DigitalGlobe, July 23, 2012), http://investor.digitalglobe.com/phoenix.zhtml?c=70788&p=irol-newsArticle&ID=1717100.

92 Ibid.

93   Ibid.

94   "DigitalGlobe: Securities and Exchange Commission Form 10K 2015" (Westminster, CO: DigitalGlobe, December 31, 2015), 6.

95   Ibid.

96   Ibid., 17.

97   "DigitalGlobe: Securities and Exchange Commission Form 10K 2014" (Longmont, CO: DigitalGlobe, December 31, 2014), 35.

98   Peter Selding, "DigitalGlobe Revenue up Despite Steep Drop in Russian Business," *Space News*, August 1, 2014, http://spacenews.com/41459digitalglobe-revenue-up-despite-steep-drop-in-russian-business/.

99   Ibid.

100 Jodi Sorensen, "Press Release: Spaceflight Successfully Launches Flock of 12 Planet Dove Satellites on India's PSLV" (Spaceflight, June 21, 2016), http://spaceflight.com/spaceflight-successfully-launches-flock-of-12-planet-dove-satellites-on-indias-pslv/.

101 Caleb Henry, "MDA Closes DigitalGlobe Merger, Rebrands as Maxar Technologies," *Space News*, October 5, 2017, https://spacenews.com/mda-closes-digitalglobe-merger-rebrands-as-maxar-technologies/.

102 "Press Release: Maxar Technologies Shareholders Approve U.S. Domestication" (Maxar Technologies, November 16, 2018), http://investor.maxar.com/investor-news/press-release-details/2018/Maxar-Technologies-Shareholders-Approve-US-Domestication/default.aspx.

103 Anselmo, "Commercial Space's Sharp New Image," 52.

104 Weber and O'Connell, "Alternative Futures: United States Commercial Satellite Imagery in 2020: Research Report for Department of Commerce and NOAA," 21.

105 Ibid.

106 "Radarsat-1," Canadian Space Agency, accessed December 26, 2016, www.asc-csa.gc.ca/eng/satellites/radarsat1/default.asp.

107 Weber and O'Connell, "Alternative Futures: United States Commercial Satellite Imagery in 2020: Research Report for Department of Commerce and NOAA," 33.

108 David S. Germroth, "Commercial SAR Comes to the U.S. (Finally!)," *ApoGeo Spatial*, May 9, 2016, http://apogeospatial.com/commercial-sar-comes-to-the-u-s-finally/.

109 Turner Brinton, "US Loosens Restrictions on Commercial Radar Satellites," *Space News*, October 8, 2009, https://spacenews.com/us-loosens-restrictions-commercial-radar-satellites/.

110 "Deutsches Zentrum Für Luft- Und Raumfahrt (National Aeronautics and Space Research Center)," National Aeronautics and Space Research Center, Federal Republic of Germany, accessed December 26, 2016, www.dlr.de/dlr/en/desktopdefault.aspx/tabid-10377/565_read-436/#/gallery/350.

111 "XpressSAR Inc. Private Remote Sensing License Public Summary" (NOAA, October 28, 2015), www.nesdis.noaa.gov/CRSRA/files/xpresssar.pdf.

112 "XPRESSSAR, INC. Selects IAI's Tecsar Technology for Its High-Resolution X-Band Satellite Constellation," Israel Aerospace Industries, November 21, 2018, www.iai.co.il/xpresssar-inc-selects-iais-tecsar-technology-its-high-resolution-x-band-satellite-constellation-0.

113 "PanGeo-Alliance," PanGeo Alliance, accessed November 15, 2018, www.pangeo-alliance.com/.

114 "UrtheCast," UrtheCast Company Webpage, November 16, 2018, www.urthecast.com/constellation.

115 "National Security Presidential Directive (NSPD) 27, US Commercial Remote Sensing Policy" (White House, April 25, 2003).

116 Turner Brinton, "NGA Solicits Proposals for Commercial Radar Imagery," *Space News*, September 11, 2009, http://spacenews.com/nga-solicits-proposals-commercial-radar-imagery/.

117 Selding, "EnhancedView Contract Awards Carefully Structured, NGA Says."

118 Douglas Loverro, Deputy Assistant Secretary of Defense for Space Policy-interview by author, Phone, January 13, 2017.

119 "Resiliency and Disaggregated Space Architectures, White Paper" (Air Force Space Command, 2013), 3.

120 Ibid.

121 Loverro, Deputy Assistant Secretary of Defense for Space Policy-interview by author.

122 Ibid.

123 "Commercial Remote Sensing Regulatory Affairs," NOAA Official Website, accessed January 28, 2017, www.nesdis.noaa.gov/CRSRA/licenseHome.html.

124 Ibid.

125 Alan Robinson to author, "NOAA Senior Licensing Officer," November 30, 2016.

126 Loverro, Deputy Assistant Secretary of Defense for Space Policy-interview by author.

127 Ibid.

128 "Resolution Restrictions Lifted," DigitalGlobe Website, accessed April 19, 2017, http://blog.digitalglobe.com/news/resolutionrestrictionslifted/.

129 "Licensing of Private Land Remote-Sensing Space Systems," 15 CFR 960, Vol. 71, No. 79 § (2006), sec. 960.11.

130 Land Remote-Sensing Commercialization Act of 1984, sec. 402a.

131 "General Conditions for Private Remote Sensing Space System Licenses" (NOAA, 2016), 1.

132 "Report for Congress: Commercial Remote Sensing by Satellite: Status and Issues" (Washington, DC: Congressional Research Service, January 8, 2002), 1.

133 Rick Heidner, "Shutter Control: An Approach to Regulating Imagery from Privately-Operated RS Satellites" (ACCRES Meeting, May 15, 2014).

134 Duncan Campbell, "US Buys up All Satellite War Images," *The Guardian*, October 17, 2001.

135 Ibid.

136 Denette L. Sleeth, "Commercial Imagery Satellite Threat: How Can U.S. Forces Protect Themselves?" (Master's Thesis, Rhode Island, Naval War College, 2004), 12.

137 "SPOT Medium Resolution Satellite Imagery," Apollo Mapping Website, https://apollomapping.com/imagery/medium-resolution-satellite-imagery/spot.

138 "Remote Sensing Services Market by Platform (Satellites, UAVs, Manned Aircraft, and Ground), End User (Defense and Commercial), Resolution (Spatial, Spectral, Radiometric, and Temporal), and Region—Global Forecast to 2022," Research Report (Markets and Markets, October 2017), www.marketsandmarkets.com/Market-Reports/remote-sensing-services-market-87605355.html.

7 Missile warning

On September 26, 1983, Soviet Lieutenant Colonel Stanislav Petrov was in command of the operations center for the USSR's Oko missile warning satellites when they detected the launch of five inter-continental ballistic missiles (ICBMs) from the US.[1] Tensions between the USSR and the US were at an all-time high. Six months previously, President Reagan had labeled the USSR the "evil empire" and just three weeks earlier a Soviet jet had shot down a civilian airliner killing 269 people including a US congressman.[2] Making a judgment call, Petrov decided that the missile tracks reported by the satellite must be false alarms and reported this to his superiors. Petrov's instincts were correct; the missile tracks were the result of reflections off the arctic snow.[3] Later lauded by many as a man who averted a nuclear war, Petrov remained self-effacing stating that "I was just doing my job."[4] Whether the incident receives more attention than it deserves is arguable, but it does highlight the strategic importance of missile warning satellites and the people who operate them.

The US currently operates a constellation of new and old missile warning satellites operating in a variety of orbits. The oldest systems on orbit are the Defense Support Program (DSP) satellites, with the newest being the Space Based Infrared System (SBIRS). The US launched the first of 23 DSP satellites in 1970, and the last satellite in 2007, all of them are well beyond their original five-year design life.[5] Currently, there are four remaining active DSP satellites which operate as a backup to the far newer and more capable SBIRS constellation of satellites that is still in the process of completion.[6] The SBIRS constellation currently consists of four hosted payloads in highly elliptical orbits (HEO) and four dedicated satellites in GEO with two more planned for GEO.[7] This constellation is supported by two Space Tracking and Surveillance System (STSS) satellites operated by the Missile Defense Agency in inclined LEO orbits.[8] The Air Force is already in the process of replacing the SBIRS constellation with a next-generation Overhead Persistent Infrared (OPIR) constellation specifically designed to be more survivable against threats from China and Russia.[9]

The total number of satellites in the missile warning architecture is small but they are vital to the nuclear missile warning mission, and individually they are among the most expensive satellites ever constructed. The SBIRS GEO constellation alone has cost $19.2 billion since inception with the purchase price of the last two satellites approaching $1.75 billion each.[10] The aggregate total cost estimate for each GEO satellite varies between $1.92 billion and $3.2 billion. Despite their small numbers relative to SATCOM or remote sensing platforms, these satellites still contribute to the security dilemma. It is not through the fear or uncertainty they generate in adversaries, but rather it is the fear of loss by the host country that drives their contribution.

The fear of loss and the development of new missile technologies, such as hypersonic missiles by Russia and China, are driving proposals for a vast expansion in the number of missile warning satellites, actions that may further drive the security dilemma in a downward spiral. Michael Griffith, the Undersecretary of Defense for Research and Engineering, has called for the creation of a new constellation of missile warning satellites in LEO leveraging the latest developments in commercial small satellite technology.[11] He argues that this mission cannot be performed by ground-based radars, but must instead come from a "proliferated space sensor layer."[12] Such a development would significantly expand the number of military satellites in orbit. To suspicious adversaries this creates opportunities to place disguised weapons and other dual-use capabilities in orbit, further contributing to the security dilemma by generating fear and uncertainty in adversaries that the current much smaller constellation does not.

Missile warning and the security dilemma

Missile warning satellites are far fewer in number than either remote sensing or SATCOM satellites. Individually the current generation of these satellites is among the most expensive ever produced, but not excessively more than other government satellite programs.[13] Despite this, they still contribute as much or more to the security dilemma in space as either SATCOM or remote sensing. Neither uncertainty of purpose nor numbers provide a missile warning contribution to the security dilemma. Their contribution is almost entirely driven by fear of loss, though efforts to address this fear of loss through the development of constellations of satellites supporting missile defense will shift their contribution toward uncertainty based on fear.

Missile warning satellites provide a capability that is critical to national defense and one that is achievable only from space. These satellites allow for global missile detection from the moment of launch, maximizing warning time. For theater ballistic missiles and ICBMs every additional second of warning is valuable, and at best radar can only partially compensate for their loss. Radar systems have limitations on range and viewing area that do not impact satellites. Attempts to overcome these shortcomings by leveraging alliances to place missile detection radars near the borders of adversary nations can ameliorate these shortcomings, but these systems make their own contribution to the security dilemma and still suffer from range and scope limitations if the adversary is as large as China or Russia. It is the irreplaceability of this capability that drives the fear of loss which is missile warning's contribution to the security dilemma even as it enjoys a relatively high threshold for attack.

Missile warning has the highest threshold for non-destructive or destructive attack of any satellite system. Its association with the nuclear deterrent mission is the source of this deterrent effect though its value is subjective. The value of missile warning's deterrence through association decreases as the mission and capabilities of missile warning satellites expand beyond simply detecting nuclear threats to the US toward characterizing events on the battlefield during limited conflicts.[14] An adversary must now weigh the possibility that the US will see attacks on these satellites as a prelude to nuclear escalation against the value of the non-missile warning information the US gathers. If proposals for a significant expansion in the number and mission set of missile warning platforms become a reality, then the deterrence value granted to missile warning satellites through their association with the nuclear deterrent will weaken further and add uncertainty of purpose to missile warning's security dilemma contribution.

Recognition of the decreasing value of the deterrent by association that the current generation of missile warning satellites has is driving the Air Force to pursue new capabilities.[15] The recent Air Force decision to shift funds away from building more SBIRS satellites toward developing a more survivable generation of satellites despite decades of expensive development is indicative of a rising fear of loss. Since the offense–defense balance favors the offense at the tactical and operational level for individual satellites and low-density capabilities, the US military's efforts to develop adequate defenses for a small number of replacement satellites will most likely be expensive and unsuccessful.

The current contribution of missile warning to the security dilemma is driven by fear of loss. However, calls by national leaders to integrate missile warning satellites with space-based missile defense will dramatically alter this dynamic. President Trump has called for the development of a space-based missile defense layer which would integrate missile warning and missile defense in space.[16] He has also proposed that every allied nation that benefits from it should help pay for it. This effort would shift the dynamic of the security dilemma associated with missile warning away from fear of loss as well as uncertainty. These satellites would instead become an existential threat to the nuclear deterrent capabilities of adversary nations. A space-based hybrid missile shield and warning system would cross a new threshold of space weaponization and accelerate the arms race in space. Exploring methods of defusing missile warning's contribution to the security dilemma and shifting away from the seemingly inevitable pivot toward space weaponization is a difficult task though one that may be possible.

Missile warning development and acquisition

The path to developing and launching the first generation of missile warning systems was not easy. While the potential benefits of a space-based early warning architecture were recognized even before the first successful satellite launch, the challenges and uncertainties meant that men would walk on the Moon before the first space-based missile detection constellation became active. Today, the early warning satellites on orbit are the most strategically vital portion of the US—or any nuclear-armed nation's—space architecture. Ensuring that these satellites remain functional, accurate, and protected is an area of intense focus for military space personnel. Studying the expensive and troubled path that led from inception to reality for space-based missile warning provides valuable insight into future options for this capability.

The development of ICBMs presented a problem for nations seeking some warning of an impending attack. Prior to the development of ICBMs, nuclear weapons delivery was limited to long-range bombers that required hours of flight time and could be spotted by existing air-defense radars. The advent of the ICBM meant that a nation could launch a missile from its own soil undetected, reach sub-orbital altitudes, evade standard air-defense radars, and arrive on target within 30 minutes of launch.[17] With flight times so short, every minute of additional warning provided additional time for national leaders to order the launch of retaliatory capabilities and for people to take shelter from the impending attack. Faster and more accurate warning subsequently contributed to nuclear stability by further deterring an opposing nation from believing it could conduct a successful first strike.[18] Additionally, the more and varied types of warning systems that a nation could employ increased the confidence it could have in its ability to be sure an attack was occurring.[19]

A space-based infrared system provided multiple advantages over a ground-based radar system. Because it could detect the heat signature of a launch it provided the maximum possible warning time, five to eight more minutes than even the most powerful radars.[20] A space-based system also could provide much broader coverage area than Ballistic Missile Early Warning System (BMEWS) radar sites that suffered from gaps in coverage areas and other limitations due to their fixed orientation.[21] In addition, Space-based systems could detect the exact launch location of ICBMs, providing definitive attribution of the source of the attack, something that no radar could provide.[22]

Despite the advantages of a space-based system, the development of a constellation of missile warning satellites was not a forgone conclusion. Desperate for the quick deployment of some type of early warning system, President Eisenhower's Science Advisory Committee on Early Warning decided not to support a joint Air Force–Lockheed proposal for the development of a satellite-based infrared detection system in 1959.[23] The advisory committee opted instead to speed up the construction of the planned BMEWS radars and deploy an airborne infrared detection capability on 50–100 U2 aircraft that would be stationed in the Arctic.[24] The committee was not blind to the advantages of a space-based sensor, but felt at the time that there was "insufficient evidence concerning the effective implementation of such a system to justify its construction."[25] At that time the Air Force–Lockheed proposal, the Missile Defense Alarm System (MIDAS), was estimated to require 20 satellites at the enormous cost of $200 to $600 million per year depending on the lifetime of the satellite, this represented over 1 percent of the entire national defense budget.[26] The technology also faced significant technical uncertainties despite several years of Air Force and Lockheed testing. As a result of these uncertainties, the committee recommended that research and technology development continue pending a future decision on implementation.

The rejected MIDAS program was born immediately following the awarding of the first contract for a photographic reconnaissance satellite in 1956.[27] A young Lockheed engineer, Joseph Knopow, proposed using an infrared radiometer mounted on a satellite to detect missile launches.[28] Knopow's idea was not entirely novel; since 1955 the US had been investigating the concept of placing sensors on high-altitude aircraft flying near an opponent's border to detect the unique signature of missile plumes during launch.[29] Knopow's idea simply expanded this concept to satellites, and by 1957 his proposal had formally become sub-system G of the DOD's broader space surveillance program, WS-117L.[30] Following the launch of Sputnik late in that year, sub-system G was transferred to the newly established Advanced Research Projects Agency (ARPA), given the name MIDAS, and made an independent program.[31]

The development program for MIDAS consisted of a series of ten test platforms launched over a period from November 1959 to May 1961.[32] These tests were designed to test the performance of an infrared sensor in orbit, the impact of the space environment, and the economic feasibility of the program. Typical of launches early in the space age, the first MIDAS test satellite failed to reach orbit and the second failed to return any data.[33] The Air Force, now back in control of the program, was under pressure to show some success from the test program and in 1961 it successfully orbited MIDAS III and MIDAS IV. MIDAS III lasted for only a few orbits, and MIDAS IV lasted less than a week, though both did manage to return some useful data.[34] Despite this partial success, a review of the program in late 1961 concluded that MIDAS presented one of the "most difficult system reliability problems ever faced by the United States," and that the "organization and management of the program has not been appropriate to the magnitude of the problem."[35]

These troubles with MIDAS put the program at risk of cancelation, and in August 1962 Secretary McNamara put the program on a limited budget with a research-only focus.[36] Most programs would have quietly faded away at this point, but MIDAS had a powerful defender in General Bernard Schriever, the head of Air Force Systems Command (AFSC).[37] Schriever lobbied hard for the continuation of the MIDAS program based on its technical and cost advantages over ground-based radar. With a significantly reduced budget and a looming threat of cancelation, the program was in desperate need of a successful proof of concept, which it finally achieved on its seventh test in May 1963.[38] The test MIDAS satellite not only easily detected the launch of land-based Minuteman and Titan II missiles in real time, it was also able to detect the launch of a sub-launched Polaris missile, albeit only after post-processing of the data.[39] The performance of the satellite was far better "than was generally believed obtainable," and the evidence that detection of submarine-launched ballistic missiles (SLBM) was possible from satellites breathed new life into the program and ensured further research would continue.[40]

With the basic technology no longer in doubt, research on the program shifted to improving detection and proving economic feasibility. Renamed project 461 and then project 949 in November 1966 by the Air Force, the new goal of the program was to demonstrate a technical base "from which cost effective multi-purpose satellite systems may be evolved during the 1970 time period."[41] In order to achieve this goal, the program needed to demonstrate increased satellite longevity and find a way to reduce the cost of an operational system. The ideal orbit for launch detection would be a highly elliptical Molniya orbit with long dwell times over the northern hemisphere and in sight of ground stations located in the US during the observation portions of its orbit. The issue with this orbit was that it required at least two satellites to maintain constant observation of the USSR and China since the satellites were not stationary relative to the Earth. At an estimated cost of at least $300 million each ($2.3 billion in 2018), the program decided instead to use a geostationary orbit that could cover the same area with a single satellite, though this came at the cost of reduced detection near the poles and the need for a ground station based on foreign soil.[42]

Testing continued throughout the latter half of the 1960s with a focus on deploying an operational constellation of missile warning satellites in 1970. By 1969, the program name had shifted again to project 647 as the result of security breaches, and the program had adopted the unclassified and non-descript cover name of the Defense Support Program (DSP).[43] The first satellite of the operational DSP constellation launched successfully on November 6, 1970, followed by three more first-generation satellites over the next two years.[44] Each of these satellites was 22 ft long, weighed 2,400 lbs, and was spin-stabilized such that it had a frame rate of ten seconds.[45] Despite the failure of DSP-1 to achieve the correct orbit, the three other first-generation DSP constellation satellites had detected 1,014 missile launches by June 1973, beginning the era of overhead persistent infrared (OPIR) monitoring.

The continued classification of the mission and the sensitivity associated with it remained a challenge for the DSP program. First, data acquisition from the satellites over Russia and China required a ground station in Australia. Posing significant political challenges for the Australian government.[46] Second, the USSR's lack of awareness regarding the system's capabilities compromised the deterrence benefits of having a secure and reliable early warning network. The US Air Force attempted to solve this problem by leaking enough information into the public domain about DSP system capabilities to create a deterrent effect, though it could never be sure if it had released enough.[47] Finally, the classification of the program created confusion between NRO missions and DSP launches.[48] The NRO actually attempted to have the DSP program declassified so that its launches could serve as cover for NRO satellites.[49]

As the DSP program matured throughout the 1970s and became an integral part of the national missile warning architecture, the military slowly became aware that DSP could be used for more than just warning of ICBM launches. Even the first-generation DSP satellites were equipped with a sensor sensitive enough to "detect a lighted match at 100 to 200 miles."[50] While operating at an altitude of 22,000 miles meant that heat sources had to be many times hotter to enable detection, the DSP satellite was still able to detect far more than ICBM launches. System operators correlated DSP detections to much smaller surface-to-air missiles as well as mid-air collisions, gas fires, and military engagements.[51] The information provided was of significant enough interest that daily summaries of detections were reported throughout the 1973 Yom Kippur War between Israel and its Arab neighbors.[52] The IR detection capability of DSP was broadly capable of providing early warning of any significant event occurring globally that produced a strong IR signature unless it was obscured by cloud cover.[53]

The evolution of DSP and the understanding of the capabilities continued throughout the 1970s and 1980s. By the launch of DSP 12 in 1984, the system sensitivity had increased with the primary short-wavelength infrared (SWIR) sensor increasing from 2,000 detectors to 6,000 and the addition of a second mid-wavelength infrared (MWIR) sensor.[54] Like other US satellite programs, DSP suffered from the dearth of US launch capability following the Space Shuttle Challenger disaster, and the Air Force struggled to keep the constellation fully capable during this period. The 1980s did allow for the first test of DSP's ability to detect smaller theater ballistic missiles (TBMs) in an operational environment. The Iran–Iraq war saw the employment of Soviet-made missiles in significant numbers and the DSP constellation was able to detect more than 191 TBM launches in real time throughout the war.[55] This experience served as excellent preparation for the Persian Gulf War.

Despite the observation of the launch of TBMs throughout the Iran–Iraq war and globally in testing scenarios, the Air Force had not developed a formalized system of distributing warning of TBM launches.[56] When Iraq invaded Kuwait on August 2, Air Force Space Command quickly recognized the need for a TBM warning system. Within a week of the invasion, efforts were underway to provide DSP warning to deploying US combat troops.[57] This warning was essential because the time of flight for TBMs was extremely short and every additional second of warning provided time for troops to don chemical protective equipment and seek shelter. It also increased the effectiveness of the Patriot missile defense system which quickly became dependent on the tipping and cueing provided by DSP.[58] While the basic Soviet SCUD missile was on the very edge of DSP's detection capability, Iraqi modifications to increase its range made DSP detection much easier.[59] DSP not only provided launch warning, it also was capable of determining the launch location and the impact point with enough accuracy that only those soldiers in the immediate impact area had to seek shelter.[60] These were invaluable contributions as DSP allowed the Air Force to locate and destroy mobile launchers and it reduced the exhaustion and complacency often created through frequent regional alarms that are irrelevant to the majority of soldiers in the combat zone.

Figure 7.1 DSP 16 deployed from STS-44 Shuttle Bay.

Source: "STS-44," NASA Science, November 1991, https://science.ksc.nasa.gov/mirrors/images/html/STS44.htm.

Note
DISCLAIMER: These photographs are available for preview and download in electronic digital color form ONLY. They are a cropped or some other electronically processed version of an original NASA negative and cannot be ordered from NASA in photograph form. No copyright protection is asserted for these photographs. If a recognizable person appears in this photograph, use for commercial purposes may infringe a right of privacy or publicity. It may not be used to state or imply the endorsement by NASA of a commercial product.

The theater missile warning mission required a rapid evolution in how missile warning was employed. When conducting the strategic mission of detecting inbound ICBMs accuracy is paramount, and false alarms are unacceptable.[61] For the theater missile warning mission, the time of flight is short, and the danger created by a false alarm is much less than the potential that operators may miss an inbound missile.[62] This required a complete shift in the mentality of the operators and systems designed to conduct the warning mission. US Space Command established a new voice warning network and practiced the shift in operating procedures during the several months that elapsed between the invasion of Kuwait in August and the launch of the first SCUD missile in January 1991.[63]

Even with all of this preparation, DSP was operating beyond its design parameters, and failed to detect the single deadliest attack on US troops during the war. On February 26, 1991, fragments of a SCUD missile struck a US barracks killing 27 and wounding 98.[64] DSP failed to detect this attack due to a combination of factors. The launch site was obscured by clouds, blocking the telltale heat signature of the launch and preventing DSP 13 from gaining enough data to observe the launch directly.[65] Another satellite, DSP 15, did detect the launch but was in off-line backup mode, while a third DSP satellite did not gather enough data to identify the launch positively.[66] Desert Storm and the TBM mission brought new attention to DSP; it was the first time DSP had ever been used in a real-world operational scenario since its inception. The conflict demonstrated that despite its success in detecting events beyond its design threshold, DSP was a less than ideal tool and a new system would be needed if tactical warning was to become a future mission.

Efforts to replace DSP began as early as 1983 and took on a new urgency following President Reagan's announcement of the SDI initiative. At the time of President Reagan's SDI announcement, the Air Force was pursuing a DSP follow-on named the Advanced Warning System (AWS). The AWS program was suffering from the common failings of military space programs, "immature technology and high costs," when the DOD transferred control over the program to the newly created Strategic Defense Initiative Organization.[67] The SDI Organization sought to combine the requirements of missile warning and strategic defense into one platform known as the Boost Surveillance and Tracking System (BSTS).[68] Technical requirements for BSTS included support for a 2.5-second scan rate, four times faster than DSP, which was believed necessary to support the missile defense mission and as many as 17 additional classified auxiliary systems. Ambitions for BSTS quickly outgrew technical capability, and the program was scaled back in 1989.[69] Plans for BSTS also moved away from the missile defense mission to the missile warning mission in order to avoid violating the 1972 Anti-Ballistic Missile (ABM) Treaty. Modifications included the elimination of an antenna designed to communicate directly with SDI's planned orbital defensive platforms and the removal of SDI computer hardware and software.[70] In 1990, after spending more than $1 billion in technology development, the program was no longer deemed essential to SDI, and Congress directed the transfer of BSTS back to the Air Force.

After reverting to its pre-SDI name of AWS, the program faced threats of cancelation almost immediately. The DOD Office of the Comptroller proposed canceling AWS due to "high costs, technical and schedule risks, and the availability of an alternative system—an enhanced DSP."[71] Representatives from within OSD and the Air Force strongly objected to plans to cancel AWS, instead proposing a scaled down version of AWS called the Follow-on Early

Warning System (FEWS). Disagreement immediately arose within Congress and the DOD over the necessity of a new system or if an upgraded DSP could meet the national requirement for missile warning.[72] Various studies conducted by the Air Force and other agencies calculated that the savings from pursuing an enhanced DSP over developing a new system would approach $3 billion; yet the Air Force continued to support FEWS.[73]

The experience of the Gulf War further complicated the disagreement over the future of the missile warning mission. Success in detecting and providing warning of TBMs made the Army and Navy eager to further exploit DSP for use in warning of TBMs and other battlefield events. The Air Force initially vigorously resisted this effort, with senior Air Force leaders arguing that nothing could distract from the primary DSP mission of strategic missile warning.[74] However, the Army and Navy were successful in breaking the Air Force monopoly on the use of DSP and developed plans for a deployable ground station for processing and distributing DSP data to tactical users. The Army and Navy's planned system was a deployable ground station named the Joint Tactical Ground Station (JTAGS) that used direct line of sight of up to three DSP satellites to generate in-theater missile warning information.[75] In a pivot away from its previous position, the Air Force began using the TBM detection mission of JTAGS to justify the development of the advanced capabilities of FEWS.[76]

The struggle over the form of the replacement system for DSP and the future of FEWS grew extremely contentious as the US defense budget contracted and the Soviet Union collapsed. The priority for space-based infrared shifted to global awareness and theater missile defense as the need for strategic attack warning faded.[77] The success of DSP during the Gulf War in the newly prioritized theater missile warning mission became the driver for a desired higher revisit rate and more sensitive sensor capabilities to increase the utility of FEWS. These requirements were understandable, but the Air Force also added many more technically challenging requirements including one for full onboard processing and another for laser crosslinks in the FEWS constellation.[78] Despite FEWS being a supposedly less expensive version of the canceled AWS program, Congress remained concerned about costs and directed the Air Force to study alternatives. The resulting study, led by the Aerospace Corporation with Air Force personnel in support, concluded that FEWS had a high potential for cost overruns, delays, and cancelation.[79] The report, "Preserving the Air Force's Options," recommended instead that the Air Force pursue an upgraded DSP system, DSP-II. The report's authors found that an updated DSP system would meet the nation's needs for a space-based infrared sensor while costing much less than FEWS and presenting far less technical risk.[80]

The fact that a congressionally mandated internal Air Force study so completely disagreed with the Air Force's chosen course of action created significant problems. Air Force leadership immediately attacked the study, General Charles Horner, the commander of US Space Command, penned a personal letter to the CEO of Aerospace Corporation telling him that "my fellas are very hot about this—they see it as a short-sighted effort of the DSP SPO to sell FEWS out."[81] Pete Aldridge, CEO of Aerospace Corp, responded to General Horner with a letter that apologized profusely, promised to withdraw the report, and discipline his staff.[82] The Air Force also initiated an investigation into the report led by the program director for space-based early warning, concluding that the report was "improper in its tone and flawed in its content and should be withdrawn" and that the "conclusions of the study are counter to the AF stated position."[83] For a brief period, these damage control efforts seemed to quiet the controversy over FEWS.

Due to the issues surrounding FEWS, Under Secretary of Defense John Deutch directed the MITRE Corporation to lead another study into FEWS and its requirements. The study concluded that stereo DSP provided "adequate near-term capability" for detecting missiles with a range greater than 300 km and that radar was a better answer for missiles with a shorter range than this.[84] Further, the assessment team determined that space-based infrared needs could be met with a simpler system than FEWS. They argued that FEWS was trying to do too much and that separating the strategic global missile warning mission from other missions could lead to "the potential for doing both [missions] better and at lower cost."[85] The fact that the findings of the MITRE led study supported those of the Aerospace Corporation group, whose authors the Air Force had refused to let the MITRE study group interview, created additional problems for the FEWS program.[86] Another issue was that the new program director for space-based early warning, Colonel Edward Dietz, supported the position that Air Force Space Command and the SMC were overstating the benefits of FEWS to General Horner and understating the benefits of a follow-on DSP option.[87] The MITRE study, along with a cascade of other reports and analyses, effectively killed FEWS but the controversy was still growing.

The FEWS controversy entered the public sphere when the House Committee on Government Operations decided to hold public hearings on FEWS. In February 1994, the House Committee heard from representatives of the Air Force as well as the authors of the suppressed Aerospace Corporation report. The report's authors, including Aerospace contractors and active Air Force officers, accused senior Air Force leaders of "unethical and perhaps illegal" conduct in their efforts to save the FEWS program.[88] According to Guido Aru, the primary author of the Aerospace report, Air Force Space Command and SMC intentionally presented Congress with misleading data on FEWS alternatives and systematically suppressed and punished dissenting analysis from a variety of sources. Aru also argued that Major General Donald Hard, the Air Force director of space and strategic defense initiatives, provided false testimony to Congress in support of FEWS and implied that he had mixed motives as he was hired immediately after retirement by Aerospace Corporation with the responsibility to lobby US Space Command on its behalf.[89] Aru's testimony, supported and amplified by Colonel Dietz and another Air Force Colonel, painted a damning picture of the Air Force and its wasteful acquisition processes. The controversy gained enough attention that CBS's *60 minutes* ran a special on the FEWS controversy featuring interviews with Aru, Dietz, and General Horner.[90] The death of the FEWS program left the Air Force with a damaged reputation and ensured that the media and Congress would carefully follow any future successor program.

Even though the Air Force canceled FEWS, the need for an eventual successor to DSP with more capability for detecting theater ballistic missiles had not disappeared. The next manifestation of AWS/BSTS/FEWS called the Alert Locate and Report Missiles (ALARM) program emerged immediately after the Air Force canceled FEWS in November 1993.[91] ALARM received the same sales pitch from the Air Force that FEWS had—a faster, better, and cheaper alternative to its predecessor.[92] Congress had significant doubts about ALARM, and within a year cost estimates had already spiraled upward, forcing the Air Force to cancel its latest effort at developing a space-based infrared system.[93]

In August 1995, SMC awarded two contracts for development work on the next missile warning system, the space-based infrared system (SBIRS).[94] The SBIRS program would have two major elements, SBIRS High which would be a direct successor to DSP, and SBIRS Low which would be a LEO system evolved from the SDI's Brilliant Eyes program.[95] SBIRS High would consist of two payloads hosted onboard another government satellite in a Molniya like HEO orbit and four dedicated satellites in GEO.[96] SBIRS Low would consist of 20–30 satellites in LEO that could provide higher resolution of battlefield events and allow for mid-course tracking of missiles in flight after the hot-burning boost phase was complete.[97] SBIRS quickly became US Space Command's number one priority as it offered a "quantum leap over DSP" with sensors capable of going beyond simply detecting missiles to detecting events as small as the signature of artillery fire on the battlefield, according to an Air Force Space Command pitch for the SBIRS platform.[98]

DOD planned to deploy the SBIRS constellation in two major phases. SBIRS High would be deployed first due to the need to replace DSP and its lower program risk. The goal for deployment of SBIRS High was fiscal year 2002.[99] SBIRS Low had higher technical risk, and so DOD planned to deploy it no earlier than 2006 following a decision to be made in 2000 after the contractor validated key technologies using two planned demonstration satellites.[100] Dissatisfied with this timeline, Congress directed that SBIRS Low would start deployment in fiscal year 2002 and obligated an additional $135 million in funds in FY1996 to make this possible.[101] The DOD protested congressional efforts to accelerate the timeline for SBIRS Low development before the flight demonstration proved the validity of the technology. In an exceedingly rare event, a General Accounting Office (GAO) study into the issue of accelerating SBIRS Low deployment agreed with the DOD that accelerating the program would create "higher risks and substantial additional funding requirements."[102] By 2001, SBIRS Low had returned to its roots in SDI and was transferred away from the Air Force to the newly established Missile Defense Agency and renamed the Space Tracking and Surveillance System (STSS).

While SBIRS Low/STSS had problems with congressional impatience, SBIRS High was not without issues of its own. Originally SBIRS High was to be fielded by 2002 at a cost of $1.8 billion, but by the fall of 2001 the DOD had identified that the SBIRS High program would exceed its planned budget by at least $2 billion.[103] This funding breach triggered a Nunn-McCurdy review, a congressionally mandated review that occurs when a DOD program exceeds its original baseline projected budget by more than 15 percent.[104] In the case of SBIRS High, it had exceeded its planned baseline by more than 50 percent, triggering a critical review that required the Secretary of Defense to certify to Congress that: "the program was essential to national security, there was no viable cost-effective alternative to the program, the new cost estimate was reasonable, and the management structure was sufficient to control additional cost growth."[105] The DOD met these requirements to congressional satisfaction and set the new baseline budget estimate for SBIRS High at $4.4 billion and delayed the launch of the first satellite until 2006.[106]

In certifying to Congress why the program was so far over budget, the DOD identified three primary issues with the SBIRS High program. First, the program was "too immature to enter the system design and development phase."[107] The Air Force had relied on overly optimistic assumptions about software development timelines and the benefits of its latest contractor driven acquisition process. Second, the complexity of the engineering problems was not well understood by those running the program. This led to contract officials underestimating the complexity and impact of the requirements and changes to them on the program. Finally, there was a "breakdown in execution and management" by overwhelmed government contracting officials.[108] Believing that it had solved the problems that led to the Nunn-McCurdy breach by adding tighter management control, improved contractor reporting, and a formal review process, the Air Force proceeded with SBIRS along the new procurement timeline.

The fixes proposed by the Air Force were ineffective, and SBIRS quickly ran into trouble and additional cost overruns. In 2004, and again in 2005, the SBIRS program triggered Nunn-McCurdy reviews.[109] Much of the cost growth was driven by software challenges and issues with the SBIRS HEO sensor which would fly as a hosted payload on a classified platform. In 2006, the estimated cost of the SBIRS program had increased from its 2002 revised estimate of $4.4 billion to $10.4 billion.[110] Each SBIRS GEO satellite was now projected to cost $3.5 billion, more than the original estimated cost of the entire program.[111] The ongoing cost increases and development problems led the DOD to pursue an alternative space-based sensor in parallel with SBIRS known as the Alternative Infrared Satellite System (AIRSS) that would potentially replace one of the planned SBIRS GEO satellites and act as competition for the SBIRS program.[112] AIRSS also was designed as insurance against SBIRS cancelation and would pursue a more conservative approach with less technical risk by replicating existing DSP capabilities. However, the GAO found in 2007 that rather than serve as a "plan B" in case of SBIRS cancelation using a more conservative approach, AIRSS had already evolved into an effort to exceed

SBIRS capabilities and serve as a next-generation replacement for SBIRS.[113] This program was renamed the Infrared Augmentation Satellite (IRAS) in 2008 and canceled in 2009 as the first SBIRS GEO satellite finally looked likely to launch.[114]

By 2006, the SBIRS project finally showed some tangible outcomes. The first SBIRS HEO hosted payload launched in June 2006, followed in 2008 by the second SBIRS HEO payload. These successes gave the program some breathing room, and despite additional cost overruns, SBIRS GEO 1 was finally launched in 2011. This satellite has been followed by three more SBIRS GEO satellites and two more HEO platforms as of 2018 with a fifth and sixth satellite planned. The current cost of the SBIRS program is estimated at $19.2 billion, many times its original cost estimate.[115] Rather than continue the SBIRS program, however, the Air Force is currently seeking a successor system for it.

Concerns over the survivability of the current SBIRS platforms are driving the Air Force pursuit of a SBIRS replacement. In 2018 the Air Force awarded Lockheed Martin a sole source $2.9 billion contract toward the development of three next-generation Overhead Persistent Infrared satellites to replace SBIRS.[116] The motivation for a new system to replace SBIRS is that the Air Force needs a "survivable missile warning capability by the mid-2020s to counter adversary advantages."[117] The originally stated timeline for the readiness of this system was 2029, but the commander of US Strategic Command, General John Hyten, called this timeline "ridiculous."[118] At a news conference, General Hyten expressed frustration with this 2029 timeline and the entire acquisition process remarking that "that's 12 years from now, Boeing will go through four generations of commercial satellites" in that time.[119] He went on to elaborate that all he wanted was an infrared sensor on orbit that would work with US ground hardware, preferably something built off a commercial bus to commercial timelines.

The existing SBIRS constellation is unsustainable, and General Hyten has remarked that "there's not enough money in the budget to keep buying billion-dollar satellites and put them up and try to defend them."[120] General Hyten's frustration is understandable, though the track record for Air Force satellite acquisition provides little hope that a new system will not suffer from the failures and cost overruns of past acquisition programs. However, the Air Force does have one existing piece of evidence that it can acquire missile warning systems cheaper, faster, and at lower risk. This example is a legacy of the canceled AIRSS program, the Commercially Hosted Infrared Payload (CHIRP).

Exploring the possibilities of commercialization: Commercially Hosted Infrared Payload (CHIRP)

When a satellite acquisition program is canceled, the money spent on that program is not always a complete loss. The new technologies and capabilities developed for that program may find their way into future systems. CHIRP was a beneficiary of legacy technology from one of the many canceled OPIR programs, in this case AIRSS. AIRSS was initially conceived as a cheaper alternative to SBIRS. It evolved into a technology development program for next-generation OPIR before being canceled after SMC determined that a single capability demonstration satellite would cost at least $500 million.[121] Under the research contract for AIRSS, Science Applications International Corporation (SAIC) developed a new OPIR sensor that used four one-quarter earth staring sensors to achieve complete hemispheric coverage.[122] Following the cancelation of AIRSS, Americom Government Services (AGS) saw an opportunity for using the SAIC-developed hardware as a hosted payload onboard a commercial communications spacecraft. This proposal would eventually become CHIRP.

AGS submitted an unsolicited proposal to the Air Force for CHIRP in January 2008.[123] This was a bold proposal as the US government at that point had never succeeded in launching a commercially hosted payload of any type. The AGS proposal involved placing one of the SAIC staring sensors on the nadir (Earth-facing) surface of a soon to launch GEO communications satellite. The proposal was accepted and AGS received a $65 million firm fixed price (FFP) sole source contract to build, integrate, launch, and operate CHIRP for one year.[124] This process sounds simple, however, because it was a sole source contract and an unsolicited proposal using hardware previously developed for the government, it required an extraordinarily complex organizational structure to avoid violating contracting and acquisition regulations.

AGS was the managing company on the CHIRP program, but there were at least three other companies involved in a complex web of contracting relationships. The parent company of AGS, Systems Engineering Solutions (SES), had a contract to develop three communications satellites with Orbital Sciences Corporation. AGS contracted Orbital Sciences to modify the third of its planned communications satellites (SES-3) to accommodate CHIRP. Orbital Sciences, in turn, subcontracted the payload integration to SAIC. However, the core of CHIRP was the SAIC-developed and Air Force Research Lab (AFRL) sponsored OPIR payload designed under the AIRSS program. To avoid contracting issues and protect proprietary data, the CHIRP sensor development team was isolated from the rest of SAIC, including the payload integration team.[125] A further complication was that the launch of the SES satellites was aboard a non-US launch vehicle, a French Ariane rocket, and the ultimate satellite owner would be SES, a non-US company based in Luxembourg.[126] Overcoming the foreign security issues that this contracting relationship generated required AGS to clear significant hurdles related to concerns with International Traffic in Arms Regulations (ITAR).

CHIRP also faced major engineering hurdles as a hosted payload. First, a hosted payload cannot interfere with the primary commercial mission of the satellite.[127] A commercial platform is expected to have years of performance life ahead of it, and AGS was only contracted to operate the CHIRP payload for one year of operations. As a result, the CHIRP payload would be required to make design sacrifices if it interfered with the primary payload. Second, as a hosted payload, CHIRP needed to be ready before the host satellite was completed and the host satellite would not make design compromises to support CHIRP's eventual integration.[128] This meant that the CHIRP integration team needed to develop the payload interface in anticipation of the host satellite's requirements. The payload integration team faced a further complication by not having direct insight into the sensor development process, which meant that they had only the most basic information on the sensor and assumptions on the host satellite requirements, a complex engineering challenge. Finally, since CHIRP was a sensitive military payload, its data could not be passed directly through the host satellite's downlink. Instead, it required a National Security Agency (NSA) approved encryption device and the leasing of a dedicated transponder on the host satellite to transmit its data.[129]

Meeting the engineering challenges of being a hosted payload under these conditions necessitated several compromises. When conducting integration, the Orbital Sciences payload team found that the solar shield in front of the telescope interfered with one of the host satellite's antennas.[130] The solar shield needed to be shortened to prevent interference, and this somewhat compromised the CHIRP sensor by decreasing its sensitivity. Also, the coolers for the infrared sensor on CHIRP were mounted in sub-optimal locations because they initially interfered with the host satellite's cooling sub-system. These changes resulted in the operating temperature of the sensor being warmer than desired when the nadir surface of the spacecraft was exposed to the sun.[131]

AGS was aware of many of the fundamental risks associated with integrating a government payload when it made the unsolicited proposal. Due to the uncertain nature of the sensor development, AGS included a "shared risk" clause in the contract that meant that both the US government and AGS shared the risk if the sensor development was delayed and the launch window was missed.[132] AGS's actions proved prescient as technical issues delayed the sensor development by one year and AGS missed its original window to integrate into the SES-3 satellite which was completed before the SES-2 satellite.[133] Unlike in a typical government to government hosting arrangement this was not catastrophic for CHIRP because SES-2 was still in development and could serve as the host for CHIRP. The delay in integration due to sensor development problems forced SMC to pay an additional $17 million in costs to AGS.[134] The final cost to integrate, launch, and operate CHIRP under the fixed price contract was $82.9 million, excluding the sensor development costs which were absorbed by AFRL under a separate contract.[135]

Ultimately, the CHIRP mission was extended beyond its original one-year timeline to 27 months. It was canceled after it became a victim of congressionally driven DOD budget cuts.[136] During that period, it collected 300 terabytes of OPIR data and detected 70 missile launches as well as 150 other significant infrared events.[137] CHIRP also cost over $400 million less than the Air Force initially estimated for a dedicated satellite to put the SAIC-developed sensor in orbit. Even with a delay of one year, the CHIRP mission took only 38 months from proposal to operational payload on orbit. Under historical Air Force satellite acquisition timelines, the requirements development and approval process would only just be approaching completion.[138]

By any measure, CHIRP was a stunning success. It demonstrated that by partnering with commercial companies, the Air Force could get a capability on orbit for a fraction of what it would cost using traditional acquisition methods. CHIRP, along with another experiment in commercially hosted payloads, proved so successful that the 2013 National Space Transportation Policy directed all US government agencies to "explore the use of hosted payload arrangements ... and other ride-sharing opportunities" in an effort to reduce costs.[139] The DOD took some steps to implement this recommendation, but since CHIRP, the DOD has launched or initiated only four additional commercially hosted payloads.[140] To help improve its hosted payload efforts a dedicated hosted payload office within the Air Force is making headway on easing the process. An example of these efforts is the development of a standard NSA approved data interface unit to carry encrypted government data generated by the hosted payload.[141] Despite the success of CHIRP and the ongoing efforts within the DOD to take advantage of commercial hosting opportunities, the US government still has significant work to do before using commercial platforms to host government payloads is normalized.

Options for the infrared missile warning mission

The history of DSP and SBIRS demonstrates that developing an infrared missile warning satellite is among the most expensive and time-consuming satellite projects that a nation can embark upon. False starts and poorly developed requirements inflated costs and timelines to the degree that if DSP or SBIRS were anything other than vital to national security, they would never have succeeded in being built. The exorbitant cost of their development means that there are relatively few of these satellites on orbit. Only China, Russia, and the US possess a space-based infrared missile warning capability. Even so, these satellites still contribute to the security dilemma. Not through any fear or uncertainty they might generate in adversaries, rather it is the fear of loss by the host country that primarily drives their contribution to the security dilemma.

Efforts to protect these small constellations of sensitive and expensive satellites are futile. The offense–defense balance at the tactical level of the individual satellite strongly favors the attacker. Further, with a small constellation of systems performing a time-sensitive mission, the offense–defense balance also favors the attacker at the operational level. It is not feasible to shift the location of SBIRS GEO satellites in a timely enough manner to fully compensate for the loss of one of these exquisite platforms. Due to their exceptionally high cost, it is also not feasible to build more of the existing platforms to create enough resiliency in the constellation that the loss of more than one satellite would not substantially degrade its missile warning capabilities. These inherent weaknesses are why Michael Griffith, the Undersecretary of Defense for Research and

Engineering, has called for the creation of a new constellation of missile warning satellites in LEO leveraging the latest developments in commercial small satellite technology.[142] While this might help lessen the fear of loss that drives the security dilemma today, the development of a vast constellation of infrared LEO satellites would contribute to the security dilemma by generating the fear and uncertainty in adversaries that the current constellation does not.[143]

The challenge of reducing the contribution of missile warning satellites to the security dilemma is further exacerbated by the fact that, unlike with electro-optical remote sensing and SATCOM, it is unlikely that the commercial market will ever have the same need for global real-time infrared data that national governments do. While communications and imagery are commodities that can be acquired on the commercial market, infrared data is not. A very small number of commercial satellites, such as Maxar's WorldView-3, do possess imagery capabilities in the infrared spectrum.[144] However, these platforms are focused on the potential commercial gains present in providing the best possible imagery of a specified area and not on conducting broad hemispheric scans of the globe in real time. It is possible that additional future commercial systems with infrared capability may someday exist, but there are no current plans to build such a system. It is also unlikely that a future commercial capability could meet the stringent real-time uninterrupted hemispheric coverage required for the strategic missile warning mission. Without a feasible commercial market available to leverage in order to subsidize costs and defray risks, many of the preferred approaches to commercialization, such as acting as an anchor tenant, applying the CRAF model, or purchasing the capability as a commodity are not possible with the missile warning mission.

What then are the options for lessening the contribution of space-based infrared to the security dilemma if full commercialization is not an option? One alternative would be to pursue a terrestrially based alternative, though it would be prohibitively expensive and much less effective. While radar technology has no doubt improved since the decision to pursue DSP was made, even today the most advanced radars are limited to ranges on the order of 1,500 km.[145] This range issue creates even more risk for a nation if it is unable to place radars beyond its borders. These radars would then provide minimal warning of inbound ballistic missiles and not allow for the independent verification of an attack that radar and space sensors working together can provide; a point that General Schriever leaned upon in his push for the continued development of DSP.[146] Returning to the days of flying hundreds of aircraft at high altitude with launch detection sensors onboard also is not feasible. Geography and physics dictate that a space-based sensor remains the best method of providing cost-effective global warning of attack.

CHIRP provided an example of the potential to utilize hosted payloads as an alternative for dedicated missile warning satellites. Hosted payloads provide a solution for the inability of commercial providers to meet the unique requirements of the missile warning mission. If the US military adopted this approach in a limited manner for the strategic missile warning mission, then it would preserve some of the advantages of signaling benign intent provided by a full commercialization approach. If instead the US aggressively pursued hosted payloads as a method of disaggregation then it would only further fuel the security dilemma, especially if it continued to maintain a dedicated constellation of missile warning platforms. The greatest difficulty with this approach is the initial implementation phase where hosted payloads would exist at the same time as SBIRS creating the perception of proliferating sensor platforms and capabilities. Careful messaging of the limited scope of a hosted payload approach would be required to avoid creating uncertainty over US intentions.

In this situation, where there is no obvious method of commercialization for reducing the number of dedicated military satellites, alternative approaches must be considered. One innovative approach that has potential is to internationalize the strategic missile warning mission.[147] While this does not initially seem like a credible approach, it does have strategic advantages. All nuclear-armed nations desire missile warning, but not all nations can afford to build and maintain a space-based missile warning constellation. This lack of assured warning harms all nations. It is to the benefit of global nuclear stability that all nuclear-armed nations possess the most unambiguous and assured methods of detecting and verifying an attack to prevent possible mistakes such as what almost occurred in the USSR in 1983.

Russia serves as a clear example that the difficulty of maintaining a space-based missile warning constellation is not limited to the US. Russia possesses the largest nuclear arsenal and is the third largest global space power, yet it had no missile warning satellites on orbit from 2014 to 2017 and even today has only the skeleton of an effective space-based warning system.[148] Following a pattern similar to the US development of SBIRS, the Russian replacement missile warning system faced years of delays and technical difficulties. It even involved a court battle between the Russian Defense Ministry and Energia Corporation which was responsible for building the next-generation Russian system.[149] If a nation with the space expertise of Russia cannot afford to maintain an active missile warning constellation, then the many smaller nuclear powers certainly cannot. As nuclear proliferation continues, the potential for an accidental nuclear exchange will increase. Without access to the data from a space-based infrared system, nations will have to rely on terrestrial systems with much narrower fields of view. This limits these nations' situational awareness and any number of incidents occurring during a time of heightened tension may be mistaken for a nuclear strike leading to a nuclear exchange.

The US already practices a limited form of internationalization with its missile warning satellites through the Shared Early Warning System (SEWS). Under the SEWS program, the US shares reporting on adversary ballistic missile launches with nine partner nations.[150] Started following the end of the Cold War, the SEWS program is maintained by the US for the benefit of these nine allied nations who have access to the US missile warning system. The names of all of the nations that participate in the SEWS program are not known, but it does include Japan and most likely other close allies who face a significant missile threat.[151] That the US has not granted more allied nations access to its missile warning data is indicative of just how sensitive the information is. Even if the US could be convinced to make its space-based infrared data freely available as it does its GPS signals, only close allies could trust that this data was unaltered. Even with less critical GPS data freely available, Russia, China and even the EU have still launched their own navigation satellites in an effort to avoid dependence on the US.

An international missile warning system would be a valuable addition to nuclear stability, but how would it be implemented? An international missile warning system would need to be controlled by an internationally recognized body that could be trusted to provide unaltered data. A UN-sponsored and operated organization is one possibility. Another might be a separate international body modeled on the IMF with ground stations in all participating nations. Funding is another issue, even a system modeled on CHIRP would require funding on the order of at least $1.5 billion to develop a global monitoring constellation and the associated ground stations.[152] The current UN budget is only $5.4 billion and wholly inadequate for its current mission even before the additional expense of managing a global missile warning constellation is added to it.[153]

Should funding be somehow secured, an additional challenge would be building the necessary level of trust in the system for it to be effective. The missile warning mission is understandably something that nations would be hesitant to trust to others. If a nation could afford to build its own satellites it would, this would then remove the wealthiest nuclear nations from the pool of potential contributors to an international system. The number of remaining nuclear nations with the financial ability to contribute to such a system is small if China, Russia, and the US, as well as its SEWS partners, are removed from the equation. Non-nuclear nations may desire a capability to warn of small theater ballistic missiles, though with the exception of Israel, few nations are threatened by non-nuclear missiles with enough range that a space-based sensor is truly necessary.[154]

Another approach to reducing the fear of loss that the expensive SBIRS satellites generate is to separate the strategic missile warning mission from tactical infrared detection tasks. In its 1993 report for the DOD on future missile warning options, MITRE obliquely mentioned the idea of separating these missions. From a technical standpoint the MITRE team argued that "separating sensor support to Global Missile Warning and other missions offers the potential for doing both better and at lower cost."[155] While they approached this from a purely technical perspective, the idea of separating the missions and developing a less expensive and less sensitive sensor makes strategic sense as well. In a limited war scenario, a nation is incentivized to avoid attacking a sensor that provides its adversary with assurance that nuclear escalation is not occurring. When the sensors that provide nuclear missile warning also provide an opponent with data that substantially supports conventional limited conflicts as well, then nations face a more difficult choice. A nation is then forced to weigh the benefits of denying that sensor to its opponent versus the risk that an attack on that sensor will be seen as a prelude to nuclear escalation. When viewed through a rational strategic lens, separating the strategic mission from other missions seems like a viable method of reducing the fear of loss without generating fear of uncertainty given the small number of satellites required to do this mission.

A counter-argument to separating the strategic and conventional missions is that that conflating them creates a deterrent effect. While that is true, it also makes any strategic capabilities using this approach a legitimate target in limited war. This complicates the risk assessment for both sides in a conflict. Adversaries must consider whether the US will over-react to an attack on strategic capabilities or shift to existing ground-based warning systems. For the US, it must weigh the real possibility that it could lose the benefits of strategic missile warning during conflict forcing it to rely entirely on BMEWS radar and other capabilities. An adversary may determine that the risk of attacking dual-use strategic satellites is acceptable if it avoids hostile acts against the strategic BMEWS radar systems. It is difficult to make an accurate assessment of how much deterrence conflating missions would create without knowing the risk tolerance of an adversary. What is known, is that pursuing this approach would continue to fuel the fear of loss that is the primary contribution of missile warning satellites to the security dilemma.

The primary issue with a separation approach is that the window for successfully implementing it may have already closed. While the legacy DSP sensor had more than enough sensitivity to detect large hot-burning ICBMs and larger theater ballistic missiles, it struggled to detect anything smaller than that.[156] Until recently DSP would still have been an adequate capability to detect nuclear threats. This changed in 2018 when Russian President Vladimir Putin unveiled nuclear-armed hypersonic weapons in a speech to the Russian Parliament.[157] Putin claimed he developed these fast and stealthy weapons in response to the US withdrawal from the ABM treaty. Detecting this new nuclear threat is a challenge. Hypersonic missiles do not need the hot-burning rockets required of ICBMs, and so a more sensitive sensor than those designed to meet existing threats is required.[158] Any sensor with enough sensitivity to detect the new nuclear hypersonic threat will no doubt gather information on other battlefield events which are useful outside of just the nuclear mission, potentially reviving the current security dilemma.

The US withdrawal from the Intermediate-Range Nuclear Forces (INF) Treaty is further driving the future need for an even more sensitive space-based sensor. The US withdrawal from the INF treaty is supposedly in response to the Russian development of its new missile systems.[159] Since Putin announced that he developed these systems in response to US withdrawal from the ABM treaty, it is clear that a negative arms spiral is already occurring. This spiral will impact the need for more sensitive space-based detection systems and place options for removing their contribution to the security dilemma in space farther out of reach.

Removing the contribution of a specialized military space capability such as missile warning satellites to the security dilemma is a much more difficult task than for other categories of satellites. Their association with the nuclear warning mission makes them a challenging asset to replace. In comparison, the other categories of military satellites: satellite communications, remote sensing, and navigation are comparatively easy to commercialize or transition away from military control. Protecting existing missile warning satellites is a nearly impossible task that contributes to the security dilemma mostly through the fear of loss generated by the sub-optimal arms race. If enough reassurance was generated through other means that the US does not intend to dominate space, that China does not intend to cripple US space architecture, or that Russia does not have malicious intent, then this fear could be eliminated. Only when the fear of loss is gone, or at least significantly reduced, will the contribution that missile warning satellites make to the security dilemma decrease.

Summary

The development of infrared space-based missile warning was not inevitable. The difficulty and expense of developing and placing these satellites on orbit required massive investment and years of perseverance. It was only the timely success of a MIDAS test satellite that kept space-based missile warning alive. Once on orbit, space-based missile warning became one of the most critical space-based capabilities the US possessed. After the Gulf War, the recognition that these infrared sensors could do much more than merely warn of impending missile attacks against the US increased their value and created demand for more sensitive systems.

Finding a replacement for the original DSP missile warning system was not an easy task. The effort to replace DSP began prior to President Reagan's SDI program as AWS before becoming an integral part of it (e.g., BSTS). Following recognition that separate sensors could better perform the missile warning and missile defense missions, the DOD returned BSTS to the Air Force as AWS. Once AWS grew too expensive, the Air Force turned to the simpler FEWS system. When this system also turned out to be too complex and expensive, the Air Force canceled it and proposed ALARM as its replacement. The Air Force then quickly replaced the short-lived ALARM system with the SBIRS program when it also grew too expensive.

The SBIRS program suffered from all of the weaknesses of its predecessor programs but benefited from the growing urgency to replace DSP. The SBIRS High program ballooned in cost from an original estimate of $1.8 billion to a final cost of $19.2 billion for just six GEO satellites and four hosted payloads in HEO.[160] SBIRS Low never recovered from early technical challenges and today has only two functional technology demonstration satellites operated by the Missile Defense Agency.

The CHIRP hosted payload that evolved from another Air Force missile warning program demonstrated that placing an infrared sensor on orbit without spending billions is possible. Using technology derived from the defunct AIRSS program, CHIRP placed a one-quarter Earth viewing infrared sensor on orbit in three years for only $82.9 million.[161] CHIRP had minimal oversight from the Air Force and used commercial processes under a fixed price contract to meet its tight timeline and cost goals. Despite CHIRP's success, the Air Force has initiated only a handful of commercially hosted payload efforts in the last decade, choosing instead to rely on traditional acquisition methods.

Reducing the contribution of missile warning satellites to the security dilemma is a challenging task. Existing satellites are expensive and small in number, yet perform the essential nuclear missile warning mission. Their role in performing this mission has historically protected these satellites from attack, but the expanding role of these platforms to other missions makes them tempting targets. No commercial alternative exits and there is no market for global real-time infrared monitoring that could lead to one. Developing an international system would contribute to global stability, though such a system would face severe political and financial hurdles. An argument exists for separating the strategic warning mission from other tasks by deliberately building a less sensitive and cheaper platform. However, recent events that are indicative of the worsening security dilemma in space may have already closed the window of opportunity to execute this approach.

Missile warning presents the most challenging case for a commercialization approach to reducing the security dilemma and one that is nearly intractable when nations are already engaged in an arms race. The US faces several dilemmas with its existing SBIRS constellation; there is no realistic approach to protecting these individually expensive satellites, yet it cannot afford to lose them. The alternative is to follow Undersecretary Griffin's approach and build a disaggregated missile warning architecture of dozens or hundreds of LEO satellites in a mega-constellation to increase resiliency.[162] This approach is likely to be even more expensive and unaffordable than the existing SBIRS constellation since the Air Force cannot seem to repeat the success of CHIRP. It will also further fuel the arms race in space as it worsens the existing security dilemma.

Notes

1 Simon Shuster, "Stanislav Petrov, the Russian Officer Who Averted a Nuclear War, Feared History Repeating Itself," *Time Magazine*, September 19, 2017, http://time.com/4947879/stanislav-petrov-russia-nuclear-war-obituary/.

2 A.D. Horne, "U.S. Says Soviets Shot Down Airliner," *Washington Post*, September 2, 1983.

3 Shuster, "Stanislav Petrov, the Russian Officer Who Averted a Nuclear War, Feared History Repeating Itself."

4 Ibid.

5 Ian Williams, "Defense Support Program (DSP)," Missile Threat, accessed November 27, 2018, https://missilethreat.csis.org/defsys/dsp/.

6 "Union of Concerned Scientists: Satellite Database" (Union of Concerned Scientists, May 1, 2018), www.ucsusa.org/nuclear-weapons/space-weapons/satellite-database#.W7eDtPZFxEZ.

7 "Space Based Infrared System," Air Force Space Command, accessed November 27, 2018, www.afspc.af.mil/About-Us/Fact-Sheets/Display/Article/1012596/space-based-infrared-system/.

8 "Fact Sheet: Space Tracking and Surveillance System" (Missile Defense Agency, March 27, 2017).

9 Sandra Erwin, "Lockheed Martin Awarded $2.9 Billion Air Force Contract for Three Missile-Warning Satellites," *Space News*, August 14, 2018, https://spacenews.com/lockheed-martin-secures-2-9-billion-air-force-contract-for-three-missile-warning-satellites/.

10 Cristina T. Chaplain, "Space Acquisitions: DOD Continues to Face Challenges of Delayed Delivery of Critical Space Capabilities and Fragmented Leadership," Testimony before US Senate Subcommittee on Strategic Forces (United States Government Accountability Office, May 17, 2017), 6; "Selected Acquisition Report: Space Based Infrared System High (SBIRS High)" (Department of Defense, March 18, 2015), 21.

11 Sandra Erwin, "U.S. Would Need a Mega-Constellation to Counter China's Hypersonic Weapons," *Space News*, August 8, 2018, https://spacenews.com/u-s-would-need-a-mega-constellation-to-counter-chinas-hypersonic-weapons/.

12 Ibid.

13 For instance the AEHF communications satellites cost $2.5 billion each including development costs compared to around $3 billion for SBIRS see Cristina T. Chaplain, Director Acquisition and Sourcing Management, "Space Acquisitions: DOD Continues to Face Challenges of Delayed Delivery of Critical Space Capabilities and Fragmented Leadership," § Testimony before the Subcommittee on Strategic Forces, Committee on Armed Services, US Senate. (2017), www.gao.gov/assets/690/684664.pdf.

14 Lockheed Martin Corp., "SBIRS Fact Sheet" (Lockheed Martin Corp., 2017), www.lockheedmartin.com/content/dam/lockheed-martin/space/photo/sbirs/SBIRS_Fact_Sheet_(Final).pdf.

15 "2019 Missile Defense Review" (Department of Defense, Office of the Secretary of Defense, 2019), vi, 19.

16 Donald J. Trump, "Remarks by President Trump and Vice President Pence Announcing the Missile Defense Review" (Pentagon, Washington, DC, January 17, 2019).

17 "Draft of a Report on MIDAS by DDR&E Ad hoc Group," November 1, 1961, II–1, https://nsarchive2.gwu.edu/NSAEBB/NSAEBB235/04.pdf.

18 Ibid., III–1.

19 Ibid., III–2.

20 Ibid., III–3.

21 Bernard Schriever to Eugene Zuckert, "DOD Program Change (4.4.040) on MIDAS (239A)," *Memorandum*, August 13, 1962, 3, Curtis LeMay Papers, Box B141 (AFSC 1962), Library of Congress, https://nsarchive2.gwu.edu/NSAEBB/NSAEBB235/06.pdf.

22 Ibid., 2.

23 "The President's Science Advisory Committee: Report of the Early Warning Panel," March 13, 1959, 11, White House Office, Office of the Special Assistant for Science and Technology, Records (James R. Killian and George B. Kistiakowsky, 1957–1961) Box 4 AICBM-Early Warning [November 1957–December 1960], Eisenhower Library.

24 Ibid., 5.

25 Ibid., 11.

26 "The President's Science Advisory Committee: Report of the Early Warning Panel," 11.

27 Jeffrey T. Richelson, *America's Space Sentinels: DSP Satellites and National Security* (Lawrence, KS: University Press of Kansas, 1999), 9.

28 Ibid., 8.

29 William W. Kellog and Sidney Passman, "Infrared Techniques Applied to the Detection and Interception of Intercontinental Ballistic Missiles," Research Memorandum (RAND Corp., October 21, 1955), ii.

30 Robert E. McClellan, "History of the Space Systems Division July–December 1965" (SAMSO Office of Information Historical Division, October 1968), 35, https://nsarchive2.gwu.edu/NSAEBB/NSAEBB235/09.pdf.

31 "Draft of a Report on MIDAS by DDR&E Ad hoc Group," IV–1.

32 Ibid.

33 Ibid., IV–2–3.

34 Ibid., IV–4.

35 Ibid., IV–10.

36 Richelson, *America's Space Sentinels: DSP Satellites and National Security*, 28.

37 Schriever to Zuckert, "DOD Program Change (4.4.040) on MIDAS (239A)," 1.

38 Richelson, *America's Space Sentinels: DSP Satellites and National Security*, 36.

39 Adam Yarmolinsky, "Memorandum For: Mr. Timothy J. Reardon Jr., Special Assistant to the President on Classified Space Project 461," *Memorandum*, May 31, 1963, 1, https://nsarchive2.gwu.edu/NSAEBB/NSAEBB235/08.pdf.

40 Ibid.

41 Robert F. Piper, "History of Space and Missile Systems Organization, 1 July 1967–30 June 1969" (SAMSO Chief of Staff Historical Division, March 1970), 214, https://nsarchive2.gwu.edu/NSAEBB/NSAEBB235/10.pdf.

42 Richelson, *America's Space Sentinels: DSP Satellites and National Security*, 49.

43 Ibid., 63.

44 "Defense Support Program: Celebrating 40 Years of Service" (Northrop Grumman, 2010), 6.

45 Ellis E. Lapin, "Surveillance by Satellite" (Aerospace Corporation, August 11, 1975), 182.

46 J.W. Plummer, "Memorandum for the Deputy Secretary of Defense, Subject: Defense Support Program" (National Reconnaissance Office, May 3, 1974), 4–5.

47 HQ ADCOM J-3 to HQ USAF, "Knowledge of the DSP System," *Memorandum*, October 10, 1980, 1–2, https://nsarchive2.gwu.edu/NSAEBB/NSAEBB235/14.pdf.

48 Plummer, "Memorandum for the Deputy Secretary of Defense, Subject: Defense Support Program," 3.

49 Ibid., 1.

50 Lapin, "Surveillance by Satellite," 174.

51 Ibid., 182–183.

52 Ibid., 183.

53 William O'Brien to Jeffrey T. Richelson, "Air Force Space and Missile Systems Center Response to: Request under the FOIA for Report on 'Preliminary Analysis of Project Hot Spot IR Signals' and 'Application of Infrared Tactical Surveillance,'" November 4, 1992, 2–3, https://nsarchive2.gwu.edu/NSAEBB/NSAEBB235/11.pdf.

54 "Defense Support Program: Celebrating 40 Years of Service," 5–6.

55 Richelson, *America's Space Sentinels: DSP Satellites and National Security*, 159.

56 "United States Space Command: Operations Desert Shield and Desert Storm Assessment," January 1992, 10, https://nsarchive2.gwu.edu/NSAEBB/NSAEBB235/25.pdf.

57 Ibid.
58 Ibid., 15.
59 Ibid.
60 Ibid., 4.
61 Ibid., 19.
62 Ibid.
63 Ibid.
64 R.W. Apple Jr., "War in the Gulf: Scud Attack; Scud Missile Hits a US Barracks, Killing 27," *New York Times*, February 26, 1991.
65 Space Systems Division, "DSP Desert Storm Summary Briefing," June 1991, 48, https://nsarchive2.gwu.edu/NSAEBB/NSAEBB235/24.pdf.
66 Ibid.
67 "Report to the Chairman, Subcommittee on Defense, Committee on Appropriations, House of Representatives; Early Warning Satellites: Funding for Follow-on System Is Premature" (Washington, DC: United States General Accounting Office/National Security and International Affairs Division, November 7, 1991), 1.
68 Ibid., 2.
69 Jeffrey T. Richelson, *America's Space Sentinels: The History of DSP and SBIRS Satellite Systems*, 2nd Edition (Lawrence, KS: University Press of Kansas, 2012), 183.
70 Ibid., 185.
71 "GAO/NSAID-92-30," 2.
72 Ibid., 3.
73 Ibid., 4.
74 Richelson, *America's Space Sentinels: The History of DSP and SBIRS Satellite Systems*, 191.
75 Ibid., 191–192.
76 Garry Schnelzer, "Air Force Space Sensor Study" (AFPEO/SP, April 12, 1993), 11, https://nsarchive2.gwu.edu/NSAEBB/NSAEBB235/20130108.html.
77 Robert R. Everett et al., "Report of the Space Based IR Sensor/Technical Support Group" (MITRE Corp, October 1993), 12, https://nsarchive2.gwu.edu/NSAEBB/NSAEBB235/37.pdf.
78 Joe Bailey, "Point Paper on DSP-II TOR" (Air Force System Program Director, Space-Based Early Warning Systems, May 24, 1993), 3, https://nsarchive2.gwu.edu/NSAEBB/NSAEBB235/35.pdf.
79 Guido W. Aru and Carl T. Lunde, "DSP-II: Preserving the Air Force's Options" (El Segudo, CA: The Aerospace Corporation, April 23, 1993), 10, https://nsarchive2.gwu.edu/NSAEBB/NSAEBB235/33.pdf.
80 Ibid., 32.
81 Charles Horner to Pete Aldridge, "Letter from CINC US SPACECOM to CEO Aerospace Corp.," May 24, 1993, https://nsarchive2.gwu.edu/NSAEBB/NSAEBB235/34.pdf.
82 Pete Aldridge to Charles A. Horner, "Aerospace Corporation Response to Gen Charles Horner CINC USSPACECOM," May 27, 1993, https://nsarchive2.gwu.edu/NSAEBB/NSAEBB235/36a.pdf.
83 Bailey, "Point Paper on DSP-II TOR," 2.
84 Everett et al., "Report of the Space Based IR Sensor/Technical Support Group," 6.
85 Ibid.
86 Richelson, *America's Space Sentinels: The History of DSP and SBIRS Satellite Systems*, 204.

87 Ibid., 203.

88 Guido W. Aru, "Statement of Mr. Guido William Aru (Project Leader, System Architecture and Integration Space-Based Surveillance Division, Aerospace Corporation) Before the House of Representatives Committee on Government Operations Legislation and National Security Subcommittee," February 2, 1994, 2, https://nsarchive2.gwu.edu/NSAEBB/NSAEBB235/38.pdf.

89 Ibid., 7.

90 Richelson, *America's Space Sentinels: The History of DSP and SBIRS Satellite Systems*, 210.

91 Harry N. Waldron, "History of the Space and Missile Systems Center: October 1994–September 1997" (Space and Missile Systems Center, March 9, 2002), 113, https://nsarchive2.gwu.edu/NSAEBB/NSAEBB235/27.pdf.

92 Richelson, *America's Space Sentinels: The History of DSP and SBIRS Satellite Systems*, 213.

93 Ibid., 214–218.

94 Waldron, "History of the Space and Missile Systems Center: October 1994–September 1997," 113.

95 Ibid., 122.

96 Ibid., 116–118.

97 Ibid., 123–125.

98 Air Force Space Command, "SBIRS Overview Brief: Combat Air Force Commander's Conference," November 16, 1998, 8–9, https://nsarchive2.gwu.edu/NSAEBB/NSAEBB235/39.pdf.

99 Homer H. Thomson et al., "National Missile Defense: Risk and Funding Implications for the Space-Based Infrared Low Component" (United States General Accounting Office, February 1997), 13.

100 Ibid.

101 Ibid., 13–14.

102 Ibid., 16.

103 "Defense Acquisitions: Despite Restructuring SBIRS High Program Remains at Risk of Cost and Schedule Overruns" (Washington, DC: General Accounting Office, October 2003), 2.

104 Moshe Schwarts and Charles V. O'Conner, "The Nunn-McCurdy Act: Background, Analysis, and Issues for Congress" (Washington, DC: Congressional Research Service, May 12, 2016), 2.

105 Ibid., 25.

106 "GAO-04-48," 9.

107 Ibid., 7.

108 Schwarts and O'Conner, "The Nunn-McCurdy Act: Background, Analysis, and Issues for Congress," 7.

109 Jeremy Singer, "DOD Notifies Congress of Higher SBIRS Cost," *Space News*, March 14, 2005, https://spacenews.com/dod-notifies-congress-higher-sbirs-cost/.

110 "Missile Detection Systems Development: SBIRS/AIRSS Briefing to Congressional Staff Preliminary Findings" (Washington, DC: General Accounting Office, March 13, 2007), 1.

111 Ibid., 19.

112 Ibid., 5.

113 Ibid., 28.

114 Richelson, *America's Space Sentinels: The History of DSP and SBIRS Satellite Systems*, 267.

115 Chaplain, "Space Acquisitions: DOD Continues to Face Challenges of Delayed Delivery of Critical Space Capabilities and Fragmented Leadership," 6.

116 Erwin, "Lockheed Martin Awarded $2.9 Billion Air Force Contract for Three Missile-Warning Satellites."

117 Sandra Erwin, "The End of SBIRS: Air Force Says It's Time to Move on," *Space News*, February 19, 2018, https://spacenews.com/the-end-of-sbirs-air-force-says-its-time-to-move-on/.

118 Valerie Insinna, "STRATCOM Head: Timeline to Field New Missile Warning Satellites Is 'Ridiculous,'" *Defense News*, December 4, 2017, www.defensenews.com/digital-show-dailies/reagan-defense-forum/2017/12/04/stratcom-head-timeline-to-field-new-missile-warning-satellites-is-ridiculous/.

119 Alistair Reign, *Reagan National Defense Forum: STRATCOM Commander Gen. Hyten Discusses Future of Space War*, Panel, 2017, sec. 23:24, www.youtube.com/watch?v=wUIHctFgTac.

120 Insinna, "STRATCOM Head: Timeline to Field New Missile Warning Satellites Is 'Ridiculous.'"

121 Carl Schueler, "Commercially Hosted Payloads: Low-Cost Research to Operations" (American Meteorological Society Third Conference on Transition of Research to Operations, Austin, TX: Orbital Sciences, 2013), 3.

122 Mark Andraschko et al., "Commercially Hosted Government Payloads: Lessons from Recent Programs" (NASA Langley Research Center, 2011), 9, https://ntrs.nasa.gov/archive/nasa/casi.ntrs.nasa.gov/20110008235.pdf.

123 Ibid.

124 Ibid.

125 Ibid.

126 Schueler, "Commercially Hosted Payloads: Low-Cost Research to Operations," 4.

127 Ibid.

128 Ibid., 5.

129 Andraschko et al., "Commercially Hosted Government Payloads: Lessons from Recent Programs," 9.

130 Schueler, "Commercially Hosted Payloads: Low-Cost Research to Operations," 5.

131 Ibid.

132 Andraschko et al., "Commercially Hosted Government Payloads: Lessons from Recent Programs," 10.

133 Carl Schueler, "Commercially-Hosted Payloads: Low-Cost Research Options" (AMS Third Conference on Transition of Research to Operations, Austin, TX, January 9, 2013), https://ams.confex.com/ams/93Annual/flvgateway.cgi/id/24205?recordingid=24205&entry_password=706714&uniqueid=Paper224173.

134 Ibid.

135 Ibid.

136 Mike Gruss, "U.S. Air Force Decision To End CHIRP Mission Was Budget Driven," *Space News*, December 12, 2013, https://spacenews.com/38628us-air-force-decision-to-end-chirp-mission-was-budget-driven/.

137 "Air Force Commercially Hosted Infrared Payload Mission Completed," US Air Force, Los Angeles Air Force Base, December 6, 2013, www.losangeles.af.mil/News/Article-Display/Article/734794/air-force-commercially-hosted-infrared-payload-mission-completed/.

138 Reign, *Reagan National Defense Forum: STRATCOM Commander Gen. Hyten Discusses Future of Space War.*

139 "National Space Transportation Policy" (Executive Office of the President of the United States, November 21, 2013), 4.

140 "Military Space Systems: DOD's Use of Commercial Satellites to Host Defense Payloads Would Benefit from Centralizing Data" (United States Government Accountability Office, July 2018), 10.

141 Ibid., 9.

142 Erwin, "U.S. Would Need a Mega-Constellation to Counter China's Hypersonic Weapons."

143 The current constellation consists of a small number of large satellites in GEO. A constellation of hundreds or thousands of satellites operating in LEO provides many operational advantages to its owner. It also provides opportunities to seed the constellation with dual-use systems or other capabilities that could be used to attack adversary satellites. At the extreme an armed mega-constellation could be used to attack any new satellite immediately after launch denying adversaries access to space.

144 "SWIR Imagery," Official Company Page, Digital Globe, accessed December 14, 2018, www.digitalglobe.com/products/swir-imagery.

145 George Lewis and Theodore Postol, "Ballistic Missile Defense: Radar Range Calculations for the AN/TPY-2 X-Band and NAS Proposed GBX Radars (September 21, 2012)," *Mostlymissiledefense* (blog), September 21, 2012, https://mostlymissilede fense.com/2012/09/21/ballistic-missile-defense-radar-range-calculations-for-the-antpy-2-x-band-and-nas-proposed-gbx-radars-september-21-2012/.

146 Schriever to Zuckert, "DOD Program Change (4.4.040) on MIDAS (239A)," 2.

147 A technical analysis of the required capabilities for an international missile warning system is discussed by Geoffrey Forden, "A Constellation of Satellites for Shared Missile Launch Surveillance" (MIT's Program on Science, Technology, and Society, September 7, 2006), 4.

148 "Russia's Soyuz Launches EKS Missile Warning Satellite, Ends Year-Long Military Launch Gap," *SpaceFlight 101*, May 25, 2017, http://spaceflight101.com/soyuz-successfully-launches-second-eks-satellite/.

149 "'Doomsday Sputnik': Russia Said to Launch New Missile-Attack Warning Satellite," RT International, July 19, 2014, www.rt.com/news/174076-early-warning-satellite-russia/.

150 Benjamin Newell, "$93 Million Contract Keeps Missile Warning System Vigilant," Air Force Global Strike Command, July 13, 2017, www.afgsc.af.mil/News/Article Display/tabid/2612/Article/1247579/93-million-contract-keeps-missile-warning-system-vigilant.aspx.

151 Masaya Kato, "Japan and US to Sharpen Missile Defense with Real-Time Data Sharing," Nikkei Asian Review, August 25, 2018, https://asia.nikkei.com/Politics/International-Relations/Japan-and-US-to-sharpen-missile-defense-with-real-time-data-sharing.

152 This estimate is based on the cost of CHIRP which used a one-quarter Earth viewing sensor. Four CHIRP like systems would be needed to provide the equivalent of "single ball" coverage over one face of the Earth. Since a missile warning constellation would need "four ball" coverage of the Earth at least 16 CHIRP like sensors would be needed.

153 "Secretary-General Unveils $5.4 Billion 2018–2019 Programme Budget to Fifth Committee, Drawing Mixed Reviews from Delegates on Funding, Staffing Cuts | Meetings

Coverage and Press Releases," October 11, 2017, www.un.org/press/en/2017/gaab4243.doc.htm.

154 Everett et al., "Report of the Space Based IR Sensor/Technical Support Group," 8.

155 Ibid., 6.

156 "United States Space Command: Operations Desert Shield and Desert Storm Assessment," 15.

157 Daniel Cebul, "Coercive Tactics? Putin Touts Russia's 'Invincible' Nuclear Weapons," *Defense News*, March 2, 2018, www.defensenews.com/smr/nuclear-triad/2018/03/01/coersive-tactics-putin-touts-russias-invincible-nuclear-weapons/.

158 Aaron Mehta, "3 Thoughts on Hypersonic Weapons from the Pentagon's Technology Chief," *Defense News*, July 16, 2018, www.defensenews.com/air/2018/07/16/3-thoughts-on-hypersonic-weapons-from-the-pentagons-technology-chief/.

159 Kingston Reif, "Trump to Withdraw U.S. From INF Treaty | Arms Control Association," November 2018, www.armscontrol.org/act/2018-11/news/trump-withdraw-us-inf-treaty.

160 Chaplain, "Space Acquisitions: DOD Continues to Face Challenges of Delayed Delivery of Critical Space Capabilities and Fragmented Leadership," 6.

161 Schueler, "Commercially-Hosted Payloads: Low-Cost Research Options."

162 Erwin, "U.S. Would Need a Mega-Constellation to Counter China's Hypersonic Weapons."

8 Competition or cooperation?

This book asserts that a strategy of cooperation centered on commercialization is the rational strategic choice in space. Despite being the rational choice, it is difficult for a state to believably signal a desire to cooperate within a security dilemma without making itself too vulnerable. The difficulties that the security dilemma generates make choosing to compete by pursuing arms an attractive alternative to cooperation, but one that will ultimately lead to tragedy given the unique environment of space. This chapter will review how the conclusion that cooperation is preferable to competition was reached before evaluating the validity of this book's hypotheses on how signaling a desire to cooperate might be made possible.

The security dilemma arises when a state's attempts to increase its security threaten other states, leading to unnecessary conflict or intensified security competition.[1] This relatively simple concept results in complex outcomes and lies at the heart of realist theories of international relations. Within realism, war is the result of either tragic competition between status quo states, as described by the security dilemma, or from the actions of greedy states.[2] Key to each realist theory's prediction of outcomes is its assumption on the nature of states as security seekers bent on power balancing, preserving the status quo, or in maximizing their power.

Assumptions on the nature of state behavior differentiate realist theories from each other and allow them to achieve a significant degree of explanatory power relying only on material measures of power as an independent variable. However, this reliance on material measures of power fails to fully explain many facets of international relations. Other theories of international relations reject security seeking as the primary driver behind state behavior and attempt to use an alternative basis for explaining state behavior, but they also have weaknesses. The theory that forms the framework for this book, Charles Glaser's theory of rational strategic choice, retains the security dilemma at its core but adds state's motivations and the information a state has about other states motivations as well as its own to the more common realist material measures of power. This theory's roots in the realist security dilemma, combined with its elements of other theories, make it well suited to explaining state behavior in space.

Including the non-material factors of motivation and information adds explanatory power as well as complexity to Glaser's theory of rational strategic choice. The additional explanatory power added by the motive and information variables allows for differentiation between insecure security seekers and purely greedy states. Lacking information on motivation, the behavior of both of these types of states will look identical to external observers. Having this information allows a state to choose whether it is best to pursue cooperative or competitive policies. The issue is that state motivation is a difficult variable to measure, and one that is easily misinterpreted. A wrong interpretation of state motivation can lead to a state choosing cooperative policies that leave it vulnerable to a greedy adversary. Alternatively, a state may choose to compete when cooperation is preferable, perpetuating the tragic competition that is at the heart of the security dilemma.

Choosing the optimal policy of competition or cooperation requires an assessment of an adversaries' strategic goals. This is challenging enough when it comes to assessing the motivations and goals of a single adversary state, but the shared geography of space dictates that an optimal policy must account for the motivations of all adversaries. This multistate dilemma in space makes a purely bipolar analysis infeasible. It also means that in space a state cannot signal benign intent and a willingness to cooperate with one adversary, while at the same time competing with another. The difficulty of achieving collective cooperation within the multistate dilemma encourages states to pursue competitive strategies even knowing that it is a sub-optimal policy. Ultimately, since space is an environmentally driven stag hunt (see Figure 2.1) choosing to compete is not the worst option for a state, as the danger of another state defecting and pursuing space weapons and holding another's space assets hostage without the ability to respond in kind is high.

Even though environmental factors in space make cooperation the ideal option, the multistate dilemma seems to make competitive policies the rational choice for states. Competition is not, however, the best rational choice for the US. The US is more dependent on its space assets to project its military power than any other nation. It also has far more space assets than any other nation. This lack of mutual dependence and vulnerability prevents the US from holding other nation's space assets in equal jeopardy to its own. If the offense–defense balance favored the defense in space, the US could overcome this weakness by building appropriate defenses, but it does not. As demonstrated earlier, the offense–defense balance in space varies based on the level of war, favoring the offense under certain conditions. Due to the nature of the balance, as the aims of the attacker become more limited the difficulty of deterrence through denial increases. Practically this means that a capable adversary can gain temporary tactical advantage in a geographic region over the US in space even if the US has invested in a robust and resilient space architecture.

The argument in favor of pursuing cooperation is further reinforced by the military aims of the US in space. US actions and rhetoric strongly support the conclusion that the US is an extremely insecure security seeker attempting to preserve the status quo in space. The US military leverages space for the ability to transfer information through it and to gather information from it. US efforts to build a more resilient space architecture are focused on preserving this capability. Investments in capabilities such as the $1.5 billion spent on the new space fence are directed at detecting threats to US satellites, not in targeting adversary satellites.[3] Further, the motivation to replace exorbitantly expensive platforms such as SBIRS is not due to dissatisfaction with their current capabilities; rather it is a desire to improve the survivability of the existing system.[4] Returning to the questions asked in Chapter 2 that determine if a state should arm or cooperate in space; does choosing to pursue an arms race in space increase military capability? Given the factors discussed above, for the US the answer is a resounding no. An arms race in space will be one where the US invests heavily in new systems merely to preserve existing capability. Any new capability of significance that is added will require further investments to preserve that capability and only increase the magnitude of the arms race.

The second of three factors that a nation must weigh when choosing to pursue arms over cooperation is whether the increased security gained by pursuing arms has a negative impact on an adversary's security. It is already clear that the US is engaged in an arms race to preserve capability and that any increase in capability will be a small ancillary benefit to the goal of protecting what it already has. Despite this, uncertainty over US motivations, actions, and capabilities generate fear and insecurity in its adversaries. The fear and insecurity that drives the US to invest billions in more resilient and protected satellites is magnified in adversaries who cannot be certain that the US is not building offensive capabilities designed to dominate space. Dominance centric US rhetoric toward space by US national leaders only reinforces uncertainty over its aims, further negatively impacting its adversaries' perceptions of their security.[5]

The final factor to consider in pursuing an arms race is whether the benefits to security created by the pursuit of arms are greater than the dangers generated by the increases to an adversaries' level of insecurity both in and out of the space domain. This seemingly redundant question hints at the dangers of an unnecessary arms race. As the US struggles to develop more resilient satellites, it will inevitably develop defensive systems that are distinguishable from offensive systems only by intent. Crossing the threshold of space-to-space weapons will encourage adversaries to pursue equivalent capabilities as well as additional offensive systems in an effort to deter a feared US attack. This tragic spiral of misperception and competition will create instability that could quickly spread beyond the space domain. While military action in space is unlikely to occur absent some terrestrial source of tension, events in space can quickly escalate outside the domain. Current US policy which affirms that attacks in space will be met with responses outside of space only increase the probability of escalation.

Since competition is a self-defeating strategy for the US, it must make every effort to encourage cooperation in space. The challenge is that the US must first escape the accelerating cycle of action and reaction typified by the security dilemma in which it is currently trapped. Escaping the competitive fear-based cycle of a sub-optimal arms race requires that a state must first convince its adversaries that it does not have malign intent. The more entrenched an adversary's view is of another state as malign, the stronger the required signal of benign intent needs to be. The issue is that both the US and its largest rivals in space are developing increasingly firm beliefs in each other's malign intentions.

This research focused on the three largest space powers and geopolitical rivals: the US, China, and Russia. These powers are not all in equal competition with each other, rather China and Russia are locked in nearly independent security competitions with the US. As the dominant space power by most material measures, the US is sensitive to the increasing capability of its rivals. While US power in space continues to increase in absolute terms, gains by China create a sense of relative decline and insecurity. Meanwhile, Russia is struggling to modernize and rebuild its space industrial base even as the fundamental weakness of its economy and disruptive technological advances leave its Soviet-era technology farther behind. Within this power dynamic, each state views itself as a security seeker and its adversaries' motives as greedy. Actions designed to address each state's insecurity only reinforce the perception that its adversary is a greedy state. Both Russia and China see the US as a greedy hegemon while the US sees its rivals as revisionist threats to the existing status quo. The nature of these misperceptions and the resulting competition are all indicative of a moderate security dilemma that is increasing in severity.

As the focal point of the security dilemma between the three largest space powers, the US is the state best positioned to break the cycle of arming. This requires that the US reassure its rivals that it does not, in fact, intend to dominate space and that its intentions within the domain are benign. Changing US rhetoric toward space would be a positive first step, but not nearly enough to change Chinese and Russian perceptions of the US as a greedy hegemon whose goals extend into space. The higher the cost in security associated with sending a signal the more likely it is to be interpreted as signaling benign intent.[6] The difficulty is that the signaling state can never be sure that its adversaries are truly security seeking status quo states.

Uncertainty over an adversary's motivations makes accepting risk by sending a signal of benign intent a problematic choice. In descending order of cost a security seeking state can indicate its benign intentions in three primary ways: through arms control agreements, adopting a defensive posture, or through unilateral disarmament.[7] Far weaker powers in terms of explicit military power, neither China or Russia can afford to send the costlier signals of adopting a defensive posture or a policy of unilateral restraint. This does not mean that they have not tried to signal benign intent; both states have actively pursued arms control agreements with the US in space for over a decade. Their proposals are not without weaknesses that the US readily highlights, but they are attempts at arms control and possible signals of benign intent. To Russian and Chinese diplomats, continued US resistance to the PPWT and the lack of a meaningful counter-proposal on space arms control can only reinforce their views that the US has malign intentions of dominating space at their expense.

It is in the strategic interest of the US to avoid an arms race in space so it must find a way of sparking cooperation. A belated attempt by the US to seriously pursue arms control in space would certainly be a signal of benign intent, albeit one that is too weak to change the entrenched perception of the US as a greedy hegemon. Adopting the costliest signal, unilateral disarmament, is difficult in the space domain as neither the US nor any other state currently acknowledges possessing space-based weapons. Disbanding US missile defenses that can act as ASAT weapons would be a form of unilateral disarmament. However, it is one that would come at the cost of making the US and its military forces vulnerable to a variety of theater and regional ballistic missiles from states that are not security seekers and instead are most likely contained greedy states, most notably Iran and North Korea.

The middle approach to signaling benign intent is unilaterally adopting a defensive posture. This method is best applied when there is a high level of distinguishability between the offense and defense and the balance between the two is neutral. Under these conditions a state has a choice of whether to adopt offensive or defensive weapons and that choice can signal intent. A greedy state is unlikely to choose defensive weapons, while a security seeking is free to choose defensive weapons since that choice does not put the state at undue risk.[8] Within the framework of the security dilemma, the US can use the type and nature of the satellites and space systems it acquires to reassure other states and signal benign intent. Since established norms in space already identify the passive military use of space as non-aggressive, some significant change is required to distinguish the status quo from a signal of benign intent.

This book proposes commercialization as an approach to signaling benign intent. It offers a blend of the reassurance strategies of unilateral disarmament and defensive posture adoption that would allow the US to reassure its adversaries without sacrificing national space power. Commercialization preserves national space power by maintaining or increasing US national space presence and therefore national space power. It also reduces the need for dedicated military and intelligence satellites which reduces the uncertainty created each time the US launches non-commercial payloads. In contrast to most national security satellites, the purpose and intent of commercial satellites is clear, and verification through use is possible by any paying customer. Also, reducing the active military presence in space would reduce the need for military space forces, leading to a reduction in their numbers. Finally, due to the rapid pace of commercial space innovation in western countries, the US is uniquely capable of adopting and benefiting from a commercialization approach. The combination of these effects and others discussed in previous chapters would signal benign intent while reducing uncertainty and fear without placing the US in unnecessary strategic jeopardy.

Commercialization as a strategy

Outside of the increasingly contentious Outer Space Treaty, national behavior in space has been largely constrained only by technical limitations and by tacitly accepted norms established between the US and the former Soviet Union rather than formal agreements. Increases in the importance of space to global economies and militaries have not been matched by the establishment of more robust systems and laws to manage behavior. This shortcoming was not readily apparent in the decade following the end of the Cold War since the US faced no real competition in orbit. As that competition has arisen and technical barriers to action decrease, the lack of agreed-upon rules is helping to fuel the security dilemma. The advantage that this lack of structure has is that there is an opportunity for China, Russia, and the US to shape it jointly. For this to happen the US must step back from fueling the security dilemma by signaling that it does not intend to dominate space at other's expense, full commercialization would send that signal.

Full commercialization

Full commercialization would send the clearest believable signal that the US does not have malign intentions toward its adversaries in space and set the conditions for the development of a jointly agreed-upon rules-based order suited for the new space age. There are significant challenges and obstacles to adopting this strategy. While it would be possible to expand existing commercial contracts to commercialize satellite communications and remote sensing, missile warning presents a more difficult challenge. In addition, the risk associated with adopting a reassurance strategy built around full commercialization rises as the space arms race progresses. As the threshold for each new degree of space militarization is crossed, a reassurance strategy built around commercialization will look less like adopting a defensive posture and more like one of adopting unilateral disarmament. While the degree of effective signaling will increase so will the associated risk. As the risk associated with a course of action increases the likelihood that it will be adopted decreases even if the potential reward increases proportionally to the risk.[9] While the theory of pursuing a reassurance strategy built upon full commercialization has merit, the practicality of achieving it is more challenging.

After subdividing the US national security space architecture into five categories this book focused on the commercialization potential present in remote sensing, communications, and missile warning. The two remaining categories of launch and PNT were excluded in favor of the other three because they contribute relatively little to the security dilemma. Also, since SpaceX broke ULA's long-running monopoly on national launch services the category is now open to competitive bidding and is already effectively commercialized. The US PNT architecture, primarily in the form of GPS, adds little to the security dilemma in space because of its unique orbit, global civilian and military use, infrequent replenishment launches, and the ease of interfering with the signal at the point of reception. In its current form, GPS could transition from a military to civilian-led organization relatively easily as it has expanded far beyond its original military purpose. The remaining three categories are the focus of this research and offer varying degrees of commercialization potential.

Variations in the degree of possible commercialization within these three categories impact the strength of the signal that the US can send. The current US approach to commercialization focuses on supplementing existing dedicated military platforms. This allows commercial platforms to act as flexible surge capacity for expensive dedicated government platforms. Any incomplete or partial commercialization effort might appear to be no more than an expansion of the existing policy of augmentation. The more complete the degree of com-mercialization the clearer and stronger the signal of benign intent is. Analyzing the practical possibility of fully commercializing satellite communications, remote sensing, and missile warning is integral to determining the validity of the first hypothesis that this research sought to prove: that full commercialization by the status quo dominant space power (the US alone) would significantly reduce the security dilemma.

Of the three categories analyzed in this research, satellite communications presented the fewest challenges to full commercialization. In Chapter 5 three broad approaches to achieving the goal of commercialization were discussed: purchasing SATCOM directly on the market, subsidizing national systems using a civil reserve space fleet (CRSF) approach, or using hosted payloads. Purchasing SATCOM directly from the market is the purest form of commer-cialization and one that is possible, though this approach had several potential weaknesses. These weaknesses stem from the relatively small market share that the US government would have as a traditional customer. As a risky client with niche demands subject to unpredictable funding and termination for conveni-ence clauses, the US government is an unattractive customer. The US military could also not be sure that it could depend on the commercial markets to fully meet sudden surges in demand during wartime. Even though Gen. Hyten, as the commander of US Strategic Command, asked why SATCOM is not just pur-chased, stating that "it's just a commodity, why don't we buy it as a commod-ity," there are significant risks with relying solely on purchased bandwidth.[10] An alternative approach that grants the US government more leverage over the commercial market and carries less risk might be more attractive as a signaling strategy.

A subsidized approach to commercialization similar to the civil reserve air fleet (CRAF) model has many advantages over relying entirely on the commer-cial market. Adapting CRAF to space could involve a variety of methods of providing subsidies to US commercial SATCOM providers in exchange for guaranteed future access. A civil reserve space fleet (CRSF) approach would address many of the shortcomings of a contract-driven approach to commercial-ization while only marginally impacting on its value as a reassurance strategy. Using a CRSF model would ensure the availability of US-flagged satellites during periods of national crisis. It would also help overcome any reluctance to carry US military signals and give the US government more leverage over the industry than it would have as a purely commercial customer. Perhaps the greatest benefit of a CRSF model is that it would further encourage the development of the domestic space industrial base, increasing US presence and therefore power in space. As attractive as a CRSF model might be, there remain fundamental technical shortcomings in the commercial market that neither contracts nor subsidies may be able to address that could result in less than full commercialization.

The challenges faced in commercializing SATCOM due to market share are not present in the remote sensing market. The remote sensing market is much smaller than the commercial SATCOM market and as a result of this US government market share and leverage is much greater than with SATCOM. Since the demand for remote sensing on the civilian market is relatively small, the US military and Intelligence community have been able to exercise significant control over the market. This market control has meant that the vagaries of US government contracts and regulations have had more impact on the fate of the domestic remote sensing market than any other factor. US remote sensing regulation has managed to stay ahead of commercial demands in the electro-optical market but stifled domestic development of other capabilities such as space-based radar. This has ceded certain aspects of the commercial market to foreign powers with less restrictive regulation. The lack of an existing robust commercial market in some fields of remote sensing is the largest challenge to a commercialization strategy.

US regulatory limitations are driven by a balancing act between promoting commercial competitiveness and preserving national security. Overriding national security concerns has only been possible when it becomes clear that domestic providers are in danger of losing competitiveness to foreign companies. At .25-meters, the current restriction on panchromatic resolution has reached the point where it is no longer a hindrance to the optical portion of the commercial remote sensing market for the foreseeable future. There is no existing market for satellite imagery of such high quality even as it approaches the limits of technical capability. Another challenge to the commercial market is that when it becomes technically and financially feasible to produce better imagery, it is likely that the commercial market will face increased public resistance over privacy concerns that are muted at the current commonly available resolutions.[11] If the national security community did desire higher-resolution imagery a shift toward commercialization, the two-tier compromise that existed under the previous .5-meter resolution limit could be used again to allow commercial providers to provide its highest-resolution imagery only to approved customers.

At the same time, that spectral resolution restrictions are receding as a significant national security issue temporal resolution issues are arising that present challenges and opportunities. The operational risk to military units from high-revisit rate commercial satellite systems is something that the national security community will have to cope with through a combination of regulatory, diplomatic, and eventually military means.[12] While these systems present a significant risk, they also present a significant opportunity. The distributed nature of these systems will make them naturally resilient to interference and make reliance on commercial systems a real possibility as long as the US regulatory environment is not overly restrictive on their development.

Restrictive regulation on imagery platforms has not driven the commercial market entirely overseas because the US military and Intelligence community are lucrative customers. Since DOD and other US government agencies are legally obligated to buy American products when available, domestic providers have an advantage.[13] The US market was attractive enough to encourage Canadian space company MDA to rebase itself in the US as Maxar Technologies following its purchase of DigitalGlobe to preserve its preferred access to the US government. This move hints at the leverage over the commercial market that the US would have under a reassurance strategy based on commercialization.

A shift toward commercialization would require significant growth in the commercial remote sensing market that would be advantageous to US space power. With global revenues for the remote sensing market in 2016 estimated at just $9.7 billion reinvesting half of the existing NRO and NGA budget toward commercial capabilities would nearly double the global market.[14] Reinforcing the link between access to US funding and US corporate presence would incentivize the bulk of the global industry to shift to the US. This shift of the industry would allow the US to consolidate the commercial industry under US regulatory rules. US government market leverage would also encourage commercial providers to innovate to meet unique military and intelligence needs for which there is otherwise not enough demand to justify the investment.

Perhaps the most significant benefit of a commercialization approach is that the market shift toward the US would grant the US government the ability to exercise some control over access to imagery during periods of crisis. The myriad methods of control that the US government currently has over the domestic remote sensing market are increasingly undercut by proliferating foreign competition. Under existing conditions, military maneuvers such as the surprise "left hook" executed by US forces during the first Gulf War that caught the Iraqi Army unprepared would be impossible to disguise.[15] While the era of large military maneuvers remaining unobserved from space is unlikely ever to return, US dominance of the market could ensure that many future niche commercial capabilities that might compromise national security remain under US regulatory control.

Commercialization of the remote sensing market is possible, though it could not be implemented quickly. It would take several years for the industry to develop the ability to meet the needs of the US national security community. A dedicated shift in the direction of commercialization by halting the development of future national systems and soliciting commercial providers would send the appropriate signal of reassurance without requiring any sudden loss of capability. With the increasingly rapid pace of innovation in the new space industry, national security needs would likely be met sooner than anticipated. Perhaps the greatest challenge for remote sensing commercialization would be developing a large enough customer base to avoid being labeled as adjunct government systems. Given the significant benefits for US national space power, control, and security that come through commercialization of the remote sensing function, it is a viable strategy for reassurance.[16]

Commercializing the functions associated with the nuclear deterrent, primarily missile warning, presents a much more challenging problem than that presented by remote sensing or SATCOM. Among the many challenges associated with surrendering direct control of missile warnings to a commercial provider is the complete lack of a commercial need for real-time global monitoring of infrared events. An alternative to commercialization that would preserve the intent of its impact on the security dilemma is pursuing internationalization. Placing national trust of such an extremely sensitive mission as warning of nuclear attack in an international body is a difficult concept to envision, though it does have strategic advantages. An international missile warning constellation would aid global nuclear stability by providing access to the same assured early warning information that only the wealthiest nations possess to all nuclear-armed nations. It would also share the burden of costs associated with building these expensive satellites while ensuring that all nations had an equal stake in keeping these platforms inviolate. An international system has many theoretical advantages, but also many practical challenges.

Despite the clear benefits of assured missile warning for all nations, there are too many practical challenges to making it a reality. Among the most significant issues would be building trust in an internationalized system. While this is a difficulty that applies to all aspects of a commercialization approach, it is one that becomes acute when considering missile warning. These platforms drive the security dilemma through the fear of their loss, not through any fear they generate in an adversary. This fear of loss is driven by the importance of these systems to national security. The history leading up to the development of SBIRS demonstrates that, at least in the US, space-based missile warning is considered essential enough to overcome even the most egregious acquisition failures. It is an unavoidable fact that while nations that do not currently have space-based missile warning capabilities would likely welcome data from an international system, the three major space powers at the center of this research already have it. That fear of loss that drives the security dilemma aspect of these systems will also prevent them from ever surrendering complete control. Asking the wealthiest space-faring nations to bear the bulk of the burden of funding an international system that they do not fully control, or fully trust, is not realistic.

Partial commercialization

The inability to fully commercialize a task as sensitive as missile warning creates an obstacle to full commercialization that presents a challenge to the first hypothesis of this research. This hypothesis, that full commercialization by the US alone would significantly reduce the security dilemma, most closely resembled the strongest form of signaling benign intent, pursuing unilateral disarmament. In theory, this approach would send the clearest possible signal of benign intent by the US, especially as it would require a shift away from the current US dominance centric rhetoric. While full commercialization would undoubtedly send the clearest signal of benign intent, this research included a second more measured approach in anticipation of this difficulty; that commercialization of all functions except those directly associated with the nuclear deterrent would somewhat reduce the security dilemma.

The less aggressive approach to commercialization would exclude those satellites involved in missile warning as well as a very small number of GEO satellites designed for low-bandwidth, high-frequency strategic communications. Preserving this relatively small number of the 334 out of 839 active US-flagged satellites that the Union of Concerned scientists tracks as government or military would have a more substantial impact on the security dilemma than raw numbers suggest. While small in number these satellites continue to fuel the dilemma through the fear of loss they generate. Even so, removing all but a few dedicated US national security platforms from orbit would shift the focus of the US security community away from direct involvement in space and instead make it primarily a consumer and regulator of space affairs. Partial commercialization may still defuse the developing arms race, but only if the US military avoids the temptation to leverage developments in the commercial market and invest in an expensive distributed constellation of platforms. Commercializing satellite communications and remote sensing while investing in a massive expansion in the number of missile warning satellites would completely remove any signaling benefit that might be gained from this approach.

The development of commercial small satellites organized into constellations and the production of satellites at scale is only just beginning. Until recently satellites were usually unique from each other or produced as small batches in parallel. The development of LEO constellations is leading to the need for true satellite production at scale along with an associated decrease in costs. No longer the leader in technology development, the US national security apparatus is already looking to benefit from the innovations in the commercial industry to build constellations of its own.[17] An explosion of military small satellites in orbit is the logical end-state of the current quest for resiliency through disaggregation and one that will further fuel the security dilemma even if these satellites are nominally only associated with the nuclear deterrent.

As US military small satellite constellations are deployed, they will initially seem to achieve the resiliency that existing military satellites do not provide. They will be individually less expensive than existing satellites and much easier to produce and launch in quantities. However, from the Russian and Chinese perspective, the launch of hundreds of US military platforms into LEO orbit will only be seen as further proof of the US quest to dominate space.

Given the US technical advantage, neither nation could be sure what the exact capabilities these small satellites have. Would the US seed this constellation with hunter-killer satellites in a quest to prevent China from launching into GEO transfer orbits or mimic its success with stealth aircraft and develop stealthy satellites hidden among small satellite launches that are designed to shadow and disable future Chinese or Russian constellations? The remarkable decreases in launch costs and technology miniaturization that are making the current revolution in space possible are also making it easier to hide the launch of satellites whose purpose goes beyond the passive military collection and transfer of information.

If the path to avoiding a future of competition and potential conflict is to be brought about by a reassurance strategy built on partial commercialization could the US pursuing this strategy alone reduce the security dilemma? As the dominant power in space in terms of presence and technical capability, a shift by the US toward commercializing most of its national security architecture in space would not go unnoticed. However, a decrease in mission sets requiring dedicated military platforms cannot be accompanied by an absolute increase in the number of military satellites in orbit. If partial commercialization is successful in reducing the intensity of the security dilemma and still avoiding an arms race, then the US could continue to rely on the handful of primarily GEO satellites that are currently associated with the nuclear deterrent. If the US were able to avoid the temptation to seek resiliency through disaggregation at the same time that it commercialized then it would be in a position to observe the impact of its actions on its two primary adversaries in space. The reaction of these adversaries, Russia and China, would likely vary for different reasons.

Russia would no doubt welcome signals of reassurance that allowed it a respite from security competition in orbit. The Russian economy is only a fraction of the US economy, and there are significant signs of strain in the formerly world-leading Russian space industry. These signs are rampant and range from the increase in the failure rate of previously reliable Soviet-era Russian rockets to innovation issues arising from the inability to attract talented young engineers.[18] Russian efforts to stop the decline by preventing its space workforce from traveling abroad or seeking higher-paying jobs in other industries preserved the existing workforce at the cost of frightening away new entrants. As the number of engineers and scientists remaining in the space workforce from before the fall of the Soviet Union declines, the signs of decay and low morale in the Russian space industry are growing. The recent controversy over a shoddily covered hole accidentally drilled in a Soyuz spacecraft is only the latest example of the decline of the Russian space industry, many other less public examples exist.[19]

Commercial space-flight contracts and the exorbitant amount that the US and its partner countries pay for access to the ISS have helped subsidize the Russian space industry until recently. Profits from these activities are in rapid decline as the Russian share of the commercial launch market is undercut by SpaceX and other emerging low-cost rivals. The cutting of the Russian crew aboard the ISS from three to just two cosmonauts in an effort to save costs and possibly attract a paying customer is yet another sign of the strain in the Russian space industry.[20] When income from NASA and existing commercial orders disappears over the next few years, the Russian state will not have the resources to replace the lost sources of external income, further reducing Russian prestige.

Russia is a security seeking state desperate to restore the previous status quo and prevent any further loss of status within the existing international structure. It lacks the material power to enforce its demands through either military or economic means conventionally, so it has relied on asymmetric strategies. The effectiveness of these asymmetric strategies has created concern within NATO and placed Russia back on the geopolitical map as a US rival. What many fail to recognize is that Russian hybrid warfare, while effective in achieving limited goals, is a strategy of weakness. It is a strategy that works only as long as it does not face active and organized resistance from NATO. Reducing the need to compete militarily in space with the US will allow Russia to husband its resources for use in other domains. Russia will no doubt continue to explore methods of denying or disrupting US and allied advantages in space, though the urgency and therefore funding available for its efforts will be dramatically reduced as it is shifted toward other security priorities. It will also likely welcome the prestige that comes from any accompanying US efforts to negotiate arms control on an equal footing with the US.

China's probable response to US commercialization is more problematic. Unlike Russia, China is not a declining power that might welcome relief from competition. In a close contest, it exceeded the number of space launches by any other nation in 2018 for the first time, albeit only placing half the mass in orbit of the US.[21] With 38 successful attempts, China successfully launched more rockets in a year than any nation since the end of the Cold War. While many of the payloads carried to orbit aboard Chinese rockets were for paying foreign commercial companies, the majority carried communications, remote sensing, and navigation satellites for the Chinese government and domestic semi-commercial companies. It is this rapid increase in Chinese space power and cap-abilities which is causing concern in the US national security community. One overlooked advantage of the increasing presence of Chinese platforms in orbit is the corresponding increase in mutual dependence. While the US remains far ahead in the number and cost of its orbital assets, China's increasing investment raises its motivation to avoid significant conflict in space. A US strategy of partial commercialization that involves a significant decrease in the number of dedicated military platforms and is accompanied by overtures for arms control will likely be met with cautious optimism by China that its space assets are not potential targets of a US quest for space dominance.

Even a properly interpreted and accepted signal that the US is not seeking armed competition will not stop China from seeking prestige in space, seemingly at US expense. The prestige the US gained in winning the race to the

Moon persists today, and is something that China seeks to emulate. Chinese General Secretary Xi Jinping has made it clear that achievements in space are "important symbols of a country's scientific and technological strength."[22] China's manned space program, as well as its unmanned scientific missions, have seen a steady march of increasing accomplishments that have brought significant attention to its growing technological prowess. In the US, China's increasing capabilities are often fearfully contrasted with the relative malaise of the US space program, which has not left Earth orbit since the last mission to the Moon. The pace of China's accomplishments feeds US perceptions of relative decline and insecurity. To overcome these perceptions, US observers must separate true security threats in space from Chinese technical accomplishments.

The difficulty of separating true security threats from Chinese technical accomplishments is complicated by the unprecedented scale and scope of the increase in China's economic power. It is the speed of this increase which the world, and especially the US, is having difficulty adjusting to. In little more than a single generation, China has gone from being a poor economically backward nation to one whose economy rivals or surpasses the US. China's corresponding increase in military and space power appears as an aggressive build-up at first glance though at least publicly China's military expenditure has remained steady relative to its GDP and remains much smaller than US defense spending in absolute and relative terms. Of ongoing concern is that the focus of China's defense budget seems to be aimed at acquiring capabilities to counter the US which is increasingly seen as a rival. Despite China's change in status and increasing rivalry with the US, there has been only relatively minor friction with the institutions and systems established in the post-war world in which it had no involvement in creating. This speaks to the resiliency and adaptability of the current world system. The challenge is that the legal structures and organizations that extend into space are relatively weak compared to their terrestrial counterparts.

In contrast to a strategy of full commercialization a partial commercialization effort that sees a substantial decrease in the number of dedicated US government platforms will need to be accompanied by additional signals to ensure it is not misinterpreted as an effort to optimize existing force structure. Abandoning dominance centric language while pursuing arms control agreements and other legal structures that work to the advantage of both the US and its adversaries would help achieve this goal. As the nation seen by both China and Russia as their primary adversary, the US is uniquely positioned to pursue a strategy of reassurance that avoids further escalating the arms race in space. Its status as the dominant space power in terms of presence and capability ensures that a substantial change in posture will not go unnoticed. However, as the degree of commercialization decreases its ability to act as a successful strategy of reassurance also decreases rapidly. A strategy of partial commercialization could still positively impact the security dilemma, but it must be conscious of the danger of seeming to be nothing more than an effort to optimize capability with limited resources.

Global commercialization

The third hypothesis proposed in this book expands the concept of commercialization to all three of the great space powers. This third hypothesis, that if Russia and China also pursued an equivalent strategy of commercialization it would effectively eliminate the security dilemma, can only come about following effective US efforts. It would not eliminate national competition, but it would remove the destabilizing arms race aspect of that competition. Commercialization in space is the first step toward a new norm in space. Global commercialization would not disrupt the norm established at the beginning of the space age that the passive military use of space is peaceful and acceptable. Instead, it would establish a new norm that doing so does not require dedicated national platforms. Future rising space powers would also be held to this new norm. Deviating from it by militarizing would clearly demonstrate the greedy motives and malign intent that existing norms do not.

The beneficial outcomes of global commercialization mask the original question of whether it would effectively eliminate the security dilemma in space. Returning to the definition of the security dilemma, that one nation's quest for security through the accumulation of power creates insecurity in other nations, it is clear that it is imbalances in power that create insecurity, not necessarily imbalances in arms. An arms race is the usual result of this insecurity and the cause of future insecurity. Since space power theory equates presence in space to power, global commercialization in space would not remove power imbalances that might lead to insecurity. Rather than explicit power in the form of military presence, power would be shifted to latent power in the form of economic power. Since a power imbalance would still exist, some form of the security dilemma must also exist.

There is a theoretical disconnect between methods of escaping the security dilemma and the causes of it. In his original formulation of the security dilemma, John Herz described how states would seek to expand individual power through economic means "in order to be self-sufficient in war."[23] He explicitly linked the security dilemma with economic power, and since then all realist theories that seek ways to measure power include a nation's economic capability. Despite this clear link between latent economic power and the security dilemma, methods of building trust through reassurance to escape the security dilemma focus on methods of disrupting the military power imbalance, not the economic one. Perhaps because it is easier to envision a nation choosing not to build arms than it is to envision a nation choosing to impoverish itself deliberately.

Among the primary advantages of a strategy of reassurance built upon commercialization is that it preserves national space power, shifting it from explicit military presence to latent commercial. Properly executed a commercialization strategy may, in fact, increase national space power by leveraging the self-reinforcing nature of economic growth. Attempting to avoid an arms race in space through commercialization cannot remove the fact that there would be real advantages to controlling the space domain in any form, no matter how dangerous or futile obtaining or attempting to obtain that status would be. If the competition in space shifts from military to economic and a power imbalance continues between the great space powers will a security dilemma still exist?

The answer to this question largely depends on the speed in which latent economic space power could be converted to explicit military space power. Economic imbalances that cannot be readily converted to explicit military power generate envy rather than fear, and it is fear that truly drives the security dilemma. Looking to the future, the ability to convert commercial capabilities to military purposes will increase as commercial technologies such as repair, refueling, and debris removal satellites are developed. Domination of one of these future industries by one nation could generate fear that a nation could use them to threaten another nation's space presence. Debate over the potential danger that these systems present is already ongoing.[24] These emerging dual-use commercial capabilities seem to present a challenge to the concept of resolving the security dilemma through commercialization.

The seeming danger of commercial capabilities transforming overnight into space weapons is magnified out of proportion to the threat. This magnification is the result of humanities lack of experience with space warfare. It is analogous to a nation unfamiliar with air warfare fearing another nation's commercial airline industry because it has the potential to be used as a weapon. It is, after all, a plane with significant range and carrying capacity. Could it not then be used to drop bombs, provide close air support, or deliver parachutists to the battlefield? Yes, it could, technically, but the level of inefficiency involved in using a commercial airliner as an effective battlefield weapon makes the idea ludicrous. The validity of this analogy is debatable, though it does serve to highlight a point. Collectively humanity has no experience with warfare in space, and while analysts can see the potential in specific capabilities, there is a substantial difference between a commercial system and a purpose-built weapon. In addition to being systems not designed to dominate space, these future commercial refueling or servicing platforms will have no reason to exist in significant enough numbers to allow a nation to use them in order to rapidly dominate another in space. Using the technology upon which these platforms are based to develop dedicated military platforms is possible, though that will always be a threat and one that is more likely to be avoided with full commercialization than without.

Avoiding the future need to develop sophisticated space weapons is the point of attempting to escape the security dilemma. Should humanity escape the current trend toward a tragic arms race in space, the focus of competition could shift toward the more usual forms of economic competition for market share and profit. The idea that armed force must follow, or even precede, economic development is challenged by the space environment which urges peaceful competition. What is still missing from this analysis is a vision of how a strategy would unfold and what the triggers could be for abandoning it.

Implementation

The path to global commercialization hints at the challenges in measuring the success of commercialization as a reassurance strategy. Whether the US commercialization effort is partial or complete, it is the reactions of other nations that will determine its success at building the trust necessary to defuse unnecessary military competition. The key limiting factor in measuring the success of this strategy is time. A commercialization effort cannot happen overnight. For a myriad of reasons ranging from systems compatibility to contracting timelines even the most aggressive shift toward commercialization would take nearly a decade. Something lacking from the existing literature on reassurance strategies is a practical analysis of the time it takes to execute them and when the signaling nation can judge whether its assumptions on its adversaries' motives are accurate.

Implementing a commercialization strategy would require careful calibration of the rapidity of execution. If it is executed too slowly the strength of the signal would be lost, it could be mistaken for an effort to optimize the existing budget and forces. Alternatively, if it is executed too quickly it could result in disruption of the space support that US armed forces depend on, leaving them vulnerable to other adversary states. Existing dedicated military satellites would not need to be immediately retired for the strategy to work, a definite halt to the development of any future replacements would be adequate. Most of the existing national security space infrastructure could complete its expected on-orbit lifetime. Doing so would soften institutional resistance to the strategy and allow adequate transition periods to commercial systems.

Executing a commercialization strategy would be a slow process and one that cannot happen in isolation. The purpose of commercialization is to signal unmistakably that the US is a benign security seeker in space. This does not mean that lesser signals of benign intent are ineffective or should be ignored. It would be at odds with the purpose of seeking to avoid a sub-optimal arms race for the US to continue to use dominance centric language toward space and avoid putting forward a counter-proposal to the PPWT. These signals should also be accompanied by immediate actions to assume a less militaristic stance in space such as transitioning GPS from military to civil control. It is following these initial changes in US posture that a preliminary reevaluation of Chinese and Russian motives in space can occur.

The relatively low-cost initial efforts at signaling that the US is a security seeker will not immediately change the entrenched mindset of Russia or China that the US is a greedy hegemon and one of the primary threats to their security. However, the assessment that both Russia and China are insecure security seekers can be evaluated on a preliminary basis by the US. These nations should greet US efforts with cautious optimism though they are unlikely to make any real changes in their existing space programs. It would be adequate at this point to see a positive reception to US willingness to negotiate and some signs that plans for future military space competition are slowing. After several years where the US launches no new dedicated military or secret intelligence platforms, a second evaluation can take place.

A mid-point evaluation is the first time where the US could abandon the proposed reassurance strategy. If after low-cost signaling efforts and several years of concrete steps toward commercialization either Russia or China have accelerated the development of space control systems and made no real progress on arms control agreements, then it may be time to pursue other approaches. The shared global nature of the space domain means that reciprocating peaceful overtures must come from both nations. Clear signals from one nation and mixed signals of intent from another may signal that one nation is not yet convinced of benign US intent so reassurance efforts should continue. A decision to continue with the reassurance strategy hinges on an assessment of the risk associated with allowing other nations to continue to arm and may be influenced by actions outside the space domain.

Events in space do not occur in isolation and reassurance efforts within the space domain may coincide with increased tensions elsewhere. This raises the concern that reassurance and cooperation in space cannot happen independently of terrestrial events. Tension in the Baltic states between the US and Russia, or over Taiwan between the US and China, or some other global event will likely occur during the interval of time it takes to pursue a strategy of commercialization. These events will further test assumptions on the motivations of all the states involved, influencing the decision to continue with a reassurance strategy in space.

Even as tensions exist in other domains, several historical and environmental factors support the idea that a reassurance strategy in space can continue. Some of the same factors that constrained US and Soviet competition in space continue to exist. First, the cost of competition continues to remain high in space. Even as launch costs fall and individual satellites get cheaper the aggregate cost of developing deterrent capabilities remains high. Second, while more clouded by other issues than in the past, the importance of preserving methods of strategic communication, reconnaissance, and missile warning to nuclear stability remains. The stabilizing influence of mutual dependence that existed in a different form during the Cold War and has been absent since it ended will also gradually return as China and Russia launch more satellites. Finally, a new factor is the rising awareness of the danger posed by persistent debris created by disabled or destroyed satellites and the impact that could have on the global economy as a result of the increase in economic dependence on space. These incentivizing factors make it possible that reassurance and cooperation can occur in space at the same time that states compete terrestrially.

Given the gradual nature of a commercialization effort, a continuous assessment will be needed once it passes the mid-point and it becomes unmistakable that the US is making a real effort to avoid an arms race. The same points of analysis mentioned for the mid-point evaluation are magnified as the effort approaches completion. Engagement in the development of, or compliance with, arms control agreements should be occurring as well as a halt to further adversary militarization of the domain. Efforts to develop refueling and maintenance platforms should be carefully monitored to ensure that they are being developed for their stated purpose and not as potential weapons, though their dual-use nature will make this a point of contention. Where possible an analysis of budgets and force structure should also occur. Increased relative spending elsewhere, a halt to increases, or even decreases in military space spending are also positive signs that an arms race has been averted.

A danger with continuous assessment is that any sign that an adversary is still arming could cause the policymakers to abandon the strategy before it can prove effective. A sustained trend line of adversary behavior that points to malign intent must be established before deciding whether to halt further pursuit of a commercialization reassurance strategy. The points when the strategy approaches fulfillment and in the period immediately following successful execution are when a judgment on success or failure can be made. Given the long lead time between initiation and completion, US policymakers should be able to accurately gauge the willingness of their principal adversaries in space to cooperate at that point.

Resuming the arms race in space should be a last resort if this strategy fails to encourage cooperation among the US and its principal adversaries in space. The difficulties associated with effectively deterring threats against space assets and the danger that an arms race will end with the deployment of destabilizing space-to-ground weapons remain. Should reassurance fail to encourage cooperation, then the US will know unequivocally which adversaries have greedy motives in space. Inability to cooperate in the space domain will also strongly point to the possibility that Chinese or Russian greedy motives extend across other domains and that a strategy of containment is the only option. The US should then aggressively use all other elements of national power to deter these states from further arming both in space and out of it.

Notes

1 Evan Braden Montgomery, "Breaking out of the Security Dilemma: Realism, Reassurance, and the Problem of Uncertainty," *International Security* 31, no. 2 (2006): 151.
2 Charles L. Glaser, "The Security Dilemma Revisited," *World Politics* 50, no. 1 (1997): 174.
3 "Selected Acquisition Report: Space Fence Ground-Based Radar System Increment 1" (Department of Defense, March 23, 2016), 21, www.esd.whs.mil/Portals/54/Documents/FOID/Reading%20Room/Selected_Acquisition_Reports/16-F-0402_DOC_24_Space_Fence_Inc_1_DEC_2015_SAR.pdf.
4 Sandra Erwin, "The End of SBIRS: Air Force Says It's Time to Move on," *Space News*, February 19, 2018, https://spacenews.com/the-end-of-sbirs-air-force-says-its-time-to-move-on/.
5 See remarks by President Trump and Vice President Pence: Mike Pence, "Remarks by Vice President Pence on the Future of the U.S. Military in Space" (The Pentagon, August 9, 2018), www.whitehouse.gov/briefings-statements/remarks-vice-president-pence-future-u-s-military-space/; Donald J. Trump, "Remarks by President Trump at

a Meeting with the National Space Council and Signing of Space Policy Directive-3" (White House East Room, June 18, 2018), www.whitehouse.gov/briefings-state ments/remarks-president-trump-meeting-national-space-council-signing-space-policy-directive-3/.

6 Andrew Kydd, "Trust, Reassurance, and Cooperation," *International Organization* 54, no. 2 (2000): 326.

7 Glaser, "The Security Dilemma Revisited," 181.

8 Evan Braden Montgomery, "Breaking out of the Security Dilemma: Realism, Reassurance, and the Problem of Uncertainty," 154.

9 Daniel Kahneman, *Thinking, Fast and Slow* (New York, NY: Farrar, Straus and Giroux, 2011), 278–288.

10 Alistair Reign, *Reagan National Defense Forum: STRATCOM Commander Gen. Hyten Discusses Future of Space War*, Panel, 2017, pt. 25:35, www.youtube.com/watch?v=wUIHctFgTac.

11 Google Earth images have raised only limited concerns at their current resolution but higher-resolution imagery is likely to give rise to the same privacy concerns that plague Google's street view, for an example see Claire Downs, "Google Maps' Street View Won't Let You Unblur Images," *The Daily Dot*, April 8, 2018, www.dailydot.com/debug/google-maps-blurring/.

12 Brad Townsend, "The Implications of Innovation in Space-Based Remote Sensing on Maneuver Warfare," *Armor*, Fall (2019): 7.

13 "Foreign Acquisition," 48 part 25 Code of Federal Regulations § (2018).

14 "Remote Sensing Services Market by Platform (Satellites, UAVs, Manned Aircraft, and Ground), End User (Defense and Commercial), Resolution (Spatial, Spectral, Radiometric, and Temporal), and Region—Global Forecast to 2022," Research Report (Markets and Markets, October 2017), www.marketsandmarkets.com/Market-Reports/remote-sensing-services-market-87605355.html.

15 Carl H. Builder, Steven C. Bankes, and Richard Nordin, "Command Concepts: A Theory Derived from the Practice of Command and Control" (RAND Corporation, 1999), www.rand.org/content/dam/rand/pubs/monograph_reports/2006/MR775.pdf.

16 Remote sensing as described in this research broadly covers the imagery centric functions of the US government. The manner and method of any other intelligence centric functions that might be performed by various US government agencies in space are outside the scope of this research.

17 See the following: Sandra Erwin, "U.S. Would Need a Mega-Constellation to Counter China's Hypersonic Weapons," *Space News*, August 8, 2018, https://spacenews.com/u-s-would-need-a-mega-constellation-to-counter-chinas-hypersonic-weapons/; Sandra Erwin, "DARPA to Begin New Effort to Build Military Constellations in Low Earth Orbit," *Space News*, May 31, 2018, https://spacenews.com/darpa-to-begin-new-effort-to-build-military-constellations-in-low-earth-orbit/.

18 Oliver Carroll, "Russian Space Programme Close to Collapse as Latest Failure Exposes Its Fragility | The Independent," *Independent*, December 6, 2017, www.independent.co.uk/news/world/europe/russia-space-programme-collapse-soyuz-2-1b-rocket-cosmo drome-launch-failure-latest-news-a8094856.html.

19 During the author's tenure at NASA there were frequent examples of poor-quality control in the Russian space industry that did not gain public attention. Examples included the failure of a Soyuz capsule during testing due to suspected gross negligence, the failure of a landing impact mitigation device to function, and skipped pre-launch checklist items by ground crew that endangered crew safety and led to delayed launches.

20 Andrey Borisov, "Полетали и Хватит," *Lenta.Ru*, July 3, 2018, https://lenta.ru/articles/2018/07/03/roscosmos/.

21 Andrew Jones, "China Had a Busy and Pioneering Year in Space in 2018," *GB Times*, December 28, 2018, https://gbtimes.com/china-in-space-2018-national-launch-record-commercial-takeoff-and-far-side-of-the-moon.

22 Tian Shaohui, "Backgrounder: Xi Jinping's Vision for China's Space Development—Xinhua | English.News.Cn," *XinhuaNet*, April 24, 2017, www.xinhuanet.com//english/2017-04/24/c_136232642.htm.

23 John H. Herz, "Idealist Internationalism and the Security Dilemma," *World Politics* 2, no. 2 (1950): 173.

24 Chow, "Op-Ed | Applying Weeden's 'Real Talk and Real Solutions' to Real Space Robotic Threats," *Space News*, January 2, 2019, https://spacenews.com/op-ed-applying-weedens-real-talk-and-real-solutions-to-real-space-robotic-threats/.

9 Conclusion

The world is approaching a critical decision point in space. Will the future of space be one centered on economic competition and exploration or one where an arms race turns space into an arena of unbounded intense military activity? Historically economic and military competition go hand-in-hand, but humanity cannot afford to follow the same pattern in space. While a limited war between space powers on Earth might have global consequences, short of nuclear conflict it will not ruin the environment. The same is not true in space. Orbital mechanics dictate that Earth orbit is a fragile shared environment where conflict cannot be geographically contained. Conflict in space could unintentionally poison Earth orbit with trapped debris or radiation, denying humanity the benefits of the domain for decades or centuries, essentially imprisoning humanity to Earth's surface. Even if one nation does succeed in achieving military dominance of space, it will only encourage that nation's adversaries to poison it intentionally. Nations are not trapped in a prisoner's dilemma in space; rather it is an environmentally driven stag hunt where cooperation benefits all parties more than competition.

How do nations that are otherwise in intense competition on Earth pursue constructive cooperation in space? There are examples that demonstrate that it is possible to compete on Earth and cooperate in space. Russia is increasingly seen as a geopolitical rival for the US, yet the US launches its Astronauts on Russian rockets, and both nations are partners in operating the ISS. There are weaknesses in this example because, despite limited signs of cooperation, the US and Russia are still locked in an intensifying arms race that is extending into space. Also, competition is far from a negative as it often leads to innovation. The space race during the Cold War resulted in rapid and stunning achievements in space, some of which have not been repeated in nearly a half-century. The challenge is finding a way for nations to cooperate on keeping competition peaceful.

The problem is that space has become increasingly vital to national military power, especially for the US. Since the end of the Cold War, the US military has become dependent on the passive exploitation of space to enable its continued military dominance. Aware of how vulnerable its space assets are, the US is aggressively trying different approaches to protecting them. Those approaches range from strategies focused on deterrence by denial to ones centered on deterrence by punishment. The issue is that strategies of denial are prohibitively expensive, and successful deterrence by punishment in space suffers from a lack of mutual dependence. Fear of losing its space-enabled capabilities is driving US actions, setting in motion the action-reaction cycle that typifies a security dilemma driven arms race.

An achievable proposal for breaking this cycle is for the US to pursue a strategy of partial commercialization of its national security space architecture to signal benign intent. Since the US is the dominant space power and sits at the center of the growing security dilemma between Russia and China it is uniquely positioned to break the cycle of arming. This strategy, when combined with other weaker signals of benign intent such as actively pursuing arms control in orbit, should encourage cooperation with the other great space powers if they are truly security seekers trapped in a security dilemma and not simply contained greedy states. If the effort to signal benign intent does not break the cycle of arming the US will have left no doubt about the true intentions of its adversaries. It will then be able to leverage its latent commercial space power to resume the undesirable yet unavoidable space arms race toward its conclusion. By pursuing a strategy of cooperation centered on commercialization the US can create the opportunity to avoid weaponizing space while preserving strategic advantage should it fail.

What the future holds

Without some significant change, current trends will eventually lead to the full weaponization of the space domain. The norms that today make this possibility seem distant will erode quickly as the major powers extend their competition for military advantage into space. Existing agreements that govern the behavior of nations in space will look anachronistic as immediate security needs and accusations of cheating override any long-term concerns about future stability.

Failing to avoid an arms race in space will ensure that it will become increasingly expensive and difficult to preserve US military access to space during times of crisis. For the foreseeable future, the US will find any limited conflict it engages in against another space power happening near its adversaries' borders rather than its own. This will mean that it is the US military that will be more dependent on its space infrastructure to support the information-centric tasks vital to success in modern war. Even though strategic advantage lies with the defense in space, a capable adversary will still be able to gain temporary tactical or operational advantages that could lead to defeat in any nuclear constrained limited war.

The resources expended on protecting existing capabilities could be better spent on expanding economic opportunities in space. Commercial innovation in space is rapidly outstripping anything that follows government timelines for development. It is economic rather than military competition that will be most effective in expanding humanity's utilization of space. The nations that successfully exploit the economic potential of space will remain global powers in the future, those that do not will decline relative to their peers. Expending resources in an unnecessary arms race only distracts from realizing the economic benefits of increasingly inexpensive access to space.

It is impossible to completely avoid some minor form of conflict extending into a domain as vital to modern warfare as space. Despite this, every effort must be made to halt the further weaponization of the domain. Should the current arms race continue unabated the world will find itself in an entirely new security paradigm. One where space-to-ground weapons orbit overhead threatening destruction in moments. These weapons will individually be extremely vulnerable to attack creating a dangerously offense-dominant environment with extremely low crisis stability. This would pressure political leaders on both sides of any crisis into a "use it or lose it situation" reminiscent of the darkest days of the Cold War.

That we have so far avoided this future is the result of both technical limitations and deliberate decision making during the initial race for space. Those technical limitations are rapidly disappearing at the same time that international cooperation seems to be breaking down. The world has shifted from a bipolar security competition to a unipolar moment and back toward a multi-polar world in the space of little more than a generation. Humanity has not witnessed war between great powers since the nuclear era began, merely conflict between proxies. Sadly, it seems increasingly likely that this situation will not hold. Containing any future conflict will be much easier should the major powers reach agreement in advance on limiting the further weaponization of space. Choosing to deliberately pursue an alternative to arming may make it possible for humanity to survive and thrive in the new space age as global power balances shift in favor of US opponents. Choosing instead to pursue the easier path of arming will likely result in short-term gains in security at the expense of long-term stability.

Bibliography

Academic papers

Aamoth, Robert J., J. Lauerent Scharf, and Enrico C. Soriano. "The Use of Remote Sensing Imagery by the News Media." *Heaven and Earth* 16 (1997).

Andraschko, Mark, Jeffrey Antol, Stephen Horan, and Neil Doreen. "Commercially Hosted Government Payloads: Lessons from Recent Programs." NASA Langley Research Center, 2011. https://ntrs.nasa.gov/archive/nasa/casi.ntrs.nasa.gov/20110008235.pdf.

Beckley, Michael. "China's Century? Why America's Edge Will Endure." *International Security* 36, no. 3 (2011): 41–78.

Dalmeyer, Dorinda, and Kosta Tsipis. "USAS: Civilian Uses of Near-Earth Space." *Heaven and Earth* 16 (1997).

Davis, James W., Bernard I. Finel, Stacie E. Goddard, Stephen Van Evera, Charles L. Glaser, and Chaim Kaufmann. "Taking Offense at Offense-Defense Theory." *International Security* 23, no. 3 (1998): 179–206.

Downs, George W., and David M. Rocke. "Tacit Bargaining and Arms Control." *World Politics* 39, no. 3 (1987): 297–325.

Finch, James P., and Shawn Steene. "Finding Space in Deterrence." *Strategic Studies Quarterly* 5, no. 4 (2011): 10–17.

Finel, Bernard I., and Kristin M. Lord. "The Surprising Logic of Transparency." *International Studies Quarterly* 43, no. 2 (1999): 315–339.

Finkelstein, S., and S.H. Sanford. "Learning from Corporate Mistakes: The Rise and Fall of Iridium." *Organizational Dynamics* 29, no. 2 (2000): 138–148.

Forest, Benjamin D. "An Analysis of Military Use of Commercial Satellite Communications." Master's Thesis, Naval Post-Graduate School, 2008.

Gao, Yangyu, and Long Ke. "New Trends in US Space Cooperation Policy." *International Research Reference, People's Liberation Army Foreign Languages Institute*, no. 6 (2014).

Glaser, Charles L. "Realists as Optimists: Cooperation as Self-Help." *International Security* 19, no. 3 (1994): 50–90.

Glaser, Charles L. "The Security Dilemma Revisited." *World Politics* 50, no. 1 (1997): 171–201.

Glaser, Charles L. "When Are Arms Races Dangerous? Rational versus Suboptimal Arming." *International Security* 28, no. 4 (2004): 44–84.

Glaser, Charles L. "A U.S.-China Grand Bargain?" *International Security* 39, no. 4 (Spring 2015): 49–90.

Glaser, Charles L., and Chaim Kaufmann. "What Is the Offense-Defense Balance and Can We Measure It?" *International Security* 22, no. 4 (1998): 44–82.

Hanwen, Ge. "'Say No to Decline' and 'American Fortification': Trump's Grand Strategy." *Journal of International Security Studies,* 国防科技大学国际关系学院国际安全研究中心 *(International Studies College, National University of Defense Technology),* no. 3 (March 11, 2018).

Hendrickx, Bart. "Naryad-V and the Soviet Anti-Satellite Fleet." *Space Chronicle* 69 (2016).

Herz, John H. "Idealist Internationalism and the Security Dilemma." *World Politics* 2, no. 2 (1950): 157–180.

Hugill, Peter J. "German Great-Power Relations in the Pages of 'Simplicissimus,' 1896–1914." *Geographical Review* 98, no. 1 (2008): 1–23.

Itzkowitz Shifrinson, Joshua R. "Deal or No Deal? The End of the Cold War and the U.S. Offer to Limit NATO Expansion." *International Security* 40, no. 4 (2016): 7–44.

Itzkowitz Shifrinson, Joshua R., and Michael Beckley. "Debating China's Rise and U.S. Decline." *International Security* 37, no. 3 (December 13, 2012): 172–181. https://doi.org/10.1162/ISEC_c_00111.

Javits, Jacob K. "Congress and Foreign Relations: The Taiwan Relations Act." *Foreign Affairs* 60, no. 1 (1981): 54–62. doi: 10.2307/20040989.

Jervis, Robert. "Cooperation Under the Security Dilemma." *World Politics* 30, no. 2 (1978): 167–214.

Jervis, Robert. *Perception and Misperception in International Politics.* Princeton, NJ: Princeton University Press, 1976.

Jervis, Robert. "Was the Cold War a Security Dilemma?" *Journal of Cold War Studies* 3, no. 1 (2001): 36–60.

Jervis, Robert. "Dilemmas About Security Dilemmas." *Security Studies* 20, no. 3 (July 2011): 416–423.

Kydd, Andrew. "Game Theory and the Spiral Model." *World Politics* 49, no. 3 (1997): 371–400.

Kydd, Andrew. "Trust, Reassurance, and Cooperation." *International Organization* 54, no. 2 (2000): 325–357.

Lambakis, Steven. "Space Control in Desert Storm and Beyond." *Orbis* 39, no. 3 (Summer 1995).

Liff, Adam P., and G. John Ikenberry. "Racing toward Tragedy?" *International Security* 39, no. 2 (2014): 52–91.

Mastro, Oriana Skylar. "The Vulnerability of Rising Powers: The Logic Behind China's Low Military Transparency." *Asian Security* 12, no. 2 (May 3, 2016): 63–81.

Merges, Robert P., and Glenn H. Reynolds. "News Media Satellites and the First Amendment: A Case Study in the Treatment of New Technologies." *Berkeley Technology Law Journal* 3, no. 1 (January 1988).

Montgomery, Evan B. "Breaking out of the Security Dilemma: Realism, Reassurance, and the Problem of Uncertainty." *International Security* 31, no. 2 (2006): 151–185.

Morrow, James D. "Arms Versus Allies: Trade-Offs in the Search for Security." *International Organization* 47, no. 2 (1993): 207–233.

Podvig, Pavel. "Did Star Wars Help End the Cold War? Soviet Response to the SDI Program." *Science & Global Security* 25, no. 1 (2017): 3–27.

Scott, David. "Conflict Irresolution in the South China Sea." *Asian Survey* 52, no. 6 (2012): 1019–1042.

Sleeth, Denette L. "Commercial Imagery Satellite Threat: How Can U.S. Forces Protect Themselves?" Master's Thesis, Naval War College, 2004.

Smith, Delbert D. "The Legal Ordering of Satellite Telecommunication: Problems and Alternatives." *Indiana Law Journal* 44, no. 3 (1969). www.repository.law.indiana.edu/ilj/vol.44/iss3/1.

Snyder, Glenn H. "'Prisoner's Dilemma' and 'Chicken' Models in International Politics." *International Studies Quarterly* 15, no. 1 (1971): 66–103.

"The South China Sea Arbitration (*The Republic of the Philippines* v. *The People's Republic of China*)." *The International Journal of Marine and Coastal Law* 31, no. 4 (November 22, 2016): 759–774.

Townsend, Brad, "The Implications of Innovation in Space-Based Remote Sensing on Maneuver Warfare," *Armor*, Fall (2019): 7.

Tung, Chen-yuan. "An Assessment of China's Taiwan Policy under the Third Generation Leadership." *Asian Survey* 45, no. 3 (2005): 343–361.

Van Evera, Stephen "Offense, Defense, and the Causes of War." *International Security* 22, no. 4 (1998): 5–43.

Zhang, ZhongXiang. "Why Are the Stakes So High?" In *Rebalancing and Sustaining Growth in China*, 2012:329–356. ANU Press, 2012.

Articles

"Air Force Commercially Hosted Infrared Payload Mission Completed." US Air Force. Los Angeles Air Force Base, December 6, 2013. www.losangeles.af.mil/News/Article-Display/Article/734794/air-force-commercially-hosted-infrared-payload-mission-completed/.

Andrews, Wilson, and Todd Lindeman. "The Black Budget." *Washington Post*, August 29, 2013. www.washingtonpost.com/wp-srv/special/national/black-budget/?noredirect=on.

Anselmo, Joseph C. "Commercial Space's Sharp New Image." *Aviation Week & Space Technology*, January 31, 2000.

Anselmo, Joseph C. "A New Image." *Aviation Week & Space Technology*, January 30, 2006.

Apple Jr., R.W. "War in the Gulf: Scud Attack; Scud Missile Hits a US Barracks, Killing 27." *New York Times*, February 26, 1991.

Artamonov, Alexander. "Americans Force Russia to Create Weapons Even More Powerful than S-500." *Pravda*, November 13, 2017. www.pravdareport.com/russia/economics/13-11-2017/139167-anti_missile_weapons-0/.

Associated Press. "Rumsfeld: China Buildup a Threat to Asia." *NBC News*, June 4, 2005. www.nbcnews.com/id/8091198/ns/world_news-asia_pacific/t/rumsfeld-china-buildup-threat-asia/#.W7edKfZFxEZ.

Berger, Eric. "In 2017, the US Led the World in Launches for the First Time since 2003: With 18 Orbital Flights, SpaceX Drove the Surge in US Missions Last Year." *ArsTechnica*, January 3, 2018. https://arstechnica.com/science/2018/01/in-2017-the-us-led-the-world-in-launches-for-the-first-time-since-2003/.

Berger, Eric. "Russia Appears to Have Surrendered to SpaceX in the Global Launch Market." *ArsTechnica*, April 18, 2018. https://arstechnica.com/science/2018/04/russia-appears-to-have-surrendered-to-spacex-in-the-global-launch-market/.

Berlocher, Greg. "Military Continues to Influence Commercial Operators." *Satellite Today*, September 1, 2008. www.satellitetoday.com/publications/via-satellite-magazine/supplement/2008/09/01/military-continues-to-influence-commercial-operators/.

Beutler, Gerhard, Rolf Dach, Urs Hugentobler, Oliver Montenbruk, George Weber, and Elmar Brockmann. "The System: GLONASS in April, What Went Wrong." *GPS*

World, June 24, 2014. http://gpsworld.com/the-system-glonass-in-april-what-went-wrong/.

Bhatia, Sanjeev. "Understanding High Throughput Satellite (HTS) Technology." June 2013. www.intelsat.com/wp-content/uploads/2013/06/HTStechnology_bhartia.pdf.

Borisov, Andrey. "Полетали и Хватит." *Lenta.Ru*, July 3, 2018. https://lenta.ru/articles/2018/07/03/roscosmos/.

Brinton, Turner. "NGA Solicits Proposals for Commercial Radar Imagery." *Space News*, September 11, 2009. http://spacenews.com/nga-solicits-proposals-commercial-radar-imagery/.

Brinton, Turner. "US Loosens Restrictions on Commercial Radar Satellites." *Space News*, October 8, 2009. https://spacenews.com/us-loosens-restrictions-commercial-radar-satellites/.

Burns, Robert. "Pentagon Chief Defends His Reversal on Space Force, Says It's the Right Thing to Do." *Military Times*, August 13, 2018. www.militarytimes.com/news/your-military/2018/08/13/pentagon-chief-defends-his-reversal-on-space-force-says-its-the-right-thing-to-do/.

Butler, Amy. "USAF Space Chief Outs Classified Spy Sat Program." *Aviation Week*, February 21, 2014. http://aviationweek.com/defense/usaf-space-chief-outs-classified-spy-sat-program.

Campbell, Duncan. "US Buys up All Satellite War Images." *Guardian*, October 17, 2001.

Carroll, Oliver. "Russian Space Programme Close to Collapse as Latest Failure Exposes Its Fragility | The Independent." *Independent*, December 6, 2017. www.independent.co.uk/news/world/europe/russia-space-programme-collapse-soyuz-2-1b-rocket-cosmodrome-launch-failure-latest-news-a8094856.html.

Caterinicchia, Dan. "NIMA Seeks 'ClearView' of World." *FCW: The Business of Federal Technology*, January 16, 2003. https://fcw.com/articles/2003/01/16/nima-seeks-clearview-of-world.aspx.

Cebul, Daniel. "Coercive Tactics? Putin Touts Russia's 'Invincible' Nuclear Weapons." *Defense News*, March 2, 2018. www.defensenews.com/smr/nuclear-triad/2018/03/01/coersive-tactics-putin-touts-russias-invincible-nuclear-weapons/.

Chan, Minnie. "'Unforgettable Humiliation' Led to Development of GPS Equivalent." *South China Morning Post*, November 13, 2009. www.scmp.com/article/698161/unforgettable-humiliation-led-development-gps-equivalent.

"CHIRP." Hosted Payload Alliance. Accessed October 25, 2018. www.hostedpayloadalliance.org/Hosted-Payloads/Case-Studies/Commercially-Hosted-Infrared-Payload-(CHIRP)-Fligh.aspx#.W9dhIfZFxEZ.

Chow. "Op-Ed | Applying Weeden's 'Real Talk and Real Solutions' to Real Space Robotic Threats." *Space News*, January 2, 2019. https://spacenews.com/op-ed-applying-weedens-real-talk-and-real-solutions-to-real-space-robotic-threats/.

Clark, Colin. "Chinese ASAT Test Was 'Successful': Lt. Gen. Raymond." *Breaking Defense*, April 14, 2015. https://breakingdefense.com/2015/04/chinese-asat-test-was-successful-lt-gen-raymond/.

Cordesman, Anthony H., and Joseph Kendall. "How China Plans to Utilize Space for A2/AD in the Pacific." *The National Interest*, August 17, 2016. https://nationalinterest.org/blog/the-buzz/how-china-plans-utilize-space-a2-ad-the-pacific-17383.

"Crimea Declares Independence, Seeks UN Recognition." *RT News*, March 17, 2014. www.rt.com/news/crimea-referendum-results-official-250/.

"Defense and Intelligence." Planet, July 18, 2018. https://planet.com/markets/defense-and-intelligence/.

"Defense Support Program: Celebrating 40 Years of Service." Northrop Grumman, 2010.

"Defense Support Program Satellites." US Air Force Space Command Public Affairs Office, November 23, 2015. www.af.mil/About-Us/Fact-Sheets/Display/Article/104611/defense-support-program-satellites/.

"Deutsches Zentrum Für Luft- Und Raumfahrt (National Aeronautics and Space Research Center)." National Aeronautics and Space Research Center, Federal Republic of Germany. Accessed December 26, 2016. www.dlr.de/dlr/en/desktopdefault.aspx/tabid-10377/565_read-436/#/gallery/350.

"'Doomsday Sputnik': Russia Said to Launch New Missile-Attack Warning Satellite." RT International, July 19, 2014. www.rt.com/news/174076-early-warning-satellite-russia/.

Downs, Claire. "Google Maps' Street View Won't Let You Unblur Images." *The Daily Dot*, April 8, 2018. www.dailydot.com/debug/google-maps-blurring/.

Du, Mingming. "China Releases White Paper on Facts and Its Position on Trade Friction with U.S." *Peoples Daily*, September 24, 2018. http://en.people.cn/n3/2018/0924/c90000-9503026.html.

Erlanger, S. "NATO Leaders Hear Putin on Russia Security Worries." *Pittsburgh Post—Gazette*, April 5, 2008.

Erwin, Sandra. "Boeing to Accelerate Production of WGS Satellites." *Space News*, April 18, 2018. https://spacenews.com/boeing-to-accelerate-production-of-wgs-satellites/.

Erwin, Sandra. "Some Fresh Tidbits on the U.S. Military Space Budget." *Space News*, March 21, 2018. https://spacenews.com/some-fresh-tidbits-on-the-u-s-military-space-budget/.

Erwin, Sandra. "The End of SBIRS: Air Force Says It's Time to Move on." *Space News*, February 19, 2018. https://spacenews.com/the-end-of-sbirs-air-force-says-its-time-to-move-on/.

Erwin, Sandra. "DARPA to Begin New Effort to Build Military Constellations in Low Earth Orbit." *Space News*, May 31, 2018. https://spacenews.com/darpa-to-begin-new-effort-to-build-military-constellations-in-low-earth-orbit/.

Erwin, Sandra. "Defense Budget Bill Creates Path for Future Network of Military, Commercial Communications Satellites." *Space News*, June 16, 2018. https://spacenews.com/defense-budget-bill-creates-path-for-future-network-of-military-commercial-communications-satellites/.

Erwin, Sandra. "U.S. Would Need a Mega-Constellation to Counter China's Hypersonic Weapons." *Space News*, August 8, 2018. https://spacenews.com/u-s-would-need-a-mega-constellation-to-counter-chinas-hypersonic-weapons/.

Erwin, Sandra. "Lockheed Martin Awarded $2.9 Billion Air Force Contract for Three Missile-Warning Satellites." *Space News*, August 14, 2018. https://spacenews.com/lockheed-martin-secures-2-9-billion-air-force-contract-for-three-missile-warning-satellites/.

Erwin, Sandra. "Army Secretary: Still Unclear What Portions of the Army Would Move to the Space Force." *Space News*, October 8, 2018. https://spacenews.com/army-secretary-still-unclear-what-portions-of-the-army-would-move-to-the-space-force/.

Erwin, Sandra. "Space Development Agency Releases Its First Solicitation," *SpaceNews.com*, July 4, 2019, https://spacenews.com/space-development-agency-releases-its-first-solicitation/.

Eurockot. "Eurockot Launch Services." Accessed October 2, 2018. www.eurockot.com/launch-services/.

"Federal Space Agency and United Rocket and Space Corporation Was Mergered in the New State Corporation—Roscosmos." *Roscosmos*, January 23, 2015. http://en.roscosmos.ru/20365/.

Ferguson, Edward, and John Klein. "The Future of War in Space Is Defensive." *Space Review*, December 19, 2016. www.thespacereview.com/article/3131/1.

Foust, Jeff. "U.S. Dismisses Space Weapons Treaty Proposal As 'Fundamentally Flawed.'" *Space News*, September 11, 2014. https://spacenews.com/41842us-dismisses-space-weapons-treaty-proposal-as-fundamentally-flawed/.

Gellman, B. "U.S. and China Nearly Came to Blows in '96; Tension over Taiwan Prompted Repair of Ties." *Washington Post*, June 21, 1998, sec. 1 of 2.

Germroth, David S. "Commercial SAR Comes to the U.S. (Finally!)." *ApoGeo Spatial*, May 9, 2016. http://apogeospatial.com/commercial-sar-comes-to-the-u-s-finally/.

Gertz, Bill. "China Conducts Test of New Anti-Satellite Missile." *The Washington Free Beacon*, May 14, 2013. https://freebeacon.com/national-security/china-conducts-test-of-new-anti-satellite-missile/.

Gertz, Bill. "China Carries Out Flight Test of Anti-Satellite Missile." *The Washington Free Beacon*, August 2, 2017. https://freebeacon.com/national-security/china-carries-flight-test-anti-satellite-missile/.

Gomez, Jim. "Duterte: China Should Temper Its Behavior in Disputed Waters." *Associated Press*, August 15, 2018. www.apnews.com/21030e69e14345bb9ce6b8bd91ce308a.

"GoodBye Earlybird and Earthwatch Too." *SpaceDaily*, January 14, 1998. www.spacedaily.com/reports/GoodBye_Earlybird__And_Earthwatch__Too.html.

Gordon, Michael R., and David E. Sanger. "The Bailout of the Kremlin: How the US Pressed the IMF." *New York Times*, July 17, 1998. www.nytimes.com/1998/07/17/world/rescuing-russia-special-report-bailout-kremlin-us-pressed-imf.html.

Grossman, Derek. "Why March 2018 Was an Active Month in Vietnam's Balancing Against China in the South China Sea." *The Diplomat*, March 23, 2018. https://thediplomat.com/2018/03/why-march-2018-was-an-active-month-in-vietnams-balancing-against-china-in-the-south-china-sea/.

Grove, T., M.R. Gordon, and J. Marson. "Putin Unveils Nuclear Weapons He Claims Could Breach U.S. Defenses; Russian Leader Sharpens Rhetoric against West, Intensifying Arms Race with the U.S." *Wall Street Journal*, March 1, 2018.

Gruss, Mike. "U.S. Air Force Decision To End CHIRP Mission Was Budget Driven." *Space News*, December 12, 2013. https://spacenews.com/38628us-air-force-decision-to-end-chirp-mission-was-budget-driven/.

Gruss, Mike. "U.S. Official: China Turned to Debris-Free ASAT Tests Following 2007 Outcry." *Space News*, January 11, 2016. https://spacenews.com/u-s-official-china-turned-to-debris-free-asat-tests-following-2007-outcry/.

Gruss, Mike. "US Air Force Award Contract for New Waveform Demonstrations." *Space News*, August 25, 2016. https://spacenews.com/u-s-air-force-awards-contracts-for-new-waveform-demonstrations/.

Guzman, Chad de. "Duterte Asserts Arbitral Ruling on South China Sea Not during His Term." *CNN*, May 21, 2018. http://cnnphilippines.com/news/2018/05/21/duterte-arbitral-ruling-south-china-sea.html.

Henry, Caleb. "MDA Closes DigitalGlobe Merger, Rebrands as Maxar Technologies." *Space News*, October 5, 2017. https://spacenews.com/mda-closes-digitalglobe-merger-rebrands-as-maxar-technologies/.

Henry, Caleb. "OneWeb Asks FCC to Authorize 1,200 More Satellites." *Space News*, March 20, 2018. https://spacenews.com/oneweb-asks-fcc-to-authorize-1200-more-satellites/.

Hill, Kashmir. "Jamming GPS Signals Is Illegal, Dangerous, Cheap, and Easy." *Gizmodo*, July 24, 2017. https://gizmodo.com/jamming-gps-signals-is-illegal-dangerous-cheap-and-e-1796778955.

Hitchens, Theresa, "DoD, Commerce Wrangle New Commercial Remote Sensing Regs," *Breaking Defense*, December 4, 2019, https://breakingdefense.com/2019/12/exclusive-dod-commerce-negotiate-commercial-remote-sensing-regs/.

Horne, A.D. "U.S. Says Soviets Shot Down Airliner." *Washington Post*, September 2, 1983.

"Ikonos Satellite." GeoImage. Accessed December 12, 2016. www.geoimage.com.au/satellite/ikonos.

Insinna, Valerie. "STRATCOM Head: Timeline to Field New Missile Warning Satellites Is 'Ridiculous.'" *Defense News*, December 4, 2017. www.defensenews.com/digital-show-dailies/reagan-defense-forum/2017/12/04/stratcom-head-timeline-to-field-new-missile-warning-satellites-is-ridiculous/.

"Intelsat Global Network." Official Company Page. Intelsat. Accessed November 5, 2018. www.intelsat.com/global-network/.

International Monetary Fund (IMF). "Components of the Gross Domestic Product (GDP) Data Series in International Financial Statistics (IFS)." Accessed August 21, 2018. http://datahelp.imf.org/knowledgebase/articles/498480-what-are-the-components-of-the-gross-domestic-prod.

"Investor Information." DigitalGlobe Website. Accessed December 23, 2016. http://investor.digitalglobe.com/phoenix.zhtml?c=70788&p=rsslanding&cat=news&id=1939027.

"Iran Jams VOA's Satellite Broadcasts." *Voice of America News*, October 5, 2012. www.voanews.com/a/iran-jams-voa-satellite-broadcasts/1521003.html.

Jones, Andrew. "Chinese Space Launch Vehicle Maker Provides Updates on Long March Launch Plans, New Rockets." *GB Times*. March 2, 2018. https://gbtimes.com/chinese-space-launch-vehicle-maker-provides-updates-on-long-march-launch-plans-new-rockets.

Jones, Andrew. "China Had a Busy and Pioneering Year in Space in 2018." *GB Times*. December 28, 2018. https://gbtimes.com/china-in-space-2018-national-launch-record-commercial-takeoff-and-far-side-of-the-moon.

Jones, Andrew. "Chinese Commercial Rocket Smart Dragon-1 Reaches Orbit with First Launch," *Space News*, August 19, 2019, https://spacenews.com/chinese-commercial-rocket-smart-dragon-1-reaches-orbit-with-first-launch/.

Jones, Sam. "Object 2014–28E—Space Junk or Russian Satellite Killer?" *Financial Times*, November 17, 2014. www.ft.com/content/cdd0bdb6-6c27-11e4-990f-00144fe abdc0#axzz3JPZDZk6I.

Kaplan, Fred. "The Pentagon A War in the Stars?; Reagan Team Presses Outer Space Arms Plan." *Boston Globe*, October 16, 1983.

Kato, Masaya. "Japan and US to Sharpen Missile Defense with Real-Time Data Sharing." *Nikkei Asian Review*, August 25, 2018. https://asia.nikkei.com/Politics/International-Relations/Japan-and-US-to-sharpen-missile-defense-with-real-time-data-sharing.

Keck, Zachary. "China Conducts Third Anti-Missile Test." *The Diplomat*, July 24, 2014. https://thediplomat.com/2014/07/china-conducts-third-anti-missile-test/.

Keck, Zachary. "China Conducted Anti-Satellite Missile Test." *The Diplomat*, July 29, 2014. https://thediplomat.com/2014/07/china-conducted-anti-satellite-missile-test/.

Kelso, T.S. "CelesTrak: Chinese ASAT Test." *CelesTrak*, June 22, 2012. https://celestrak.com/events/asat.php.

Klotz, Irene. "US Air Force Reveals 'neighborhood Watch' Spy Satellite Program." *Reuters*, February 22, 2014. www.reuters.com/article/us-space-spysatellite-idUSBRE A1L0YI20140222.

"Landsat Science: Landsat 5." NASA.gov. Accessed February 8, 2017. http://landsat. gsfc.nasa.gov/landsat-5/.

Lee Myers, Steven. "As NATO Finally Arrives on Its Border, Russia Grumbles." *New York Times*, April 3, 2004.

Lewis, George, and Theodore Postol. "Ballistic Missile Defense: Radar Range Calculations for the AN/TPY-2 X-Band and NAS Proposed GBX Radars (September 21, 2012)." *Mostlymissiledefense* (blog), September 21, 2012. https://mostlymissiledefense. com/2012/09/21/ballistic-missile-defense-radar-range-calculations-for-the-antpy-2-x-band-and-nas-proposed-gbx-radars-september-21-2012/.

Lowe, Christian. "Georgia War Shows Russian Army Strong but Flawed." *Reuters*, August 20, 2008. www.reuters.com/article/us-georgia-ossetia-military/georgia-war-shows-russian-army-strong-but-flawed-idUSLK23804020080821.

Mathewson, Samantha. "India Launches Record-Breaking 104 Satellites on Single Rocket." *Space.Com*, February 15, 2017. www.space.com/35709-india-rocket-launches-record-104-satellites.html.

Mehta, Aaron. "3 Thoughts on Hypersonic Weapons from the Pentagon's Technology Chief." *Defense News*, July 16, 2018. www.defensenews.com/air/2018/07/16/3-thoughts-on-hypersonic-weapons-from-the-pentagons-technology-chief/.

Mellow, Craig. "The Rise and Fall of Iridium." *Air and Space Magazine*, September 2004. www.airspacemag.com/space/the-rise-and-fall-and-rise-of-iridium-5615034/.

Mizokami, Kyle. "China Proposes Orbiting Laser to Combat Space Junk." *Popular Mechanics*, February 20, 2018. www.popularmechanics.com/military/weapons/a18240128/china-orbital-laser-space-junk/.

Moreno-Davis, Laura. "UltiSat Named Awardee on $2.5 Billion Complex Commercial SATCOM Solutions (CS3) IDIQ Contract." *Globe Newswire*, August 28, 2017. https:// globenewswire.com/news-release/2017/08/28/1101038/0/en/UltiSat-Named-Awardee-on-2-5-Billion-Complex-Commercial-SATCOM-Solutions-CS3-IDIQ-Contract.html.

Neilan, Terrence. "Bush Pulls Out of ABM Treaty; Putin Calls Move a Mistake." *New York Times*, December 13, 2001. www.nytimes.com/2001/12/13/international/bush-pulls-out-of-abm-treaty-putin-calls-move-a-mistake.html.

Newell, Benjamin. "$93 Million Contract Keeps Missile Warning System Vigilant." Air Force Global Strike Command, July 13, 2017. www.afgsc.af.mil/News/ArticleDisplay/tabid/2612/Article/1247579/93-million-contract-keeps-missile-warning-system-vigilant. aspx.

Nyirady, Annamarie. "Iridium Awarded $44 Million DISA Contract Extension." *Via Satellite*, October 15, 2018. www.satellitetoday.com/government-military/2018/10/15/iridium-awarded-44-million-disa-contract-extension/.

Obermueller, N. "Protesters Expect Russia Will Intervene." *USA Today*, February 7, 2014.

"Orbit Fab: Fuel Supply for Satellites." Orbit Fab. Accessed September 22, 2018. www. orbitfab.space/.

"PanGeo-Alliance." PanGeo Alliance. Accessed November 15, 2018. www.pangeo-alliance.com/.

Patranobis, Sutirtho. "China Plans to Reduce Satellite Launch Prices, ISRO Says We Can Do That Too." *Hindustan Times*, November 14, 2017. www.hindustantimes.com/world-news/india-china-in-race-to-reduce-rocket-launch-prices/story-mF7X9RwS5ai1rCjU DYb2zH.html.

Pixalytics ltd. "Earth Observation Satellites in Space in 2017?" Pixalytics Ltd., November 29, 2017. www.pixalytics.com/eo-sats-in-space-2017/.

"Press Release: DigitalGlobe and GeoEye Agree to Combine to Create a Global Leader in Earth Imagery and Geospatial Analysis." *DigitalGlobe*, July 23, 2012. http://investor. digitalglobe.com/phoenix.zhtml?c=70788&p=irol-newsArticle&ID=1717100.

"Press Release: IKONOS Imaging Satellite Achieves 15 Years of On-Orbit Operation." Lockheed-Martin Corp., September 24, 2014. www.lockheedmartin.com/us/news/ press-releases/2014/september/0924-space-IKONOS.html.

"Press Release: Maxar Technologies Shareholders Approve U.S. Domestication." Maxar Technologies, November 16, 2018. http://investor.maxar.com/investor-news/press-release-details/2018/Maxar-Technologies-Shareholders-Approve-US-Domestication/ default.aspx.

"Putin: Unilateral US Withdrawal from ABM Treaty Pushing Russia toward New Arms Race." *RT News*, June 19, 2015. www.rt.com/news/268345-putin-west-russia-relations/.

"Radarsat-1." Canadian Space Agency. Accessed December 26, 2016. www.asc-csa. gc.ca/eng/satellites/radarsat1/default.asp.

Reed, S.R. "Soviets Offer to Pull Arms from Space." *Philadelphia Inquirer*, August 19, 1983.

Reif, Kingston. "Trump to Withdraw U.S. From INF Treaty | Arms Control Association," November 2018. www.armscontrol.org/act/2018-11/news/trump-withdraw-us-inf-treaty.

"Resolution Restrictions Lifted." DigitalGlobe Website. Accessed April 19, 2017. http:// blog.digitalglobe.com/news/resolutionrestrictionslifted/.

Reuters. "Iridium Declares Bankruptcy." *New York Times*, August 14, 1999.

"Russian Deputy PM Sees No Reason for Competing with Musk on Launch Vehicles Market." *TASS*. April 17, 2018, sec. Science and Space. http://tass.com/science/ 1000229.

"Russia's Soyuz Launches EKS Missile Warning Satellite, Ends Year-Long Military Launch Gap." *SpaceFlight 101*, May 25, 2017. http://spaceflight101.com/soyuz-successfully-launches-second-eks-satellite/.

Sawyer, Kathy. "NASA Wants to Bail out the Russian Space Agency." *Washington Post*, September 21, 1998. www.washingtonpost.com/wp-srv/national/longterm/station/ stories/russia.htm.

Scharre, Paul. "The US Military Should Not Be Doubling down on Space." *Defense One* (blog), August 1, 2018. www.defenseone.com/ideas/2018/08/us-military-should-not-be-doubling-down-space/150194/.

Scott, Walter. "U.S. Satellite Imaging Regulations Must Be Modernized, Op-Ed by Digital Globe Founder." *Space News*, August 29, 2016. http://spacenews.com/op-ed-u-s-satellite-imaging-regulations-must-be-modernized/.

"Secretary-General Unveils $5.4 Billion 2018–2019 Programme Budget to Fifth Committee, Drawing Mixed Reviews from Delegates on Funding, Staffing Cuts | Meetings Coverage and Press Releases," October 11, 2017. www.un.org/press/en/2017/ gaab4243.doc.htm.

Selding, Peter. "DigitalGlobe Revenue up Despite Steep Drop in Russian Business." *Space News*, August 1, 2014. http://spacenews.com/41459digitalglobe-revenue-up-despite-steep-drop-in-russian-business/.

Selding, Peter. "EnhancedView Contract Awards Carefully Structured, NGA Says." *Space News*, September 10, 2010. http://spacenews.com/enhancedview-contract-awards-carefully-structured-nga-says/#sthash.7K7GA9Q4.dpuf.

Shaohui, Tian. "Backgrounder: Xi Jinping's Vision for China's Space Development— Xinhua | English.News.Cn." *XinhuaNet*, April 24, 2017. www.xinhuanet.com// english/2017-04/24/c_136232642.htm.

Shevchenko, Vitaly. "'Little Green Men' or 'Russian Invaders'?" *BBC*, March 11, 2014. www.bbc.com/news/world-europe-26532154.

Shuster, Simon. "Stanislav Petrov, the Russian Officer Who Averted a Nuclear War, Feared History Repeating Itself." *Time Magazine*, September 19, 2017. http://time.com/4947879/stanislav-petrov-russia-nuclear-war-obituary/.

Singer, Jeremy. "War on Terror Supersedes 2001 Space Commission Vision." *Space News*, June 29, 2004. https://spacenews.com/war-terror-supersedes-2001-space-commission-vision/.

Singer, Jeremy. "DOD Notifies Congress of Higher SBIRS Cost." *Space News*, March 14, 2005. https://spacenews.com/dod-notifies-congress-higher-sbirs-cost/.

"SIPRI Military Expenditure Database." Database. Stockholm International Peace Research Institute. Accessed September 12, 2018. www.sipri.org/databases/milex.

Smith, Marcia S. "SASC Adds Funds for ORS, STP and Commercial Imagery Purchase." *Space Policy Online*, May 25, 2012. www.spacepolicyonline.com/news/sasc-adds-funds-for-ors-stp-and-commercial-imagery-purchase.

Smith, Marcia S. "House Passes Final Version of NDAA, Goes Home for Now—UPDATE." *Space Policy Online*, December 21, 2012. www.spacepolicyonline.com/news/house-passes-final-version-of-ndaa-goes-home-for-now-update.

Socor, Vladimir. "Russia Moves toward Open Annexation of Abkhazia, South Ossetia." *Eurasia Daily Monitor*, April 18, 2008. https://jamestown.org/program/russia-moves-toward-open-annexation-of-abkhazia-south-ossetia/.

Sorensen, Jodi. "Press Release: Spaceflight Successfully Launches Flock of 12 Planet Dove Satellites on India's PSLV." *Spaceflight*, June 21, 2016. http://spaceflight.com/spaceflight-successfully-launches-flock-of-12-planet-dove-satellites-on-indias-pslv/.

"Soyuz Data Sheet." Space Launch Report. Accessed August 23, 2018. www.spacelaunchreport.com/soyuz.html.

"Space Based Infrared System." Air Force Space Command. Accessed November 27, 2018. www.afspc.af.mil/About-Us/Fact-Sheets/Display/Article/1012596/space-based-infrared-system/.

"Space Systems Bandwidth." Global Security. Accessed December 30, 2016. www.globalsecurity.org/space/systems/bandwidth.htm.

"SPOT Medium Resolution Satellite Imagery." Apollo Mapping Website, https://apollomapping.com/imagery/medium-resolution-satellite-imagery/spot.

"SPOT Satellite Pour Observation Terre." GIS Geography, July 30, 2016. http://gisgeography.com/spot-satellite-pour-observation-terre/.

Squeo, Anne Marie. "The Assault on Iraq: U.S. Bombs Iraqi GPS-Jamming Sites." *Wall Street Journal*, March 26, 2003.

Stearns, Ron. "ClearView Contract and Greater Imagery Availability Move U.S. Satellite Commercial Imaging Market Forward." *Frost & Sullivan*, February 21, 2003. www.frost.com/sublib/display-market-insight.do?id=GLEN-5JYVUJ.

Stewart, Phil, and Ben Blanchard. "Xi Tells Mattis China Won't Give up 'even One Inch' of Territory." *Reuters*, June 26, 2018. https://uk.reuters.com/article/uk-china-usa-defence/xi-tells-mattis-china-wont-give-up-one-inch-of-territory-idUKKBN1JN06O.

"Strait of Malacca Key Chokepoint for Oil Trade." *The Maritime Executive*, August 27, 2018. www.maritime-executive.com/article/strait-of-malacca-key-chokepoint-for-oil-trade.

"STS-44." NASA Science, November 1991. https://science.ksc.nasa.gov/mirrors/images/html/STS44.htm.

"SWIR Imagery." Official Company Page. Digital Globe. Accessed December 14, 2018. www.digitalglobe.com/products/swir-imagery.

"The World's Most Powerful Rocket: Falcon Heavy." SpaceX. Accessed September 12, 2018. www.spacex.com/falcon-heavy.

Tirpak, John A. "Operation Burnt Frost 'Historic.'" *Air Force Magazine*, February 22, 2008. www.airforcemag.com/Features/security/Pages/box022208shootdown.aspx.

Tritten, Travis J. "Jim Mattis Originally Opposed Space Force over Sequester, Deputy Says." *Washington Examiner*, August 10, 2018. www.washingtonexaminer.com/policy/defense-national-security/jim-mattis-originally-opposed-space-force-over-sequester-deputy-says.

Twing, Shawn L. "U.S. Bans High-Resolution Imagery of Israel." *Washington Report on Middle East Affairs*, September 1998. www.wrmea.org/1998-september/u.s.-bans-high-resolution-imagery-of-israel.html.

"Ukraine Crisis: Why Russia Sees Crimea as Its Naval Stronghold." *Guardian*, March 7, 2014. www.theguardian.com/world/2014/mar/07/ukraine-russia-crimea-naval-base-tatars-explainer.

Underwood, Kimberly. "Air Force Pursues SMC 2.0 Effort." *Signal Magazine*, October 30, 2018. www.afcea.org/content/air-force-pursues-smc-20-effort.

"Union of Concerned Scientists: Satellite Database." Union of Concerned Scientists, May 1, 2018. www.ucsusa.org/nuclear-weapons/space-weapons/satellite-database#.W7eDt PZFxEZ.

"UrtheCast." UrtheCast Company Webpage, November 16, 2018. www.urthecast.com/constellation.

Whalen, David J. "Communications Satellites: Making the Global Village Possible." NASA History Division. Accessed November 13, 2018. https://history.nasa.gov/satcom history.html.

Wilkie, Christina. "Trump Floats the Idea of Creating a 'Space Force' to Fight Wars in Space." *CNBC*, March 13, 2018. www.cnbc.com/2018/03/13/trump-floats-the-idea-of-creating-a-space-force-to-fight-wars-in-space.html.

Williams, Ian. "Defense Support Program (DSP)." Missile Threat. Accessed November 27, 2018. https://missilethreat.csis.org/defsys/dsp/.

Wu, Chengliang. "People's Daily Slams South China Sea Arbitration Tribunal for Being Political Tool." *Peoples Daily*, July 14, 2016. http://en.people.cn/n3/2016/0714/c90000-9085753.html.

"XpressSAR Inc. Private Remote Sensing License Public Summary." NOAA, October 28, 2015. www.nesdis.noaa.gov/CRSRA/files/xpresssar.pdf.

Zissis, Carin. "China's Anti-Satellite Test." Council on Foreign Relations, February 22, 2007. www.cfr.org/backgrounder/chinas-anti-satellite-test.

Books

Cheng, Dean. *Cyber Dragon*. Denver. CO: Praeger Publishers, 2017.

Clausewitz, Carl. *On War*, ed. Michael Howard and Peter Paret. Princeton, NJ: Princeton University Press, 1984.

Corbett, Julian S. *Some Principles of Maritime Strategy*. The Perfect Library, 2016.

Dolman, Everett C. *Astropolitik: Classical Geopolitics in the Space Age*. London: Routledge, 2001.

Freedman, Lawrence. *Deterrence*. Cambridge: Polity Press, 2004.

Glaser, Charles L. *Rational Theory of International Politics*. Princeton, NJ: Princeton University Press, 2010.

Gray, Colin S. *Weapons Don't Make War: Policy, Strategy, and Military Technology.* Lawrence, KS: University Press of Kansas, 1993.

Handberg, Roger. *Seeking New World Vistas: The Militarization of Space.* West Port, CT: Praeger Publishers, 2000.

Ito, Atsuyo. *Legal Aspects of Satellite Remote Sensing.* Boston, MA: Marinus Nijhoff Publishers, 2011.

Kahneman, Daniel. *Thinking, Fast and Slow.* New York, NY: Farrar, Straus and Giroux, 2011.

Klein, John J. *Space Warfare: Strategy, Principles and Policy.* London; New York, NY: Routledge, 2006.

Krepinevich, Andrew and Watts, Barry. *The Last Warrior: Andrew Marshall and the Shaping of Modern American Defense Strategy.* New York, NY: Basic Books, 2015.

Lefebvre, Jean-Luc. *Space Strategy.* Hoboken, NJ: John Wiley & Sons, 2017.

Lupton, David E. *On Space Warfare.* Maxwell Air Force Base. Alabama, AL: Air University Press, 1988.

McDougall, Walter A. *The Heavens and the Earth: A Political History of the Space Age.* Baltimore, MD: Johns Hopkins University Press, 1985.

Mahan, Alfred Thayer. *The Influence of Sea Power Upon History 1660–1783.* New York, NY: Dover Publications, 1987.

Mearsheimer, John J. *The Tragedy of Great Power Politics.* New York, NY: W. W. Norton & Company, 2014.

Mieczkowski, Yanek, ed. *Eisenhower's Sputnik Moment, The Race for Space and World Prestige.* Ithaca, NY: Cornell University Press, 2013.

Morgan, Forrest E. *Deterrence and First-Strike Stability in Space: A Preliminary Assessment.* Santa Monica, CA: RAND Project Air Force, January 2010.

Oberg, James E. *Space Power Theory.* Colorado Springs, CO: USAFA Government Printing Office, 1999.

Richelson, Jeffrey T. *America's Space Sentinels: DSP Satellites and National Security.* Lawrence, KS: University Press of Kansas, 1999.

Richelson, Jeffrey T. *America's Space Sentinels: The History of DSP and SBIRS Satellite Systems.* 2nd Edition. Lawrence, KS: University Press of Kansas, 2012.

Stares, Paul B. *The Militarization of Space: U.S. Policy, 1945–1984.* Ithaca, NY: Cornell University Press, 1985.

Van Evera, Stephen. *Causes of War.* Ithaca, NY: Cornell University Press, 1999.

Vance, Ashlee. *Elon Musk: Tesla, SpaceX, and the Quest for a Fantastic Future.* New York, NY: Harper-Collins, 2015.

Wagner, R. Harrison, ed. "The Theory of International Politics." In *War and the State*, 1–52. The Theory of International Politics. Ann Arbor, MI: University of Michigan Press, 2007.

Waltz, Kenneth. *Theory of International Politics.* Long Grove, IL: Waveland Press, 2010.

Ziarnick, Brent. *Developing National Power in Space: A Theoretical Model.* Jefferson, NC: McFarland, 2015.

Briefings/memos/messages/point papers

Air Force Space Command. "SBIRS Overview Brief: Combat Air Force Commander's Conference." presented at the Combat Commander's Conference, November 16, 1998. https://nsarchive2.gwu.edu/NSAEBB/NSAEBB235/39.pdf.

Aldridge, Pete. Letter to Charles A. Horner. "Aerospace Corporation Response to Gen Charles Horner CINC USSPACECOM," May 27, 1993. https://nsarchive2.gwu.edu/NSAEBB/NSAEBB235/36a.pdf.

Bailey, Joe. "Point Paper on DSP-II TOR." Air Force System Program Director, Space-Based Early Warning Systems, May 24, 1993. https://nsarchive2.gwu.edu/NSAEBB/NSAEBB235/35.pdf.

Berger, Samuel L. Memorandum to the President of the United States. "Subject: US Policy on Foreign Access to Remote Sensing Space Capabilities." *Memorandum*, March 3, 1994.

Broad Agency Announcement: Blackjack HR001118S0032." Defense Advanced Research Agency Tactical Technology Office, May 25, 2018, 6–7, www.darpa.mil/attachments/HR001118S0032-Amendment-01.pdf.

"Brussels Summit Declaration: Issued by the Heads of State and Governments Participating in the Meeting of the North Atlantic Council in Brussels 11–12 July 2018." North Atlantic Treaty Organization, July 11, 2018. www.nato.int/cps/en/natohq/official_texts_156624.htm.

"Commercial Remote Sensing: A Historical Chronology." Digital Globe, April 9, 2010.

"Data Bank." World Bank. Accessed August 15, 2018. http://databank.worldbank.org/data/home.aspx.

"Draft Position Paper for UN Ad Hoc Committee on Peaceful Uses of Outer Space: Legal Problems Which May Arise in the Exploration of Space." White House Office of the Staff Secretary: Records, 1952–1961, April 22, 1959. Box 24, Space Council (7). Eisenhower Library.

Eisenhower, Dwight D. "Letter from President Eisenhower to Soviet Premier Bulganin," January 12, 1958. White House Office of the Special Assistant for Disarmament (Harold Stassen): Records, 1955–1958, Box 7–8 Eisenhower-Bulganin Letters. Dwight D. Eisenhower Library, Abilene, Kansas.

Eisenhower, Dwight D. "Letter from President Eisenhower to Soviet Premier Bulganin," February 15, 1958. White House Office of the Special Assistant for Disarmament (Harold Stassen): Records, 1955–1958, Box 7–8 Eisenhower-Bulganin Letters. Dwight D. Eisenhower Library, Abilene, Kansas.

"Fact Sheet: Space Tracking and Surveillance System." Missile Defense Agency, March 27, 2017.

"Fiscal Year 2012 Commercial Satellite Communications Usage Report: In Response to Chairman of the Joint Chiefs of Staff Instruction 6250.01E." US Strategic Command, April 6, 2015.

Forden, Geoffrey. "A Constellation of Satellites for Shared Missile Launch Surveillance." MIT's Program on Science, Technology, and Society, September 7, 2006.

Heidner, Rick. "Shutter Control: An Approach to Regulating Imagery from Privately-Operated RS Satellites." ACCRES Meeting, May 15, 2014.

Hemberger, Phillip H. "The Initiatives for Proliferation and Prevention Program: Goals, Projects, and Opportunities." Los Alamos National Laboratory, 2001. www.osti.gov/servlets/purl/788224.

Horner, Charles. Letter to Pete Aldridge. "Letter from CINC US SPACECOM to CEO Aerospace Corp.," May 24, 1993. https://nsarchive2.gwu.edu/NSAEBB/NSAEBB235/34.pdf.

HQ ADCOM J-3. Memorandum to HQ USAF. "Knowledge of the DSP System." *Memorandum*, October 10, 1980. https://nsarchive2.gwu.edu/NSAEBB/NSAEBB235/14.pdf.

Kellog, William W., and Sidney Passman. "Infrared Techniques Applied to the Detection and Interception of Intercontinental Ballistic Missiles." Research Memorandum. RAND Corp., October 21, 1955.

Killian, James R. "Minutes from the Meeting of the President's Science Advisory Committee: Public Information Handling of Reconnaissance Satellites." President's Science Advisory Committee, January 19, 1960. Box 15, White House Office of the Special Assistant for Science and Technology (James R. Killian and George B. Kistiakowsky): Records, 1957–1961 Pre-Accession and Accession 76–16. Eisenhower Library.

Lessor, Marc M. Memorandum to GeoEye Imagery Collection Systems Inc. "Subject: EnhancedView Contract HM0210-10-C-0003." *Memorandum*, June 22, 2012.

Lessor, Marc M. Memorandum to GeoEye Imagery Collection Systems Inc. "Subject: EnhancedView Other Transaction For Prototype Project (OTFPP) Agreement HM0210-10-9-0001." *Memorandum*, June 22, 2012.

Lockheed Martin Corp. "SBIRS Fact Sheet." Lockheed Martin Corp., 2017. www.lock heedmartin.com/content/dam/lockheed-martin/space/photo/sbirs/SBIRS_Fact_Sheet_ (Final).pdf.

"Long March-5 (LM-5)." China Aerospace Science and Technology Corporation. Accessed September 12, 2018. http://english.spacechina.com/n16421/n17215/n17269/ n19031/c125250/content.html.

McClellan, Robert E. "History of the Space Systems Division July–December 1965." SAMSO Office of Information Historical Division, October 1968. https://nsarchive2. gwu.edu/NSAEBB/NSAEBB235/09.pdf.

"Memorandum of Understanding Among the Departments of Commerce, State, Defense, and Interior, and the Office of the Director of National Intelligence, Concerning the Licensing and Operations of Private Remote Sensing Satellite Systems." Department of Commerce, April 25, 2017.

"Minutes of Advisory Committee on Commercial Remote Sensing (ACCRES)." NOAA, September 30, 2002. www.nesdis.noaa.gov/CRSRA/accresMinutes.html.

NASA GSFC Landsat/LDCM EPO Team. "Landsat Image Gallery." Accessed December 6, 2016. http://landsat.visibleearth.nasa.gov/view.php?id=40535.

O'Brien, William. Letter to Jeffrey T. Richelson. "Air Force Space and Missile Systems Center Response to: Request Under the FOIA for Report on 'Preliminary Analysis of Project Hot Spot IR Signals' and 'Application of Infrared Tactical Surveillance,'" November 4, 1992. https://nsarchive2.gwu.edu/NSAEBB/NSAEBB235/11.pdf.

O'Neill, Jim, Pablo Morra, Humberto Medina, Richard Crump, Lesya Karpa, Nicolas Sobczak, Thomas Mayer, et al. "Goldman Sachs Economic Research Group," no. 66 (2001): 16.

Patel, Samira. "NOAA's Commercial Remote Sensing Regulatory Affairs." ACCRES Meeting, April 3, 2017.

Permanent Representative of China to the Conference on Disarmament. "Letter from the Permanent Representative of China to the Conference on Disarmament and the Charge d'affaires a.i. of the Russian Federation Addressed to the Secretary-General of the Conference Transmitting the Comments by China and the Russian Federation Regarding the United States of America Analysis of the 2014 Updated Russian and Chinese Texts of the Draft Treaty on Prevention of the Placement of Weapons in Outer Space and of the Threat or Use of Force against Outer Space Objects (PPWT)," September 11, 2015. https://documents-dds-ny.un.org/doc/UNDOC/GEN/G15/208/38/PDF/G1520838. pdf?OpenElement.

Piper, Robert F. "History of Space and Missile Systems Organization, 1 July 1967–30 June 1969." SAMSO Chief of Staff Historical Division, March 1970. https://nsarchive2.gwu.edu/NSAEBB/NSAEBB235/10.pdf.

Plummer, J.W. "Memorandum for the Deputy Secretary of Defense, Subject: Defense Support Program." National Reconnaissance Office, May 3, 1974.

Poblete, Yleem. "Remarks on Recent Russian Space Activities of Concern." US Department of State, August 14, 2018. www.state.gov/t/avc/rls/285128.htm.

"President Donald J. Trump Is Unveiling an America First National Space Strategy." The White House, March 23, 2018. www.whitehouse.gov/briefings-statements/president-donald-j-trump-unveiling-america-first-national-space-strategy/.

"Press Release: US Commercial Remote Sensing Policy." White House, April 25, 2003. www.whitehouse.gov/files/documents/ostp/press_release_files/fact_sheet_commercial_remote_sensing_policy_april_25_2003.pdf.

"Progress Report on the US Scientific Satellite Program (NSC 5520)." White House Office, National Security Council Staff, October 3, 1956. Box 38 Outer Space (1)-(6). Eisenhower Library.

"Proposed News Release from the National Academy of Sciences Regarding Soviet Plans to Launch an Earth Satellite as Part of the International Geophysical Year Program." National Academy of Sciences—S.D. Cornell, June 10, 1957. Box 625, OF 146-F-2 Outer Space, Earth-Circling Satellites. Eisenhower Library.

"Reaction to the Soviet Satellite—A Preliminary Evaluation." White House Office of the Staff Research Group, October 16, 1957. Box 35, Special Projects: Sputnik, Missiles and Related Matters. Eisenhower Library.

Reign, Alistair. *Reagan National Defense Forum: STRATCOM Commander Gen. Hyten Discusses Future of Space War*. Panel, 2017. www.youtube.com/watch?v=wUIHctFgTac.

"Russia's Accusations-Setting the Record Straight." Fact Sheet. North Atlantic Treaty Organization, April 2014. www.nato.int/nato_static/assets/pdf/pdf_2014/20140411_140411-factsheet_russia_en.pdf.

Samuel, Ron, Kay Sears, Tip Osterthaler, Daniel S. Goldberg, and Phillip Harlow. "Seven Ways to Make the DOD a Better Buyer of Commercial SATCOM," January 14, 2013. https://ses-gs.com/press-release/seven-ways-to-make-the-dod-a-better-buyer-of-commercial-satcom/.

Schriever, Bernard. Memorandum to Eugene Zuckert. "DOD Program Change (4.4.040) on MIDAS (239A)." *Memorandum*, August 13, 1962. Curtis LeMay Papers, Box B141 (AFSC 1962). Library of Congress. https://nsarchive2.gwu.edu/NSAEBB/NSAEBB235/06.pdf.

Schueler, Carl. "Commercially Hosted Payloads: Low-Cost Research to Operations." Austin, TX: Orbital Sciences, 2013.

Schueler, Carl. "Commercially-Hosted Payloads: Low-Cost Research Options." presented at the AMS Third Conference on Transition of Research to Operations, Austin, TX, January 9, 2013. https://ams.confex.com/ams/93Annual/flvgateway.cgi/id/24205?recordingid=24205&entry_password=706714&uniqueid=Paper224173.

"Space Force Fact Sheet," accessed February 8, 2020, www.spaceforce.mil/About-Us/Fact-Sheet.

Space Systems Division. "DSP Desert Storm Summary Briefing." June 1991. https://nsarchive2.gwu.edu/NSAEBB/NSAEBB235/24.pdf.

Spires, David N., and Rick W. Sturdevant. "From Advent to Milstar: The U.S. Air Force and the Challenges of Military Satellite Communications." NASA History. Accessed November 13, 2018. https://history.nasa.gov/SP-4217/ch7.htm.

Tarr, Jeffrey R. Shareowner Letter. "DigitalGlobe CEO: Letter to Investors." Shareowner Letter, 2015.

Tenent, George J. Memorandum to Director, National Imagery and Mapping Agency. "Director Central Intelligence Agency: Expanded Use of US Commercial Space Imagery." *Memorandum*, June 7, 2002.

"The Distributed Tactical Communications System: Fact Sheet." Defense Information Systems Agency, n.d. www.disa.mil/~/media/Files/DISA/Services/DTCS/DTCS-Overview.pdf.

"United States Space Command: Operations Desert Shield and Desert Storm Assessment," January 1992. https://nsarchive2.gwu.edu/NSAEBB/NSAEBB235/25.pdf.

US National PNT Advisory Board. "Russia Undermining Confidence in GPS." May 17, 2018. www.gps.gov/governance/advisory/meetings/2018-05/goward.pdf.

Waldron, Harry N. "History of the Space and Missile Systems Center: October 1994–September 1997." Space and Missile Systems Center, March 9, 2002. https://nsarchive2.gwu.edu/NSAEBB/NSAEBB235/27.pdf.

Yarmolinsky, Adam. Memorandum. "Memorandum For: Mr. Timothy J. Reardon Jr., Special Assistant to the President on Classified Space Project 461." *Memorandum*, May 31, 1963. https://nsarchive2.gwu.edu/NSAEBB/NSAEBB235/08.pdf.

Government documents

"1996 US National Space Policy." The White House National Science and Technology Council, September 19, 1996.

"2006 US National Space Policy." The White House National Science and Technology Council, August 31, 2006.

"2010 National Space Policy of the United States of America," June 28, 2010.

"2010 US National Security Space Strategy, Unclassified Summary." Department of Defense and the Director of National Intelligence, January 2011.

"2019 Missile Defense Review." Department of Defense, Office of the Secretary of Defense, 2019.

"AFSPACECOM Desert Shield/Desert Storm Lessons Learned." Air Force Space Command, July 12, 1991.

"Air Force Doctrine Document 1 (AFDD-1)." US Air Force Curtis E. Lemay Center for Doctrine Development and Education. Accessed October 12, 2018. www.au.af.mil/au/awc/awcgate/afdc/afdd1-chap3.pdf.

"China's National Defense in 2010" (Information Office of the State Council of the People's Republic of China, March 2011).

"China's Military Strategy 2015" (State Council Information Office of the People's Republic of China, May 2015).

"Civil Reserve Airfleet Allocations." US Department of Transportation. US Department of Transportation. Accessed November 13, 2018. www.transportation.gov/mission/administrations/intelligence-security-emergency-response/civil-reserve-airfleet-allocations.

"Commercial Remote Sensing Regulatory Affairs." NOAA Official Website. Accessed January 28, 2017. www.nesdis.noaa.gov/CRSRA/licenseHome.html.

Communications Satellite Act of 1962, Pub. L. No. 87–624, § 102a, HR11040 (1962).

"Complex Commercial SATCOM Solutions (CS3): CS3 Customer Ordering Guide." US General Services Administration, October 2017.

Federal Acquisition Regulation Part 49-Termination of Contracts, § 49.202 a (n.d.).

Federal Acquisition Regulation Part 49-Foreign Acquisition, § 48 part 25 (2018).

"General Conditions for Private Remote Sensing Space System Licenses." NOAA, 2016.

"GPS Space Segment." US government. Accessed October 23, 2018. www.gps.gov/systems/gps/space/.

H.R. 2809, American Space Commerce Free Enterprise Act, Pub. L. No. 2809, Title 51 (2018).

Joint Publication 3–14: Space Operations. US Department of Defense, 2009.

Joint Publication 3–14: Space Operations. US Department of Defense, 2018.

Kelley, Kerry. "Fiscal Year 2015 Commercial Satellite Communications Usage Report: In Response to Chairman of the Joint Chiefs of Staff Instruction 6250.01E." US Strategic Command, September 7, 2017.

Kyl-Bingaman Amendment—Licensing of Private Land Remote-Sensing Space Systems, 15 CFR 960, Vol. 71, No. 79 § (2006).

Land Remote-Sensing Act of 1992, Pub. L. No. 102–588, 15 USC 5623 (1992).

Land Remote-Sensing Commercialization Act of 1984, Pub. L. No. 98–365, 15 USC 4201 (1984).

Licensing of Private Land Remote-Sensing Space Systems, 15 CFR 960, Vol. 71, No. 79 § (2006).

NASA. "National Aeronautics and Space Administration Fiscal Year 1998 Estimates," n.d.

"National Aeronautics and Space Administration FY 2018 Budget Estimates." NASA, 2017. www.nasa.gov/sites/default/files/atoms/files/fy_2018_budget_estimates.pdf.

National and Commercial Space Programs, Pub. L. No. 111–314, 51–3328 (2010).

National Defense Authorization Act for Fiscal Year 1997, Pub. L. No. 104–201, 15 USC 5621 (1996).

National Defense Authorization Act for Fiscal Year 2013, Pub. L. No. 112–239, H.R. 4310–14 (2013).

"National Defense Authorization Act for Fiscal Year 2014." Washington, DC: US Senate, 2013.

National Defense Authorization Act for Fiscal Year 2017, Pub. L. No. 114–328 (2016).

"National Oceanic and Atmospheric Administration FY 2018 Budget Summary." Budget Summary. NOAA, 2017. www.corporateservices.noaa.gov/nbo/fy18_bluebook/FY18-BlueBook-508.pdf.

"National Security Council Report, NSC-68, 'United States Objectives and Programs for National Security,'" April 12, 1950. Truman Library. www.trumanlibrary.org/whistlestop/study_collections/coldwar/documents/pdf/10-1.pdf.

"National Security Decision Directive 85: Eliminating the Threat from Ballistic Missiles." White House, March 25, 1983. www.hsdl.org/?view&did=463005.

"National Security Presidential Directive (NSPD) 3, US Commercial Space Guidelines." White House, February 1991.

"National Security Presidential Directive (NSPD) 27, US Commercial Remote Sensing Policy." White House, April 25, 2003.

"National Space Transportation Policy." Executive Office of the President of the United States, November 21, 2013.

NATO. "Bucharest Summit Declaration," April 3, 2008. www.nato.int/cps/en/natolive/official_texts_8443.htm.

OECD. "Space budgets of selected OECD and non-OECD countries in current USD, 2013," in *Readiness factors: Inputs to the space economy*. OECD Publishing, Paris.

"Pathfinder 3 Request for Information: Solicitation Number 16–076." Air Force Space Command, May 20, 2016.

"Presidential Decision Directive 23: US Policy on Foreign Access to Remote Sensing Space Capabilities," March 9, 1994.

"Report by the Chair of the Group of Governmental Experts on Further Practical Measures for the Prevention of an Arms Race in Outer Space." United Nations office for Disarmament Affairs, January 31, 2019, https://s3.amazonaws.com/unoda-web/wp-content/uploads/2019/02/oral-report-chair-gge-paros-2019-01-31.pdf.

Strom Thurmond National Defense Authorization Act for Fiscal Year 1999, Pub. L. No. 105–261, 22 USC 2278 (1998).

"Remote Sensing Regulatory Reform ACCRES Recommendations," NOAA, July 15, 2019, 2, www.nesdis.noaa.gov/CRSRA/pdf/NOAA_Proposed_Remote_Sensing_Regs-FINAL_ACCRES_Recommendations_7-15-19.pdf.

Taiwan Relations Act, Pub. L. No. 96–98, 3301 22 (1979).

"The Diversified Employment of China's Armed Forces." Information Office of the State Council of the People's Republic of China, April 16, 2013. www.china.org.cn/government/whitepaper/node_7181425.htm.

"Treaty Between the United States of America and the Union Of Soviet Socialist Republics On The Limitation Of Anti-Ballistic Missile Systems," May 26, 1972. www.state.gov/t/avc/trty/101888.htm#text.

"Treaty on Principles Governing the Activities of States in the Exploration and the Use of Outer Space, Including the Moon and Other Celestial Bodies." Resolution Adopted by the UN General Assembly 2222 XXI, December 19, 1966.

Personal communications—interviews/e-mails

Koller, Joseph. Letter to author. "Office of the Under-Secretary of Defense for Space Policy—Civil Reserve Air Fleet," January 23, 2017.

Loverro, Douglas. Deputy Assistant Secretary of Defense for Space Policy-interview by author. Phone, January 13, 2017.

Robinson, Alan. Letter to author. "NOAA Senior Licensing Officer," November 30, 2016.

Reports

"Annual Report to Congress: Military and Security Developments Involving the People's Republic of China." Office of the Secretary of Defense, May 16, 2018.

Aru, Guido W., and Carl T. Lunde. "DSP-II: Preserving the Air Force's Options." El Segudo, CA: The Aerospace Corporation, April 23, 1993. https://nsarchive2.gwu.edu/NSAEBB/NSAEBB235/33.pdf.

Bedrosian, E., E. Cesar, J. Clark, G. Huth, K. Poehlmann, and P. Propper. "Tactical Satellite Orbital Simulation and Requirements Study." RAND Study. Santa Monica, CA: RAND Corp., 1993.

Bolkcom, Christopher. "Civil Reserve Air Fleet." Congressional Research Service, October 18, 2006.

Builder, Carl H., Steven C. Bankes, and Richard Nordin. "Command Concepts: A Theory Derived from the Practice of Command and Control." RAND Corporation, 1999. www.rand.org/content/dam/rand/pubs/monograph_reports/2006/MR775.pdf.

"Challenges to Security in Space," Defense Intelligence Agency, January 2019.

Cheng, Dean. "Evolving Chinese Thinking About Deterrence: What the United States Must Understand about China and Space." Heritage Foundation, March 29, 2018.

"China Armed Forces Estimate." Janes IHS, 2018.

Coats, Daniel R. Worldwide Threat Assessment of the US Intelligence Community, § US Senate Select Committee on Intelligence (2018).

"Commission to Assess United States National Security Space Management and Organization." US government, January 11, 2001.

Cooper, William H. "Russia's Economic Performance and Policies and Their Implications for the United States." Congressional Research Service, June 29, 2009. https://fas.org/sgp/crs/row/RL34512.pdf.

"Defense Acquisitions: Despite Restructuring SBIRS High Program Remains at Risk of Cost and Schedule Overruns." Washington, DC: General Accounting Office, October 2003.

"Defense Satellite Communications: DOD Needs Additional Information to Improve Procurements." Washington, DC: United States General Accounting Office, July 2015.

"DigitalGlobe: Securities and Exchange Commission Form 10K 2014." Longmont, CO: DigitalGlobe, December 31, 2014.

"DigitalGlobe: Securities and Exchange Commission Form 10K 2015." Westminster, CO: DigitalGlobe, December 31, 2015.

"Draft of a Report on MIDAS by DDR&E Ad hoc Group," November 1, 1961. https://nsarchive2.gwu.edu/NSAEBB/NSAEBB235/04.pdf.

Everett, Robert R., Penrose Albright, Roy C. Evans, David V. Kalbaugh, William Z. Lemnios, Antonio F. Pensa, and John M. Ruddy. "Report of the Space Based IR Sensor/Technical Support Group." MITRE Corp, October 1993. https://nsarchive2.gwu.edu/NSAEBB/NSAEBB235/37.pdf.

"GeoEye: Securities and Exchange Commission Form 10K." Dulles, VA: GeoEye, 2009.

"Global Trends: Paradox of Progress." National Intelligence Council, 2018. www.dni.gov/index.php/the-next-five-years/space.

Harrison, Todd, Kaitlyn Johnson, and Thomas G. Roberts. "Space Threat Assessment 2018." A Report of the CSIS Aerospace Security Project. Center for Strategic and International Studies, April 2018.

Johnson, Dana J., Scott Pace, and C. Bryan Gabbard. *Space: Emerging Options for National Power*. Santa Monica, CA: RAND Corporation, 1998.

Kelly, Terrence K., Anthony Atler, Todd Nichols, and Lloyd Thrall, eds. "Land-Based Anti-Ship Missiles in the Western Pacific." In *Employing Land-Based Anti-Ship Missiles in the Western Pacific*, 1–20. RAND Corporation, 2013.

Kent, Glenn A., and David E. Thaler. "First-Strike Stability and Strategic Defenses: Part II of a Methodology for Evaluating Strategic Forces." RAND Corporation, October 1990.

Kessler, Christian J. "Leadership in the Remote Sensing Satellite Industry: Report Prepared for US Department of Commerce and NOAA." North Raven Consulting, 2009.

Lapin, Ellis E. "Surveillance by Satellite." Aerospace Corporation, August 11, 1975.

"Military Space Systems: DOD's Use of Commercial Satellites to Host Defense Payloads Would Benefit from Centralizing Data." United States Government Accountability Office, July 2018.

"Missile Detection Systems Development: SBIRS/AIRSS Briefing to Congressional Staff Preliminary Findings." Washington, DC: General Accounting Office, March 13, 2007.

"Orbimage: Securities and Exchange Commission Form 10K." Orbimage, 2004.

"Orbimage: Securities and Exchange Commission Form 10K." Orbimage, 2005.

"Orbimage Securities and Exchange Commission Form 10-Q." Orbimage, September 30, 2004.

"Remote Sensing Services Market by Platform (Satellites, UAVs, Manned Aircraft, and Ground), End User (Defense and Commercial), Resolution (Spatial, Spectral, Radiometric, and Temporal), and Region—Global Forecast to 2022." Research Report. Markets and Markets, October 2017. www.marketsandmarkets.com/Market-Reports/remote-sensing-services-market-87605355.html.

"Report for Congress: Commercial Remote Sensing by Satellite: Status and Issues." Washington, DC: Congressional Research Service, January 8, 2002.

"Report to the Chairman, Subcommittee on Defense, Committee on Appropriations, House of Representatives; Early Warning Satellites: Funding for Follow-on System Is Premature." Washington, DC: United States General Accounting Office/National Security and International Affairs Division, November 7, 1991.

"Report to the Secretary of Defense: Taking Advantage of Opportunities for Commercial Satellite Communications Services." Defense Business Board, January 2013.

"Resiliency and Disaggregated Space Architectures." Air Force Space Command, April 14, 2016.

"Resiliency and Disaggregated Space Architectures, White Paper." Air Force Space Command, 2013.

"Satellite Communications: Strategic Approach Needed for DOD's Procurement of Commercial Satellite Bandwidth." Washington, DC: United States General Accounting Office, December 10, 2003.

"Satellite Communications Strategy Report: In Response to Senate Report 113–44 to Accompany S.1197 NDAA for FY14." Washington, DC: Department of Defense Office of the Chief Information Officer, August 4, 2014.

Schnelzer, Garry. "Air Force Space Sensor Study." AFPEO/SP, April 12, 1993. https://nsarchive2.gwu.edu/NSAEBB/NSAEBB235/20130108.html.

Schradin, Ryan. "Overcoming the Largest Threats to Military Satellites and Increasing Resiliency." *The Government Satellite Report*, April 4, 2018. https://ses-gs.com/govsat/defense-intelligence/overcoming-largest-threats-military-satellites-increasing-resiliency/.

Schwarts, Moshe, and Charles V. O'Conner. "The Nunn-McCurdy Act: Background, Analysis, and Issues for Congress." Washington, DC: Congressional Research Service, May 12, 2016.

"Securities and Exchange Commission Form 10-K." DigitalGlobe, December 31, 2012.

"Selected Acquisition Report: Space Based Infrared System High (SBIRS High)." Department of Defense, March 18, 2015.

"Selected Acquisition Report: Space Fence Ground-Based Radar System Increment 1." Department of Defense, March 23, 2016. www.esd.whs.mil/Portals/54/Documents/FOID/Reading%20Room/Selected_Acquisition_Reports/16-F-0402_DOC_24_Space_Fence_Inc_1_DEC_2015_SAR.pdf.

"State of the Satellite Industry Report, Satellite Industry Association." Tauri Group, September 2016.

"Taking Advantage of Opportunities for Commercial Satellite Communications Services: Report FY 13-02." Defense Business Board, January 2013.

"The President's Science Advisory Committee: Report of the Early Warning Panel," March 13, 1959. White House Office, Office of the Special Assistant for Science and Technology, Records (James R. Killian and George B. Kistiakowsky, 1957–1961) Box 4 AICBM-Early Warning [November 1957–December 1960]. Eisenhower Library.

Thomson, Homer H., James A. Elgas, Steve Martinez, Arthur Gallegos, and Marciela Camarena. "National Missile Defense: Risk and Funding Implications for the

Space-Based Infrared Low Component." United States General Accounting Office, February 1997.

"United States Securities and Exchange Commission Form 10-K: Iridium Communications Inc." SEC 10-k. Iridium Communications Inc., March 16, 2010.

"United States Securities and Exchange Commission Form 10-K: Iridium Communications Inc." SEC 10-k. Iridium Communications Inc., December 2015.

"United States Securities and Exchange Commission Form 10-K: Iridium Communications Inc." SEC 10-k. Iridium Communications Inc., February 22, 2018.

Weber, Robert A., and Kevin M. O'Connell. "Alternative Futures: United States Commercial Satellite Imagery in 2020: Research Report for Department of Commerce and NOAA." Washington, DC: Innovative Analytics and Training, November 2011.

Weeden, Brian. "Space Situational Awareness Fact Sheet." Secure World Foundation, May 2017. https://swfound.org/media/205874/swf_ssa_fact_sheet.pdf.

Weeden, Brian. "Through a Glass, Darkly: Chinese, American, and Russian Anti-Satellite Testing in Space." Secure World Foundation, March 17, 2014.

Testimony/hearings

Aru, Guido W. "Statement of Mr. Guido William Aru (Project Leader, System Architecture and Integration Space-Based Surveillance Division, Aerospace Corporation) Before the House of Representatives Committee on Government Operations Legislation and National Security Subcommittee," February 2, 1994. https://nsarchive2.gwu.edu/NSAEBB/NSAEBB235/38.pdf.

Chivvis, Christopher S. Understanding Russian "Hybrid Warfare," And What Can Be Done About it, § House Armed Services Committee (2017). www.rand.org/content/dam/rand/pubs/testimonies/CT400/CT468/RAND_CT468.pdf.

Chaplain, Cristina T. "Space Acquisitions: DOD Continues to Face Challenges of Delayed Delivery of Critical Space Capabilities and Fragmented Leadership." Testimony before US Senate Subcommittee on Strategic Forces. United States Government Accountability Office, May 17, 2017.

Costello, Jerry F. Hearing on the economic viability of the Civil Reserve Air fleet Program, Pub. L. No. 111–130, § House Subcommittee on Aviation (2009).

Cristina T. Chaplain, Director Acquisition and Sourcing Management. Space Acquisitions: DOD Continues to Face Challenges of Delayed Delivery of Critical Space Capabilities and Fragmented Leadership: § Testimony before the Subcommittee on Strategic Forces, Committee on Armed Services, US Senate. (2017). www.gao.gov/assets/690/684664.pdf.

Goldfien, David L. Military Space Organization, Policy, and Programs, § Subcommittee on Strategic Forces (2017). www.armed-services.senate.gov/imo/media/doc/17-46_05-17-17.pdf.

Hyten, John E. Hearing to Receive Testimony on United States Strategic Command in Review of the Defense Authorization Request for Fiscal Year 2019 and the Future Years Defense Program, § Committee on Armed Services (2018).

Hyten, John E. Statement of John E. Hyten, Commander United States Strategic Command, Before the Senate Committee on Armed Services, § Senate Committee on Armed Services (2018).

McNabb, Duncan J. Hearing on the economic viability of the Civil Reserve Air fleet Program, Pub. L. No. 111–130, § House Subcommittee on Aviation (2009).

Speeches

Pence, Mike. "Remarks by Vice President Pence on the Future of the U.S. Military in Space." The Pentagon, August 9, 2018. www.whitehouse.gov/briefings-statements/remarks-vice-president-pence-future-u-s-military-space/.

Putin, Vladimir. "Meeting of Ambassadors and Permanent Representatives of Russia." Office of the President of Russia, July 19, 2018. http://en.kremlin.ru/events/president/news/58037.

Putin, Vladimir. "Meeting with Defense Ministry Senior Officials: The President Met with Defense Ministry Senior Officials in the First of a Series of Meetings on the Development of the Armed Forces." Office of the President of Russia, May 15, 2018. http://en.kremlin.ru/events/president/transcripts/57477.

Putin, Vladimir. "Presidential Address to the Federal Assembly." Office of the President of Russia, March 1, 2018. http://en.kremlin.ru/events/president/news/56957.

Reagan, Ronald. "Address to the Nation on Defense and National Security." March 23, 1983. www.atomicarchive.com/Docs/Missile/Starwars.shtml.

Schriever, Bernard. "ICBM-A Step Toward Space Conquest." Address presented at the Astronautics Symposium, San Diego, CA, February 19, 1957.

Trump, Donald J. "Remarks by President Trump at a Meeting with the National Space Council and Signing of Space Policy Directive-3." White House East Room, June 18, 2018. www.whitehouse.gov/briefings-statements/remarks-president-trump-meeting-national-space-council-signing-space-policy-directive-3/.

Index

Page numbers in *italics* denote figures.

accumulation of power 12, 22, 214
active defense, concept of 61, 100, 102
Advanced Research Projects Agency (ARPA) 142, 174
Advanced Warning System (AWS) 178
Advisory Committee for Commercial Remote Sensing (ACCRES), US 147
aerial reconnaissance 141, 162
Aerospace Corporation 179–180
Aerospace Science and Technology Corporation (CASC), China 53
air-defense system: Ground-Based Midcourse Defense (GMD) system (US) 73; PL19 Nudol (Russia) 73; S-500 system (Russia) 72–73; THAAD system (US) 72, 109
Air Force Research Lab (AFRL), US 184
Air Force Space Command (AFSPACE) 45, 120, 126, 176, 180–181
airpower 89
Aldridge, Pete 180
Alert Locate and Report Missiles (ALARM) program 181, 191
Alternative Infrared Satellite System (AIRSS) 182–184, 191
American dominance in space 48
American Space Commerce Free Enterprise Act (2018), US 148
Americom Government Services (AGS) 184–185; "shared risk" clause 185
anchor tenancy 144
Andropov, Yuri 71
Anti-Access/Area Denial (A2/AD) strategy 57, 61
Anti-Ballistic Missile (ABM) treaty (1972) 2, 24, 44–45, 87, 178; US withdrawal from 46, 73–75

anti-satellite (ASAT) test 33; China 4, 64–66, 73; overuse of 16; Soviet proposal on 71
anti-satellite (ASAT) weapons 44, 98; in China 61; directed energy 61; Istrabitel Sputnikov (IS) system 71; Naryad-V ASAT 71–72; research and development 44; restrictions on ground-based 63; in Russia 71–72; in United States 204
ArianeGroup 72
arming decisions, quality of *21*
arms control agreements 17, 24–26, 65, 77, 107–108, 203, 213, 218, 222; Chinese position on 61, 65–66, 217; to limit offensive weapons 23; motivation to pursue 25; offense–defense balance in 25; and policy of unilateral disarmament 26; Russian position for 63, 217; in space 33, 75; US space policy and 47, 75
arms race 21, 113; due to security dilemma 214; economic burden of pursuing 24; imbalances in 214; interest of the US to avoid 204; naval arms race 90; in space 202, 222; between US, Russia, and China 111
Astrium Services (France) 151

Baker, James 68
balance of power 24–25, 102, 200; ability to project military power and capability 30–31; between China and the US 55; in space 43; Sputnik's impact on 25; technical capability and training 31; between USSR and the US 25
Ball Aerospace 149
ballistic missile defense 73

Ballistic Missile Early Warning System
(BMEWS) 173, 190; radars for 173
bandwidth 116, 119–123, 126–130,
206, 210
bankruptcy 133, 135
battlespace awareness 96
Beranger, Eric 151
Blackjack program (US) 142
Blue Origin 3, 53, 110
Boost Surveillance and Tracking System
(BSTS) 178, 181, 191
Brilliant Eyes program 181
Bulganin, Nikolai 25
Bush, George W. 46; on expansion of
NATO 69; US National Space Policy
(NSP) 46
buy-to-deny strategy 162

celestial bodies, militarization of 3, 44
celestial lines of communications 90, 91
Central Command (CENTCOM), US 120
checkbook shutter control 162
Chernobyl disaster (1986) 140
chicken, game of 15–16
Chilton, Kevin 73
China: active satellites on orbit 52;
Aerospace Science and Technology
Corporation (CASC) 53; Anti-Access/
Area Denial (A2/AD) strategy 53, 61;
ASAT program 66; balance of power
with the US 55; challenge to US in
space 42; claims over South China Sea
56; conflict with Taiwan 57; defense
white papers 61–62; disagreements with
US over the future of Taiwan 56;
economic well-being 57; efforts at arms
control in space 61; emerging security
concerns 60; energy security and trade
57; global power projection 60;
informationized military 60; interference
with Taiwan's sovereignty 56; as leader
in space launches 53; Long March (LM)
rockets 53; military budget of 66;
military expenditures 58, 59; military
strategy 61; national thresholds for
security threats and responses 61;
opposition to US withdrawal from ABM
treaty 46; Peoples Liberation Army
(PLA) 57, 65; per capita GDP 55, 58;
response to US commercialization of
space power 212–213; rise of 54–55, 58;
security concerns in the South China
Sea 57; South China Sea dispute *see*
South China Sea; space launch industry

in *see* Chinese space launch industry;
space programs in 54–67
Chinese space launch industry 52; growth
of space program 57; manned space
program 213; number of space launches
per year 55; satellite infrastructure 57
Civil Reserve Air Fleet (CRAF) 123,
130–131, 187, 206
Civil Reserve Space Fleet (CRSF) 131,
136, 206–207
ClearView contract 146, 150; purchasing
of imagery under 150
Coats, Dan 73
coercion in space: informational means of
91; through diplomatic means 91;
through economic means 91
Cold War 1, 3, 7, 9, 37, 51, 63–64, 67,
73–74, 98, 188, 212, 217, 223; end of
68, 205; space race during 221; between
US and USSR 68; US attitude toward
control of space 43
Colussy, Dan 134
commercial imagery: Astrium Services
(France) 151; demand for 150;
government funding 150; market for
150; in United States 149; *see also*
satellite imagery
commercialization, of space power: benefit
of 208; China's response to 212–213;
electro-optical market 207; full
commercialization 205–209; global
commercialization 214–215; impact on
arms race 210; implementation of
216–218; innovations in 210; launch
market 211–212; in missile warning
market 209–210; partial
commercialization 209–213;
possibilities of 133–135; purpose of
216; reassurance strategy based on
208–209, 211, 214; regulation on
imagery platforms 208; in remote
sensing market 140, 207–210; Russian
response to 211–212; SATCOM market
206, 207, 209–210; security dilemma in
209, 213; space-flight contracts 211;
space innovations and 204, 210; strategy
of 205–218; subsidized
commercialization 130–131; subsidized
CRSF style approach to 131; by United
States 204, 208, 211
Commercially Hosted Infrared Payload
(CHIRP) 111, 183–186, 191–192
commercial satellite communications
(COMSATCOM) 116–117, 123, 127,

COMSATCOM *continued*
163; alternative approaches to acquisition of *124*; Communications Satellite Corporation (COMSAT) 118; conditions for acquiring 125; cost of 125; development of 123; history of acquiring 120–127; hosted payloads 132–133; indefinite-delivery/indefinite-quantity (IDIQ) leases 123; subsidized commercialization of 130–131; Syncom III satellite 118; and US military 118–120

commercial satellites 93, 101, 107, 121, 123, 127–128, 135, 140, 145, 162, 183, 187, 204, 207; capabilities of 109; commercial satellite communications *see* commercial satellite communications (COMSATCOM); communications packages 132; in orbit 9, 52, 96; resolution of 159, 161; technological development 156; US military investment in 130

Communications Satellite Act (1962), US 118, *119*

Communications Satellite Corporation (COMSAT) 118

communications satellites 20, 33, 100–102, 111, 116, 118–120, 128, 184

competition in space 1; fear-based cycle of 203; military aspects of 216–217; offense–defense balance 201; policy of 201; between Russia and US 211; as self-defeating strategy for the US 202–203

cooperation in space 5, 20, 36–37, 42, 202, 217, 221; strategy of 200

Corbett, Julian: *Some Principles of Maritime Strategy* (1911) 90; treatise on sea power 90; work of military theory 90

cyber-attacks 70–71

Daimler-Benz Aerospace (DASA) 72
data collection 143, 145, 161
debris removal satellites 215
decision making, process of 22, 26, 61, 104, 223
DeConcini, Dennis 155
Defense Advanced Research Projects Agency (DARPA), US: Blackjack program 142
Defense Information Systems Agency (DISA), US 120–123; "termination for convenience" clause 122

defense spending 31, 51, 213
Defense Support Program (DSP) satellites 170, 175–176, 182; ability to detect theater ballistic missiles (TBMs) 176–177, 181; Advanced Warning System (AWS) 178; contributions in Gulf War 176, 179; detection capability of 176; evolution of 176; failure to detect SCUD missile 178; in-theater missile warning information 179; as missile warning system 64; replacement system for 179; US Air Force 178

Defense Working Capital Fund 122
defensive space control (DSC) 100
Delta Four Heavy rocket system 53
de Madariaga, Salvador 32
denied-parties screening 163
deterrence in space 102–107
Developing National Power in Space (Ziarnick) 89
DigitalGlobe 145–146, 149–151, 153, 160, 208; emerging competition 153; government business 164; image showing Russian military units within Ukraine *154*; merger with GeoEye 152, 154; purchase by MDA 153; WorldView-3 satellite 187
directed energy anti-satellite weapons 61
Distributed Tactical Communications System (DTCS) 134
Dolman, Everett 6; *Astropolitik* 6
Downs, George 28
dual-use commercial systems 102
Dunford, Joseph 48
Duterte, Rodrigo 56

Early Bird satellite 118, 149
Earth Observation Satellite Corporation (EOSAT) 145
EarthWatch 146, 149
economic expansion, into space 1
Eisenhower, Dwight D. 25, 43; Science Advisory Committee on Early Warning 173
electronic warfare 61
electro-optical imaging satellites 155
EnhancedView contract 151–152
Eurockot 72

Falcon Heavy rocket 53
Federal Space Agency, Russia 53–54
firm fixed price (FFP) 184
fixed price contracts 134–135, 185, 192
Follow-on Early Warning System (FEWS) 178–180, 191; cancellation of 181;

controversy over 180–181; future of 179; public hearings on 180; requirements of 180; utility of 179
freedom in space 2

Galileo constellation 110
General Accounting Office (GAO), US 121, 123, 181
GeoEye: construction of GeoEye-2 152; funding under the EnhancedView SLA 152; loss of government funding 152; merger with DigitalGlobe 152, 154; *see also* Orbimage
geosynchronous orbit (GEO) 32, 36, 53, 170, 211
geosynchronous (GEO) satellites 116; associated with the nuclear deterrent 211; communications satellites 184
Geosynchronous Space Situational Awareness Program (GSAP) 67
Glaser, Charles 5–6, 18, 200; power, definition of 20; theory of international politics 5, 19; theory of rational strategic choice 200–201
global communications 52, 118
global missile warning constellation 189
Global Positioning System (GPS) 2, 52, 98, 106, 110; Russian jamming systems 72; transitioning from military to civil control 216
global power projection, by China 60
global security system 73
GLONASS system 52, 72
Google Earth 8, 92
Gorbachev, Mikhail 68
Gray, Colin 94
Griffith, Michael 171, 186
Gross Domestic Product (GDP) 29, 58
ground-based lasers 33, 65, 100
Ground-Based Midcourse Defense (GMD) system (US) 73
ground-based radars 171, 174
ground-to-space weapons 98, 102
Gulf War (1991) 45, 120, 130, 162, 176, 179, 191, 208

Hard, Donald 180
Hayes-Ryan, Karyn 156
Herz, John 12, 14, 214
highly elliptical orbits (HEO) 131, 170, 181
high-resolution satellites, construction of 150
high-throughput satellites (HTS) 128–129; anti-jam capability 129; footprint compared to a traditional transponder *129*
Hong, Lei 64
Horner, Charles 179–181
hosted payloads 127, 132–133, 136, 170, 182–188, 191–192
Hu, Jintao 57
hunter-killer satellites 101, 211
hybrid missile shield and warning system 172
hybrid warfare: asymmetric methods of 70; key characteristic of 70; Russia's strategy of 70, 212
hypersonic missiles 171; nuclear-armed 190
Hyten, John 49, 65, 183, 206

Ikenberry, John 22
indefinite-delivery/indefinite-quantity (IDIQ) leases 123
Influence of Sea Power Upon History, The (Mahan) 89
information gathering, from space 95, 101–102
informationized military 60
Infrared Augmentation Satellite (IRAS) 183
infrared missile warning mission 186–191
infrared radiometers 174
Intelligence, Surveillance, and Reconnaissance (ISR) satellites 106, 142, 204
inter-continental ballistic missiles (ICBMs) 28, 98, 170, 177; development of 173
Intermediate-Range Nuclear Forces (INF) Treaty: US withdrawal from 190
International Geophysical Year (IGY) 44
International Monetary Fund 67
international politics, theory of 13, 17
international relations, theories of 17, 200
International Space Station (ISS) 5, 68, 211, 221
International Telecommunications Satellite Consortium (INTELSAT) 116, 118–119
International Traffic in Arms Regulations (ITAR) 147, 184
Iran–Iraq war 176
Iridium communications 133–135
Istrabitel Sputnikov (IS) ASAT system 71

jamming systems 100, 110, 129
Jervis, Robert 14–16, 18–19, 33
Joint Tactical Ground Station (JTAGS) 179

joint ventures 72, 150
Jomini, Antoine-Henri 90–91

KB Salyut 72
Khrunichev Space Center, Russia 72
Khrushchev, Nikita 25, 27
kinetic-kill missiles 65, 71, 105
Klein, John 90–91, 93–94
Knopow, Joseph 174
Korean war 60

Land Remote Sensing Commercialization
    Act (1984), US 142, 145, 157
Land Remote Sensing Policy Act (1992),
    US 144–145, 147, 163
Landsat program (US) 140, 142; Earth
    Observation Satellite Corporation
    (EOSAT) 145; privatization of
    143, 145
laser weapons, ground-based 65
launch and satellite design, technological
    revolution in 1
launch market 52; Chinese leadership in
    53; technological innovations in 53
Lavrov, Sergei 68
Lefebvre, Jean-Luc 91–93
licenses, for remote sensing satellites 143;
    application process for 158; authority to
    grant 145; conditions for operation 143;
    impact on commercial industry
    157–161; license-security-review
    process 158; memorandum of
    understanding (MOU) 148, 159;
    notification to NOAA 158; to operate a
    high-resolution commercial satellite
    145; process of approving 148; for
    sharing the imagery 150; WorldView
    Imaging Corporation 149
Liff, Adam 22
Lockheed Martin 146, 150, 173–174;
    development of Overhead Persistent
    Infrared satellites 183
Long March (LM) rockets 53
Loverro, Douglas 158–160
low-Earth orbiting (LEO) satellites 3,
    52–53, 170, 181, 187, 192; development
    of constellations of 210; global
    constellation of 116; US military
    platforms 210
Lupton's theory of space power 93

MacDonald, Dettwiler and Associates
    Ltd. (MDA) 153
McNamara, Robert 118–119, 174

Mahan, Alfred Thayer 89–90; *Influence of
    Sea Power Upon History, The* 89; theory
    of sea power 89–91
manned space program: in China 213; in
    Russia 5; in United States 5
maritime communications 90
market subsidies 131
Mastro, Oriana 65–66
Mattis, James 49
Maxar Technologies *see* DigitalGlobe
Mearsheimer, John 13–14, 17, 36
Mediasat, development of 143–145
Medium Earth Orbit (MEO) 64
memorandum of understanding (MOU)
    148, 159
Merrill Lynch 150
mid-wavelength infrared (MWIR)
    sensor 176
militarization of space 97; ground-to-space
    weapons 98; military satellite
    communications (MILSATCOM) *see*
    military satellite communications
    (MILSATCOM); space-to-ground
    weapons 98; space-to-space weapons
    98; threat to space stability *99*
military and intelligence satellites 204,
    207–208
military capability 27, 30–31, 109, 125,
    132, 202; arms race in space 38; of
    China 65–66; meaning of 29–30;
    offensive 31, 99; and policy of unilateral
    disarmament 23; process of increasing
    22; *versus* purely military assets 20; of
    Soviet Union 28, 97; in space 38, 102;
    of United States 46
military coercion 15
military power: balance of 24; definition of
    30; measurement of 31; in space *see*
    military space power
military satellite communications
    (MILSATCOM) 106–107, 111,
    116–117, 120, 123–125, 127–128, 132;
    anti-jam capability 129; bandwidth
    availability limitations 126; for
    dedicated military use 116; satellite
    constellation 119; usage in Operation
    Desert Storm 119–120; use of GEO
    orbits to achieve maximum coverage
    128; US military dependence on 126
military space competition 216–217
military space power: application of 88;
    comparison of material measures of 51;
    measurement of 31; organization and
    management of 46

Military Strategic and Tactical Relay (Milstar) 116
military use, of space 1, 52, 98–99, 109–110, 116, 123, 131, 204–205, 214
military utility, of space 95, 98
Missile Defense Alarm System (MIDAS) 173–174
missile defense and warning systems 113; Ballistic Missile Early Warning System (BMEWS) 173, 190; Boost Surveillance and Tracking System (BSTS) 178; commercialization of 111, 209–210; constellations of satellites supporting 171; Defense Support Program (DSP) satellites 170, 175–178; deterrence through association 172; development and acquisition of 172–183; Eisenhower's Science Advisory Committee on Early Warning 173; Follow-on Early Warning System (FEWS) 178–179; hybrid missile shield and warning system 172; Missile Defense Alarm System (MIDAS) 173–174; missile detection constellation 173; options for 186–191; overhead persistent infrared (OPIR) monitoring 175; proliferation of 87; satellites involved in 210; security dilemma and 171–172; space-based capabilities 172, 209; US military's efforts to develop 172
missile warning satellites 105, 171; association with the nuclear deterrent 172; deterrence value granted to 172; mission and capabilities of 172; operational constellation of 175; US Air Force 172
Mitchell, William 89
MITRE Corporation 180, 189
mobile laser system 74
mobile phone technologies 133
mobile satellite services 95, 122
Motorola Corporation 133–134
mutual defection: cost associated with 15; risk of 15; security dilemma of 15
mutual defense pacts, formation of 22
Mutually Assured Destruction (MAD) 45

Naryad-V (Briz) ASAT system 71–72
National Aeronautics and Space Administration (NASA), US 2, 51, 68, 212; Syncom III satellite 118
National Defense Authorization Act (NDAA), US 46, 125, 151; Kyl-Bingaman amendment 146

National Geospatial Intelligence Agency (NGA), US 51, 141, 150; FAR 49.202a 151; Service-Level-Agreement (SLA) 150; Termination for Convenience Article 151
National Imagery and Mapping Agency (NIMA), US 146
National Oceanic and Atmospheric Administration (NOAA), US 51, 143, 144–149, 158, 161, 163
National Reconnaissance Office (NRO), US 2, 51, 141
National Security Agency (NSA), US 185
national security space architecture 46, 110, 205, 222
national sovereignty: concept of 62; violation of 48
national space supremacy 1
National Space Transportation Policy (2013), US 186
naval arms race, between Britain and Germany 90
naval warfare, object of 90
New Glenn Rocket 53
NextView contract 150
non-aggressive military activities 44
non-kinetic weapons 105
North Atlantic Treaty Organization (NATO) 153, 212; expansion of 68, 74; membership for the Ukraine or Georgia 69; mutual defense clause 70
North Korea 49, 87, 109, 204
nuclear accidents: Chernobyl disaster (1986) 140
nuclear delivery systems 45
nuclear deterrence 172; GEO satellites associated with 211; logic of 24; between Soviet Union and US 45
nuclear missile warning mission 170, 189, 192
nuclear proliferation 188
nuclear warning mission 105, 191
nuclear weapons: modes of delivery of 173; use in orbit 16

Oberg, Jim 92; theory of space power 92–93
offense–defense balance 7, 17, 20, 22, 25, 35, 87, 102, 104, 172, 186; definition of 31; degree of differentiation between 32; distinguishability between 97–102; factors influencing 31–32; influence of geography and technology on 32; methods of measuring 32; reassurance

offense–defense balance *continued*
  and vulnerability of 27; in space 93–97,
  201; state's power in 29
offensive realism 13–14, 17, 36, 55
offensive space control (OSC) 100
offensive space weapons, tests of 34
Oko missile warning satellites (USSR) 170
OneWeb 117
on-orbit repair and maintenance systems,
  deployment of 33
Operation Burnt Frost (2008) 73, 87
Operation Desert Storm (1991) 45,
  119–120, 178
Operation Enduring Freedom (2001)
  120, 162
Operation Iraqi Freedom (2004) 120, 130
Orbimage 146, 150
Orbital ATK 53
orbital repair and refueling, technology
  of 65
Orbital Sciences Corporation 184
orbiting space robots 65
OrbView-5 satellite, construction of 150
organizational innovation 46
outer space: material balance of power in
  43; Outer Space Treaty (1967) 44, 205;
  warfare and associated actions 43
Overhead Persistent Infrared (OPIR)
  constellation 170, 175, 183–184
Overseas Contingency Operation (OCO)
  funds 125

PanGeo Alliance 156
peaceful use of space 44, 46–47, 97–98
Pence, Mike 48
Petrov, Stanislav 170
Philippines: approaches to handling
  Chinese power 57; Court of Arbitration
  case against China 56; dispute with
  China over South China Sea 56
PL19 Nudol air-defense system
  (Russia) 73
Planet Labs (satellite imagery provider)
  95–96, 158; contract with the NGA 153;
  "flock" of Dove CubeSats 153; threat to
  DigitalGlobe's business model 153
power, definition of 20
power imbalance, perception of 17,
  214–215
Powers, Gary 141
Pravda (Russian state news site) 72–73
precision targeting 45
"Preserving the Air Force's Options"
  report 179

Presidential Decision Directive 23
  (PDD-23), US 145–146, 161
Prevention of an Arms Race in Outer
  Space (PAROS) 63
prisoner's dilemma model, for social
  interaction 14–15, 36
Program Objective Memorandum
  (POM) 123
proliferated space sensor layer 171
Protected Tactical Waveform (PTW) 128
public–private partnerships 130, 133; in
  space launches 53
purchasing of imagery, under ClearView
  contract 150
purchasing power parity 58
Putin, Vladimir 68–75, 190

QuickBird 2 satellite 149

Radarsat-1 (Canada) 155
Radio-Television News Directors
  Association (RTNDA) 143, 145
rational strategic choice, theory of 6,
  18–20, 22, 29, 32, 35–38, 200–201
Raymond, Jay 64
Raytheon 150
Reagan, Ronald 2, 45, 63, 71, 170; SDI
  initiative 178, 191
reassurance strategy, based on space power
  97, 107–112, 208–209, 211, 214
reconnaissance, space-based 44, 97; space
  age reconnaissance satellites 107
remote sensing market 8, 111, 113, 117,
  142, 191, 207–210, 15 CFR 960 (2006)
  version 147, 161; Advisory Committee
  for Commercial Remote Sensing
  (ACCRES) 147; under Clinton
  Administration 146; commercial-
  imagery providers in US 145;
  commercializing of 8, 140, 207–210;
  Commercial Remote Sensing Policy
  (US) 146; conditions for investment in
  143; conditions for operation 143;
  Consolidated Federal Regulations
  (CFR), US 144; efforts to reform
  regulation of 148; evolution of 149–154;
  expansion of 148; impact of the
  licensing process on 157–161; Land
  Remote Sensing Commercialization Act
  (1984), US 142, 145, 157; Land Remote
  Sensing Policy Act (1992), US 144–145,
  147, 158; Landsat program (US) 142,
  145; law on land remote sensing 144;
  licenses for commercial remote sensing

satellites *see* licenses, for remote sensing satellites; National Security Presidential Directive 3 (NSPD-3), US 144; Presidential Decision Directive 23 (PDD-23) 145–146; satellite-based remote sensing 141; security dilemma and 141–142; Space Policy Directive-2 (SPD-2), US 148; US imaging satellites 141; US policy on 142–149

remote sensing satellites 117, 141, 143, 146, 155, 158

remote sensing technology: restrictions on the export of 145; satellite revolution 147; Synthetic-Aperture Radar (SAR) 155–157

resiliency, definition of 158

resolution limits, on satellite imagery: buy-to-deny strategy 162; denied-parties screening 163; diplomacy, option of 162; on higher-resolution imagery 160; other methods of control 161–163; relaxation of restrictions on 163; shutter control 161–162; spectral resolution restrictions 160; US-based commercial-imagery providers 159–161

risk assessment 190

Rocke, David 28

Rogozin, Dmitry 54

rogue state missile attacks, protection against 46

Roscosmos (Russia) 54

Rumsfeld, Donald 46, 58

Russia: active satellites on orbit 52; air-defense system 72; anti-satellite (ASAT) weapons 71; asymmetric methods of warfare 70; bailouts from the International Monetary Fund 67; challenge to US in space 42; collapse of launch industry 54; competition in space with US 211; development of systems designed to counter US space capabilities 72; economic reforms in 67; Federal Space Agency 53–54; invasion of Georgia 69; opposition to US withdrawal from ABM treaty 46; as powerbroker in US conflicts in the Middle East 67; relation with US 67; Rodnik military communications satellites 101; Roscosmos 54; seizure of Crimea 69; self-defense groups 70; space launch industry in 53–54; space programs in 67–75; space weapons development 71; space weapons programs 74; strategy of "hybrid warfare" 70–71; support for the PPWT space treaty 74; United Rocket and Space Corporation (USRC) 54; *see also* Soviet Union

Russian Space Agency 68, 211

S-500 air-defense system (Russia) 72–73

satellite-based observation 140

satellite communications (SATCOM) 8, 45, 98, 111, 113, 125, 136, 142, 164n7, 206, 209; bandwidth for immediate operational purposes 125; categories of 116; commercializing of 207, 209–210; commercial market for 127; commercial reliance approach for commercialization of 127–130; pathways for commercialization 127–133; SATCOM solution 3 (CS3) contract 126; satellite constellations 210; threshold for attacking 117; US government use of 116

satellite constellations 95, 141–142, 210; Galileo constellation 110; of LEO communications satellites 116; for military communications 119

satellite imagery: aerial reconnaissance and 141; American Space Commerce Free Enterprise Act (2018), US 148; availability of commercial imagery 146; ClearView contracts 146, 150; competition in 153; Consolidated Federal Regulations (CFR) for 144; embargo on sales to Iraq 162; EnhancedView contract 151; French SPOT satellite 140, 144; licensing of 143; Mediasat 143–144; news media access to 143; NextView contract 150; quality of 140; QuickBird 2 satellite 149; resolution limitations on 159–161; resolution of 140; US imaging satellites 141; US Landsat program 140, 144; Worldview-1 satellite 147

satellite reconnaissance 141

satellite revolution 147

satellites, launch of 211; ability to launch multiple satellites 101; by Indian space agency 101; LEO constellations 117; by Russian space agency 101

Schriever, Bernard 1, 174, 187

Science Applications International Corporation (SAIC) 184

Scott, Walter 145

SCUD missiles 176, 178

sea lines of communication 57

sea power: Corbett's treatise on 90; Klein's theory of 90–91; Mahan's theory of 89–91; theory of 89–90

Securities and Exchange Commission (SEC), US 152

security dilemma 67, 75, 102–112, 127; approach to measuring 35; arms race due to 214; in commercialization of space power 209, 214; concept of 12–13, 15, 22; cooperation and reassurance in 23–28; deterrence in space 102–107; Herz's formulation of 19; hypothesis of 111–112; key determinant of the severity of 16; measuring of 29–35; missile warning and 171–172; multistate dilemma in space 35–37; of mutual cooperation and mutual defection 15; offense–defense balance in 32; primary driver of 15; psychological aspects of 18; remote sensing and 141–142; severity of 20; and space power 93; state's efforts at seeking security 16; uncertainties of 23

Senate Armed Services Committee (SASC), US 49, 151–152

Service-Level-Agreement (SLA) 150

Shanahan, Pat 49

Shared Early Warning System (SEWS) 188

Shelton, William 100

short-wavelength infrared (SWIR) sensor 176

shutter control: checkbook shutter control 162; concept of 145; to restrict the operation of a satellite 161–162

social interaction: prisoner's dilemma model of 14–15; stag hunt model of 14–15

South China Sea: Chinese claims over 56; construction of artificial islands 56; Court of Arbitration verdict on China's territorial claims in 56

Soviet military capability: ICBM (intercontinental ballistic missile) force 28; Oko missile warning satellites 170; Sputnik, launch of 28

Soviet Union: agreement to German reunification 68; ASAT technology 72; effectiveness of nuclear deterrence 45; fall of 2, 68; nuclear delivery systems 45; proposal on ASAT testing 71; research and development of space weapons systems 71–72; retrenchment in Eastern Europe 68; SCUD missile 176, 178; space-flight technology 75; SS-19 ICBMs 72

Soyuz spacecraft 211; launching of 32

space apparatus inspector 74

space arms race 2, 66, 77, 111, 130, 205, 222

space-based defenses 45

space-based detection systems 190

Space Based Infrared System (SBIRS) satellites 170, 181, 188; advantages over a ground-based radar system 173; capabilities of 183; constellation of 64, 202, 209; deployment of 181–182; GEO satellite 183; Infrared Augmentation Satellite (IRAS) 183; next-generation replacement for 183; survivability of 183; of US Air Force 172, 182; US Air Force–Lockheed proposal for development of 173, 188

space-based interceptor missiles 44

space-based laser 101

space-based navigation 45; Global Positioning System (GPS) 52; GLONASS system 52

space-based weapons 16, 63, 101, 108, 204

space budget 51–52, 67

space capabilities, negation of adversary 100

space control systems, development of 217

space debris, removal of 4, 33, 101

space-enabled conventional warfare 45

space-flight contracts 211

space hegemony 62, 66

Space Imaging 146, 150, 154; contract under NextView 154; Ikonos 162

space industrial base, of US 30, 42, 51–52, 72, 93, 131, 135, 203, 207

space launch industry: in China 52, 54–67; public–private partnerships in 53; in United States 43–54

space mines 71

Space Pearl Harbor 46

Space Policy Directive-2 (SPD-2), US 148

space power 87–93; commercialization of see commercialization, of space power; definition of 88–91; functioning theory of 93; Klein's theory of 93; Lupton's theory of 93; military strategies 88; Oberg's theory of 92–93; offense–defense balance in 93–97, 208; reassurance strategy based on 97, 208–209, 211, 214; security dilemma in 93; theory of 88, 92–93, 95; US military doctrine for 88

Space Shuttle Challenger disaster 176

space systems, vulnerability of 3

space technology, proliferation of 2, 51, 91, 118, 135
space-to-ground weapons 98; categorization of 100; development of 112
space-to-space weapons 98, 202
Space Tracking and Surveillance System (STSS) satellites 170, 181
space tracking networks 52, 101
space warfare 43, 45; principles of 92; Russian strategy of 71
space weapons: Andropov's proposal to ban 71; development of 71
SpaceX 3, 52, 53, 117, 205, 211; Falcon Heavy rocket 53
SPOT satellite (France) 140, 144–145, 162; Astrium Services 151
Sputnik satellite, launch of (Soviet Union) 28, 43; impacts on US security as a result of 43; public outcry in response to 43
SS-19 ICBMs (Soviet Union) 72
stag hunt model, for social interaction 14–15, 36, 201
Star Wars 45
stealth aircraft 211
Strait of Malacca 57
Strategic Defense Initiative (SDI) Organization 178; Brilliant Eyes program 181
submarine-launched ballistic missiles (SLBM) 175
Syncom III satellite 118
Synthetic-Aperture Radar (SAR): challenge to national security 155; development of 155–157; Radarsat-1 (Canada) 155; regulations on 155; TerraSar-X (Germany) 155; X-band SAR imagery 156; XpressSAR 156
Systems Engineering Solutions (SES) 184

Taiwan Strait 57
Taiwan–US relationship: on Chinese interference with Taiwan's sovereignty 56; mutual defense treaty 56; Taiwan Relations Act (TRA) 56
technological innovations 133; in space launches 53
technological revolution, in launch and satellite design 1
"termination for convenience" clause 122, 125, 151, 206
TerraSar-X (Germany) 155
terrestrial radio interference 95

terrestrial warfighting capabilities 4
terrestrial weapons systems 108
territorial integrity, concept of 61–62
THAAD theater missile defense system (US) 72, 109
theater ballistic missiles (TBMs) 64, 109, 171, 176–178, 181, 189–190
Thompson, John 126
Treaty on Prevention of the Placement of Weapons in Outer Space and of the Threat or Use of Force Against Outer Space Objects (PPWT) 63–64, 108, 216; Russian support for 74; US resistance to 204
Trump, Donald: America first strategy 47; desire for a Space Force 49; on development of space-based missile defense layer 172; National Space Policy (NSP) 47–48, 50; Space Policy Directive-3 (SPD-3) 48
two-state solutions 36

U-2 spy plane, downing of 141
unilateral disarmament, policy of 23, 26, 28, 77, 107–110, 112, 203–205, 210
United Launch Alliance (ULA) 52–53, 110; Delta Four Heavy rocket system 53
United Nations (UN) 23; Committee on the Peaceful Use of Outer Space (COPUOS) 63
United Rocket and Space Corporation (USRC), Russia 54
UrtheCast (Canada) 156
US Air Force 1, 49, 100; ability to image satellites 101; Advanced Warning System (AWS) 178; Defense Support Program (DSP) satellites 178–179; Initial Defense Communications Satellite Program 119; missile warning satellites 172; "Preserving the Air Force's Options" report 179; pursuit of a SBIRS replacement 183; satellite-based infrared detection system 173; Space and Missile Systems Center (SMC) 126; Space Command 45; *US Air Force Doctrine Document 1* (AFDD-1) 88
US Airlines 130
US Defense Business Board (DBB) 123, 125
US Department of Commerce 143, 145, 155, 163
US Department of Defense (DOD) 2, 9, 44, 120, 123, 125–126, 130, 134, 162; Civil Reserve Air Fleet (CRAF) model 123;

US DOD *continued*
  deployment of SBIRS constellation 181;
  establishment of unified US Space
  Command 45; investment in Iridium
  135; on nuclear deterrent against Soviets
  45; Office of the Comptroller 178;
  Overseas Contingency Operation (OCO)
  funds 125; purchasing model of 123;
  SATCOM solution 3 (CS3) contract
  126; Strategic Defense Initiative (SDI)
  45; strategies against Soviet missile
  threat 45; WS-117L (space surveillance
  program) 174
US Department of Energy (DOE):
  Initiatives for Proliferation Prevention
  (IPP) program 68
US Department of Transportation 130
US National Security Space Strategy
  (NSSS) 47
US National Space Council 48, 148
US National Space Policy (NSP): under
  Bush Administration 46; under Clinton
  Administration 46; on confidence-
  building measures 47; on defense of
  space assets 47; under Obama
  Administration 47; Space Policy
  Directive-3 (SPD-3) 48; support for the
  peaceful use of space 46; toward China
  and Russia 49; toward space activities 47;
  under Trump Administration 47–48, 50
US National Space Strategy (2018) 61;
  military pillars of 102
US–Soviet relationship 28; agreement to
  German reunification 68; Anti-Ballistic
  Missile (ABM) treaty (1972) 2, 24,
  44–45; on expansion of NATO 68; on
  nuclear deterrence 45
US Space Command 3, 178–179;
  employment of space assets 45;
  establishment of 45; National Security
  Decision Directive 85 (NSDD 85) 45
US Space Commission: establishment of
  46; report on assertion of US rights in
  space 47; Space Pearl Harbor 46
US Space Development Agency (SDA) 116
US Space Force (USSF) 3; actions and
  rhetoric in space 43–51; brief history of
  43–51; budgetary concerns 49;
challenges from Russia and China 42;
  concern over organizational competition
  49; degree of dominance 51;
  establishment of 48, 50; military power
  projection 51; need for 49; NSC 5520
  recommendation on launching satellites
  44; offensive doctrines 50; power
  projection 42; on protecting and
  defending US space assets 50; reaching
  for space dominance 51–54; space
  budget 52; space warfare and associated
  actions 43; sub-optimal policy 42–43
US space industry, capability gap with
  China and Russia 54
US Space Policy Directive 48, 148–149
USSR *see* Soviet Union
US Strategic Command 206

Van Evera, Stephen 17, 20, 65
Vietnam: approaches to handling Chinese
  power 57; dispute with China over South
  China Sea 56; military ties with India
  and the US 56; Vietnam War 56, 119
Voice of America (VOA) 91
von Clausewitz, Carl 90

Waltz, Kenneth 13, 17, 55
Washington, George 89–90
Washington Naval Treaty of 1922 24
Washington treaty 70
weaponization of space 10, 44, 98, 223;
  by China 65
weapons of mass destruction 3, 44
weapons systems, categories of 32
Wideband Global SATCOM Satellites
  (WGS) 116
Wilhelm, Kaiser 90
WorldView Imaging Corporation 145,
  149; license to operate high- resolution
  imagery 149; Worldview-1 satellite 147
WS-117L (space surveillance program) 174

Xi, Jinping 62, 213
XpressSAR 156, 159

Yom Kippur War (1973) 176

Ziarnick, Brent 89, 91–92